Textbook of Soil Science

Textbook of Soil Science

Edited by **Brian Bechdal**

R CALLISTO
REFERENCE

New York

Published by Callisto Reference,
106 Park Avenue, Suite 200,
New York, NY 10016, USA
www.callistoreference.com

Textbook of Soil Science
Edited by Brian Bechdal

International Standard Book Number: 978-1-63239-686-0 (Hardback)

Printed in the United States of America.

Contents

Preface

Soil science is the study of soil as a resource of earth. It deals with studying the various concepts and properties related with formation and classification of soil. It also encapsulates the study of management of soil. Some branches of soil science like pedology and edaphology are also glanced at in this book. The study of soil has raised awareness about soil preservation and its optimum utilization. This book presents researches and studies performed by experts across the globe. The topics covered in this book offer the readers new insights in the field of soil science. It will help new researchers by foregrounding their knowledge in this field along with providing interesting topics for further research.

All of the data presented henceforth, was collaborated in the wake of recent advancements in the field. The aim of this book is to present the diversified developments from across the globe in a comprehensible manner. The opinions expressed in each chapter belong solely to the contributing authors. Their interpretations of the topics are the integral part of this book, which I have carefully compiled for a better understanding of the readers.

At the end, I would like to thank all those who dedicated their time and efforts for the successful completion of this book. I also wish to convey my gratitude towards my friends and family who supported me at every step.

Editor

Competitive Sorption Behavior of Arsenic, Selenium, Copper and Lead by Soil and Biosolid Nano- and Macro-Colloid Particles

Jessique Ghezzi, Anastasios Karathanasis, Chris Matocha, Jason Unrine, Yvonne Thompson

Department of Plant and Soil Sciences, University of Kentucky, Lexington, KY, USA
Email: akaratha@uky.edu

Abstract

Limited information exists on natural nanocolloid sorption behavior of As, Se, Cu and Pb in the environment. They are expected to have variable competitive sorption characteristics depending on size and composition and may transport elevated contaminant loads into surface and ground waters. A comprehensive characterization of their interactions with contaminants could provide a better understanding of the risks they pose to the environment. This study evaluated the sorption behavior of soil and biosolid nano- and macro-colloids with different mineralogical compositions for As, Se, Cu, and Pb contaminants. Single- and multi-contaminant Freundlich isotherms were used to evaluate sorption affinity for the contaminants among the different colloid sizes and compositions. Sorption trends based on size indicated greater affinity for As and Cu by the smectitic and kaolinitic nanocolloids, greater affinity for Pb by the kaolinitic nanocolloids, and greater affinity for As, Se and Pb by bio-nanocolloids over corresponding macrocolloid fractions. Both, single- and multi-contaminant isotherms indicated sorption preferences for cation over anion contaminants, but with somewhat contrasting sequences depending on size and composition. Multi-contaminant isotherms generally predicted greater sorption affinities likely due to bridging effects, particularly for anionic contaminants. Surface properties such as zeta potentials, cation exchange capacity (CEC), surface area (SA), organic carbon (OC), and OC:SA significantly but variably affected sorption characteristics among the differing colloid sizes and compositions. Colloid zeta potential and pH shifts in the presence of different contaminant loads suggested prevalence of inner sphere bonding mechanisms for sorption of cation contaminants by mineral colloids and outer sphere sorption for cation and anion contaminants by bio-colloids.

Keywords

Sorption Affinity, Single-Contaminant Isotherms, Multi-Contaminant Isotherms, Size Effects,

Composition Effects

---

1. Introduction

Major environmental concerns have developed within the past decade in regard to large-scale contamination of natural resources. The vast devastation experienced from catastrophic events such as hurricanes Sandy (2012) and Katrina (2005) have highlighted an ongoing water quality issue, which is the transport or mass influx of contaminants into groundwater supplies following storm events. Remediators need information on contaminant interactions at the soil-water interface and how these interactions affect contaminant plume movement and contaminant transport in surface and ground waters. One potential vector of contaminant transport that should be further investigated is that of naturally occurring environmental nanoparticles, such as those derived from soils or biosolids [1]-[4]. Soil nanoparticles include humic substances, clay minerals/colloids, and metal hydroxides [4]-[6]. Large volumes of biosolid nanoparticles are introduced into the environment from human and animal wastes applied to land as fertilizers [7]. More recently, the development and application of engineered nanoparticles for remediation of contaminant plumes has also raised concerns about their mobility and biotoxicity in the environment [8] [9]. Research on naturally occurring environmental nanoparticles is limited with regard to their behavior in environmental media, both as potential contaminant transport vectors and as models for manufactured nanoparticles [1]-[4] [10].

Current findings suggest that soil and biosolid nanocolloids may possess larger surface area and increased reactivity than macrocolloid fractions and therefore, a greater potential to sorb and transport larger quantities of heavy metals to groundwater [3] [5] [11]. Both mineral and biosolid-derived nanocolloids can form inner- and outer-sphere complexes with heavy metals via surface siloxane, aluminol, carboxylic, and phenolic groups [12] [13]. Drops in pH or zeta potential with increasing contaminant loads have been associated with inner sphere sorption due to proton release after the exchange of the cation contaminant, or due to hydrolysis/precipitation of the metals [14]. The type of bonding between nanocolloid surfaces and contaminants also dictates the likelihood of re-suspension of the contaminant in different ionic or pH environments, further demonstrating the need for a better understanding of the solid-solution interface reactions between nanocolloids and potential contaminants [11] [12].

Elevated concentrations of contaminants such as As, Se, Cu, and Pb in water supplies have raised considerable environmental concerns [12] [13]. Increased levels of these contaminants are attributed to both soluble and particulate sources [5] [11]. Both As and Se are considered as metalloids, and are usually present as oxy-anions in soil environments while Cu and Pb are cationic metals [15] [16]. Studies of their interactions with various clay minerals suggest that their sorption behavior is controlled by the competition between ions in solution and the clay surface, pH, ionic strength, and mineralogy [12] [17]. Arsenic and Se tend to form outer sphere complexes with variable charge minerals and phyllosilicate edges and inner sphere complexes directly with Fe-oxyhydroxides or through bridging with Fe/Al hydrolytic species associated with organic functional groups [18] [19]. On the other hand, the interaction of Cu and Pb with colloids is mainly controlled by ion exchange processes, with electrostatic surface bonding and chemisorption to surface SiOH and AlOH groups [19]. Since most phyllosilicate mineral surfaces are negatively charged, they are expected to repel oxy-anions like As and Se, and attract cations such as Cu and Pb. However, the presence of organic functional groups and/or Fe/Al-oxyhroxides as discrete phases or coatings on clay surfaces may cause significant surface charge alterations and drastic contaminant sorption behavior changes in colloidal fractions albeit with their mineralogical composition [20]-[22]. While the sorption behavior of As, Se, Cu, and Pb contaminants with clay sized fractions of different mineralogical composition has been extensively studied, very little information exists on their interactions with nano-sized particles under single- and multi-contaminant solution environments.

The objectives of this study were to evaluate the sorption affinity of soil and biosolid nano- and macro-colloids of diverse composition for As, Se, Cu, and Pb under single- and multi-contaminant solution environments.

2. Methods and Materials

2.1. Colloid Generation

The Bt horizons of three Kentucky soils of differing mineralogy were used to generate the mineral colloids: Ca-least-variant (fine, smectitic, mesic mollic Hapludalf), Tilsit (fine-silty, mixed, mesic Typic Fragiudult), and Trimble (fine-loamy, siliceous, mesic Typic Paleudult), referred to herein as smectitic, mixed, and kaolinitic, respectively. Biosolid colloids were fractionated from an aerobically digested municipal sewage sludge obtained from Jessamine County, Kentucky. Centrifuge fractionations using Stokes law allowed separation of the two size classes (nanocolloids < 100 nm and macrocolloids 100 - 2000 nm) using a Centra GP8R Model 120 centrifuge (Thermo IEC). Clay fractions were separated from bulk soils by centrifugation at 107 RCF for 3.5 minutes, as calculated using a rotor radius of 170 mm, 107 RCF, a density difference of 1650 kg·m^{-3}, and viscosity of 0.0008904 Pas. Nanocolloids were then separated from corresponding macrocolloids at 4387 RCF for 46 minutes, as calculated using a rotor radius of 170 mm, a speed of 4387 RCF, a density difference from water of 1650 kg·m^{-3}, and viscosity of 0.0008904 Pas [5] [23]. The colloids were generated with de-ionized water (resistivity of 1 μΩ·cm at 25°C).

2.2. Sorption Isotherms

Nano- and macro-colloid affinities for the four contaminants were evaluated with single- and multi-contaminant adsorption isotherms using duplicate colloid suspensions of 50 mg colloid L^{-1} in de-ionized water spiked with 0, 0.2, 0.5, 1, 2, 5 and 10 mg·L^{-1} of As, Se, Cu, and Pb. Equilibrium aqueous solutions of Pb, Cu, As and Se were prepared from PbCl$_2$ (98% purity, Aldrich Chemicals, Milwaukee, WI), CuCl$_2$ (>99% purity, Sigma Chemical Company, St. Louis, MO), arsenic acid Na$_2$HAsO$_4$•7H$_2$O (98% purity, Sigma Chemical Company, St. Louis, MO), and sodium selenate decahydrate Na$_2$SeO$_4$•10H$_2$O (99.9% purity, Sigma Chemical Company, St. Louis, MO). Multi-contaminant adsorption isotherms were generated with multi-contaminant equilibrium solutions in which the sum of the four contaminant concentrations was 0, 0.2, 0.5, 1, 2, 5, and 10 mg, respectively. MINEQL$^+$ speciation of the equilibrium solutions suggested the following predominant species: 99.9% Pb^{2+}, 99.9% Cu^{2+}, 99.7% SeO$_4^{-2}$ (selenate, VI), and 98.6% AsO$_4^{3-}$ (arsenate, V). Isotherm samples were equilibrated by shaking for 24 hours at room temperature (25°C) in polyethylene tubes with pH measurements taken at 0 and at 24 hours. After shaking, 0.025 μm nitrocellulose filters were used to separate the supernatant from the colloidal fraction. Supernatant fractions were preserved with 1% nitric acid, stored in polyethylene vials, and analyzed within 24 hours via inductively coupled plasma mass spectroscopy (ICP-MS).The mass of contaminant sorbed per mass of colloid was calculated and plotted using the Freundlich equation:

$$q = \frac{V \times (Cin - Co)}{M} = \mu mol \cdot kg^{-1},$$

where q is the mass of contaminant sorbed per mass of colloid, V is the solution volume used in the sorption isotherm experiment, C_{in} is the amount of contaminant added in solution, C_o is the amount of contaminant measured at equilibrium and M is the mass of the sorbent. The log version of this equation ($Logq = NLogC_{eq} + LogK_f$), yields a straight line with N slope, $LogC_{eq}$ the x-variable and $LogK_f$ (Freundlich sorption coefficient) the y-intercept [24]. Isotherms were also normalized for colloid surface area and OC content but did not produce statistically significant trends [25].

2.3. Physico-Chemical and Surface Characterizations

All analyses were performed on suspensions of 50 mg colloid L^{-1} in de-ionized water. A Malvern Instruments Zetasizer Nano ZS (Malvern, United Kingdom) measured suspensions for intensity weighted mean particle hydrodynamic diameters (z-average diameter) using dynamic light scattering (173° backscatter analysis method). Nano- and macro-colloid crystallite sizes were determined using transmission electron microscopy (TEM; JEOL 2010F, Tokyo, Japan) and scanning electron microscopy (SEM; Hitachi S-4300, Tokyo, Japan), respectively. ImageJ software was used to calculate average minimum diameters (ImageJ 1.46r, Wayne Rasband, National Institutes of Health, USA). Surface area was measured using the Ethylene Glycol Monoethyl Ether (EGME) method. Electrical conductivity and pH were measured on a Denver Instruments Model 250 pH*ISE*electrical conductivity meter (Arvada, CO). Cation exchange capacity was determined using an adapted version of the ammonium acetate method and reported as a sum of the base cations Ca^{2+}, Mg^{2+}, K$^+$, and Na$^+$, as measured with a Varian Spectr AA 50B atomic absorption spectrometer. Organic carbon was measured on a Flash EA 1112 Series NC Soil Analyzer (Thermo Electron Corporation) with a Mettler Toledo MX5 microbalance. Zeta potentials in the presence of 0 and 2 mg·L^{-1} Pb, Cu, As, and Se were converted from electrophoretic mobility measurements using the Smoluchowski approximation on a Malvern Zetasizer Nano ZS (Malvern, United Kingdom).

2.4. Mineralogical Characterization

Mineralogical characterizations were completed using X-ray diffraction (XRD) and Thermogravimetric analysis (TG) on a Phillips PW 1840 diffractometer and PW 1729 X-ray generator (Mahwah, NJ), and a Thermal Analyst 2000 (TA Instruments) equipped with a 951 Thermogravimetric Analyzer (DuPont Instruments), respectively [23].

2.5. Statistical Analysis

The standard accepted error level for all duplicate and triplicate samples was 15%. Mean differences in sorption Freundlich coefficients ($LogK_f$) and changes in the isotherm pH were calculated using the general linear model (PROC GLM). Mean differences (overall and based on mineralogy, size, and contaminant) were developed using Fisher's protected least significant difference test (LSD) in SAS using probability levels of 0.05, unless otherwise noted. Competitive sorption relationships were analyzed between colloid properties and sorption coefficients using multiple regression analysis with probability levels of $\alpha < 0.05$ and $\alpha < 0.01$ in SAS 9.3 (SAS Institute Inc., Cary, NC, USA).

3. Results and Discussion

3.1. Single-Contaminant Isotherms

The data conformed well to the Freundlich equation, showing R^2 values between 0.93 and 0.99 with differing sorption trends for each contaminant. The majority of 1/n values were <1.0, suggesting uniform sorption surface coverage, but As sorption by biosolid colloids and Se sorption by smectitic colloids indicated the potential for multiple coverage mechanisms with 1/n values >1.0. Generally, As sorption coefficients were the lowest among the contaminants studied (**Figure 1**). Macrocolloid sorption affinities for As ranged from 1.67 to 3.48, with the highest representing the mixed mineralogy and the lowest the biosolid colloids (**Figure 1**). Sorption affinities for As within the nanocolloids ranged from 1.85 to 3.60, with the highest affinity associated with the kaolinitic and the lowest with the biosolid colloids (**Figure 1**). Statistically significant trends among compositions for As affinity by macrocolloids followed the sequence: Mixed (A) > Smectitic (B) > Kaolinitic (C) > Biosolid (D); and by nanocolloids Kaolinitic (A) > Smectitic (B) > Mixed (C) > Biosolid (D) ($\alpha < 0.05$). Typically, Fe-oxy-hydroxides and kaolinitic clays have demonstrated higher sorption capacities for arsenate [26] [27] than for illite or smectite [19], especially in the presence of humic acid surface coatings [21]. However, Fe-hydroxide and organic coatings may increase As sorption by phyllosilicate minerals via inner sphere complex formation [18] [19]. The biosolid colloids showed lower overall As affinity than the mineral colloids (**Figure 1**), contrary to studies showing increased arsenate sorption with increasing organic matter composition [21] [27]. This could be explained by the higher pH (**Table 1**) and phosphate content of the bio-colloids compared to the mineral colloids. Arsenate sorption has been shown to decrease with increasing pH and phosphate has a tendency to displace As [18] [28]. A colloid size comparison for As affinity suggested greater sorption for nano-colloids than macro-colloids in all compositions, except for the mixed colloids ($\alpha < 0.05$; **Figure 1**). Ideally, the nano-fractions due to their higher negative zeta potentials (**Table 2**) were expected to have lower Kf vaues for anionic contaminants, but the presence of OC surface coatings and bridging effects with Fe/Al hydrolytic species (**Table 1**) may have enhanced their affinity for As [19] [29].

Selenium also exhibited low sorption coefficients compared to those of Cu and Pb (**Figure 1**). Sorption affinities for Se within the macrocolloids showed no particular preference for composition, ranging from 3.09 to 3.49. The highest affinity was observed with the mixed mineralogy and the lowest with the smectitic colloids (**Figure 1**). The wider range was observed within the nanocolloids (2.06 to 3.53) with the highest affinity represented by the mixed mineralogy and the lowest the smectitic composition (**Figure 1**). The higher affinity of the mixed colloids for Se over the smectitic could be associated with contributions from its higher kaolinite content and from the mica minerals that exhibit similar to kaolinite Se sorption trends at low pH (**Table 1**) [30] [31]. Statistically significant sorption affinity for Se by macrocolloids followed the sequence: Mixed (A) > Kaolinitic = Biosolid (B) > Smectitic (C); and by nanocolloids Mixed (A) > Biosolid (B) > Kaolinitic (C) > Smectitic (D) ($\alpha < 0.05$; **Figure 1**). Negative linear correlations between Se sorption affinity and smectite content are consistent

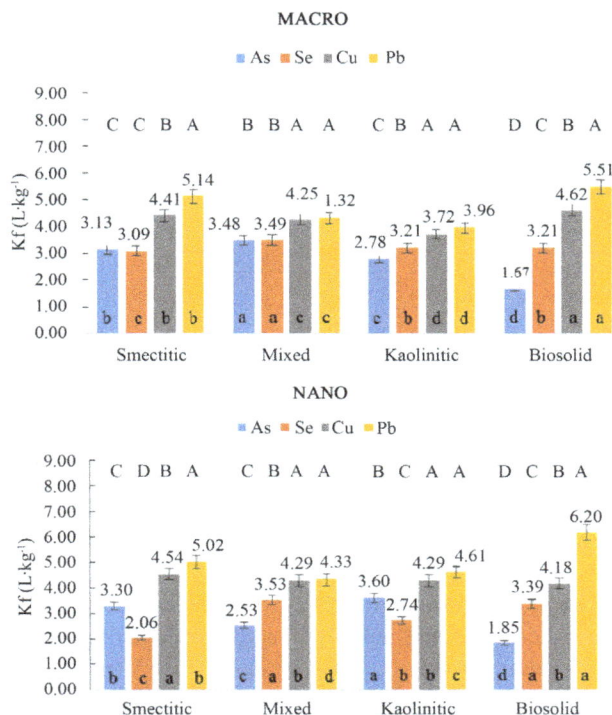

Figure 1. Single-contaminant isotherm Kf values for As, Se, Cu, and Pb for macro- and nano-colloid fractions of different composition (capital letters portray trends across all four contaminants within each composition, while lower case letters display trends for each contaminant across all four compositions).

Table 1. Selected physical and chemical characteristics of nano- and macro-colloid fractions.

Size Fraction	Colloid Composition							
	Smectitic		Mixed		Kaolinitic		Biosolid	
	Macro	Nano	Macro	Nano	Macro	Nano	Macro	Nano
DLS[†] Mean Hydrodynamic Diameter (d_h) ±SD[‡] (nm)	487 ± 10	181 ± 3	596 ± 21	205 ± 4	545 ± 25	187 ± 4	4456 ± 599	353 ± 8
SEM/TEM [¶]Mean Smallest Particle Size ±SD[‡] (nm)	328 ± 144	37 ± 13	549±394	7 ± 5	288 ± 184	41 ± 19	363 ± 338	50 ± 19
Surface Area (SA, $m^2 \cdot g^{-1}$) ±SD[‡]	708 ± 137	879 ± 76	420 ± 105	466 ± 10	333 ± 37	389 ± 44	1674 ± 70	1303 ± 63
Ionic Strength (IS, $mol \cdot L^{-1}$)[§]	4.99×10^{-5}	7.71×10^{-5}	3.70×10^{-5}	3.92×10^{-4}	3.64×10^{-5}	4.83×10^{-5}	1.97×10^{-4}	5.96×10^{-4}
Natural pH	4.92	5.12	5.07	4.92	4.91	5.38	5.39	5.25
CEC ($cmol_c \cdot kg^{-1}$)[#]	35.0 ± 12.8	42.2 ± 15.1	8.9 ± 1.6	10.5 ± 1.7	6.9 ± 1.8	13.1 ± 2.8	37.6 ± 14.8	71.0 ± 23.0
OC ($mg \cdot kg^{-1}$)[‡‡]	658	897	645	774	430	647	1300	16000
OC:SA	0.93	1.02	1.54	1.66	1.29	1.66	0.78	12.28
Mineralogy (%)[*]	K_{29}, Ge_7, Q_6, M_{10}, Sm_{48}	K_{30}, Ge_9, Q_4, M_6, Sm_{51}	K_{42}, Ge_5, Q_5, MVI_7, M_{31}, HIV_{10}	K_{46}, Ge_7, Q_3, M_{30}, MVI_7, HIV_7	K_{52}, Ge_{12}, Gi_5, Q_4, M_3, HIV_{24}	K_{55}, Ge_{15}, Gi_6, Q_2, M_3, HIV_{19}	NA	NA

[†]DLS = Mean intensity weighted hydrodynamic diameter (d_h) determined by Dynamic Light Scattering. [‡]SD = Standard Deviation of duplicate or triplicate measurements. [§]Ionic Strength (IS) = 0.0127 × Electrical Conductivity (millimhos·cm^{-1}). [¶]SEM = Scanning Electron Microscopy, TEM = Transmission Electron Microscopy. [#]CEC = Cation Exchange Capacity by sum of cations. [‡‡]OC = Total Organic Carbon-Dissolved Organic Carbon (inorganic carbon contributions were assumed to be 0 due to low pH). [*]Mineralogy (K = Kaolinite, Ge = Geothite, Gi = Gibbsite, Q = Quartz, M = Mica, Sm = Smectite, MVI = Mica-Vermiculite Interstratified, HIV = Hydroxy-Interlayered Vermiculite, NA = Not Applicable).

Table 2. Zeta potential and pH for macro- and nano-colloid fractions of different composition in the absence and in the presence of equal amounts of As, Se, Cu, and Pb, totaling 0, 2 and 10 mg·L^{-1}.

Colloid Composition	Colloid Size Fraction			
	Macro		Nano	
	Zeta potential with 0, 2, and 10 mg/L contaminant additions	pH with 0, 2, and 10 mg/L contaminant additions	Zeta potential with 0, 2, and 10 mg/L contaminant additions	pH with 0, 2, and 10 mg/L contaminant additions
	(mV)		(mV)	
Smectitic	−27, −18, −11	4.60, 4.67, 4.38	−28, −25, −18	5.11, 5.02, 4.56
Mixed	−34, −29, −26	5.50, 4.45, 4.24	−39, −38, −22	4.84, 4.61, 4.33
Kaolinitic	−34, −30, −26	4.98, 4.96, 4.40	−38, −31, −26	4.90, 4.79, 4.55
Biosolid	−19, −21, −25	5.11, 5.42, 5.18	−11, −24, −29	5.25, 5.79, 5.69

with the anionic behavior of this contaminant ($R^2 = -0.42^*$) [30]. Additionally, the Se affinity of kaolinite may have been enhanced by contributions from goethite and gibbsite (**Table 1**) through formation of inner- and outer-sphere complexes [32] [33]. With the exception of the mixed colloids, Se sorption affinity was influenced by size, showing greater sorption by the smectitic and kaolinitic macro-fraction and an opposite trend by the bio-colloid nano-fraction (**Figure 1**). Higher macrocolloid affinity for this anionic contaminant can be probably attributed to the lower pH of the smectitic and kaolinitic macro-fractions (**Table 1**) and the increased repulsion of Se by the nanocolloids due to more negative surface charge induced by OC surface coatings (**Table 1** and **Table 2**) [13] [20]. In contrast, the bio-colloids exhibited greater sorption affinity for Se in the nano-fraction-most likely due to the lower pH as was the case with As in spite of higher surface charge, OC, and OC:SA ratios (**Figure 1**, **Table 1** and **Table 2**). The bio-colloid (macro- and nano-fractions) sorption affinity for Se was significantly higher than that for As (**Figure 1**; $\alpha < 0.05$) in spite of the lower shared charge [34] in agreement with other findings in waste-amended soils [35].

As would be expected for cationic contaminants, Cu sorption affinity was significantly higher than As and Se in all colloid fractions (**Figure 1**; $\alpha < 0.05$). Sorption coefficients within the macro-fractions ranged from 3.72 to 4.62, with the highest affinity representing the biosolid composition and the lowest the kaolinitic (**Figure 1**). Copper has been shown to dominantly associate with mineralizable biosolid fractions [36], forming multi-ligand complexes with a variety of organic surface functional groups [19]. In contrast, within the nanocolloid fractions, the highest affinity was associated with the smectitic (4.54) and the lowest (4.18) with the biosolid composition (**Figure 1**). Statistically significant differences for Cu sorption affinity within the macrocolloids followed the sequence: Biosolid (A) > Smectitic (B) > Mixed (C) > Kaolinitic (D); and for the nanocolloid fraction: Smectitic (A) > Mixed = Kaolinitic (B) > Biosolid (C) ($\alpha < 0.05$; **Figure 1**). The differing trends for Cu sorption affinity observed in the biosolid macro- vs. nano-fractions may be explained by the greater OC:SA ratios of the bio-nanocolloids that could potentially induce aggregation and block surface sorption surface sites or by the presence of less reactive organic functional groups (**Table 1**). This is corroborated by the lower initial zeta potential values of the biocolloid nano-fractions (**Table 2**). In addition, the lower pH maintained by the biosolid macrocolloids compared to the nanocolloid fractions after the addition of different Cu concentrations may have enhanced dissociation and surface organic complexation reactions [37] [38]. Other than the biocolloids, statistically greater Cu sorption affinity based on size was also demonstrated by the kaolinitic and smectitic nanocolloids over the macrocolloids (**Figure 1**; $\alpha < 0.05$). The greater cation sorption affinities of these nanocolloids are likely due to higher negative zeta potentials and greater surface area as compared to that of the macrocolloids (**Table 1** and **Table 2**). There were no significant differences in sorption of Cu based on size for the mixed colloids (**Figure 1**).

Sorption affinities for Pb were the highest compared to other contaminants. Within the macrocolloids the highest coefficient (5.51) was associated with the biosolid composition and the lowest (3.96) with the kaolinitic (**Figure 1**). Within the nanocolloids Kf values for Pb ranged from 4.33 to 6.20, with the highest value representing the biosolid composition and the lowest the mixed (**Figure 1**). Based on composition, macrocolloid affinity for Pb followed the sequence: Biosolid (A) > Smectitic (B) > Mixed (C) > Kaolinitic (D); and Biosolid (A) > Smectitic (B) > Kaolinitic (C) > Mixed (D) for nanocolloids ($\alpha = 0.05$). These trends are comparable to those reported elsewhere, indicating preferential sorption of Pb by biosolids over other cation metals like Cu or

Zn [23]. The greater cation sorption affinities of the nanocolloids are likely due to greater attraction of cation contaminants to their higher negative zeta potentials and greater surface area availability for sorption as compared to that of the macrocolloids (**Table 1** and **Table 2**). Increased affinity for Pb by smectitic minerals has been attributed to their high permanent charge and ability to form inner sphere complexes with exposed surface –OH groups [19] [39]. Only the kaolinitic and biosolid colloids showed size sorption preferences, with the nano-fractions exhibiting greater Kf values than the macro-size fractions, apparently due to higher CEC, SA and OC content. Statistical correlations (R^2, $\alpha < 0.05$) of Pb sorption coefficients with the above parameters were 0.78, 0.92, and 0.56, respectively.

Single-contaminant isotherm data normalized by surface area and organic carbon did not provide significant differences from mass based isotherms in colloid size or composition trends. However, all isotherm normalization methods indicated preferential sorption of cation-(Cu, Pb) over anion-(As, Se) contaminants [12] [23] [25]. Overall, single-contaminant isotherms showed the following contaminant sorption preference: Pb (A) > Cu (B) > As (C) = Se (C) ($\alpha < 0.05$). Greater sorption of Pb over Cu is likely due to the lower hydrolysis constant of Pb [19] [22] [38]. The similar sorption affinities for As and Se contradict other findings showing preferential sorption of arsenate over selenate [34] but may be explained by the higher bio-colloid preference for Se, which has overshadowed the expected differences in anion sorption. Sorption trends based on colloid size varied, with the smectitic and kaolinitic nanocolloids demonstrating greater affinities for As and Cu contaminants than corresponding macrocolloid fractions, but showing opposite trends for Se. The kaolinitic nano-fraction indicated a greater affinity for Pb over the macro-fraction. The bio-nanocolloids showed greater affinity for As, Se, and Pb than for the corresponding macrocolloids, but showed the opposite trend for Cu. Finally, the mixed colloids exhibited size-based sorption trends only for As, showing greater affinity in the macro- than the nano-fraction (**Figure 1**; $\alpha < 0.05$).

Published research with relatively pure minerals suggests that kaolinitic clays typically have a greater affinity to sorb anions (including oxy-anions like Se and As) than illitic or smectitic clays [19]. Also, surface silanol and aluminol groups exposed on the interlayer surfaces of 2:1 minerals tend to preferentially sorb cation contaminants such as Cu and Pb [19]. The lack of consistent sorption differences by colloid size and even composition in this study may be explained by the diverse natural mixture of minerals present in each colloid fraction as well as the nature and extent of Fe, Al, and organic moieties coating their surfaces. It also demonstrates the unpredictable complexities of modeling contaminant sorption and transport behavior of natural environmental nano- and macro-colloidal fractions [5] [11] [33].

3.2. Multi-Contaminant Isotherms

Competition among As, Se, Cu and Pb sorption was established through multi-contaminant isotherms for each colloid-size fraction. The data conformed reasonably well to the Freundlich equation with lower R^2 values (0.84 ± 0.12) and more 1/n values >1 than the single contaminant sorption isotherms implying that multiple sorption mechanisms were triggered by the presence of all four contaminants. Multi-contaminant isotherm Kf values by colloid size and composition are shown in **Figure 2**. Overall, within the same composition multi-contaminant Kf values were significantly higher than single-contaminant K_f values for both sizes for all contaminants except As (**Figure 1** and **Figure 2**). This is probably the result of bridging effects induced by anionic and cationic contaminants in the mixture. The mixed macro- and nano-colloid fractions indicated equal affinity of Pb and Cu, but greater affinity for Se than As (**Figure 2**). The kaolinitic and biosolid macrocolloids indicated preferential sorption for Pb over Cu, with no significant differences between the anionic contaminants (**Figure 2**). These trends are consistent with other findings indicating preferential sorption of Pb over Cu by kaolinite [22] and similar affinities for Cu and Pb by humic acid substances, and Fe/Mn-oxides [19]. Opposing trends were displayed by the smectitic macrocolloids and kaolinitic nanocolloids, portraying greater Cu sorption over Pb, and more affinity for Se than As (**Figure 2**). Finally, the smectitic and biosolid nanofractions shared the following sorption affinity sequence: Pb (A) > Cu (B) > Se (C) > As (D) (**Figure 2**).

Overall, counting all colloids, the multi-contaminant isotherms showed the following sorption preference trends: Pb (A) = Cu (A) > Se (B) > As (C) ($\alpha < 0.05$); with cation contaminants indicating more competitive sorption affinity compared to the anion contaminants as was the case with the single-contaminant isotherms (**Figure 1** and **Figure 2**). However, two main differences are manifested in the similar sorption affinity for Cu and Pb compared to Pb > Cu and the greater affinity for Se over As compared to Se = As shown by the sin-

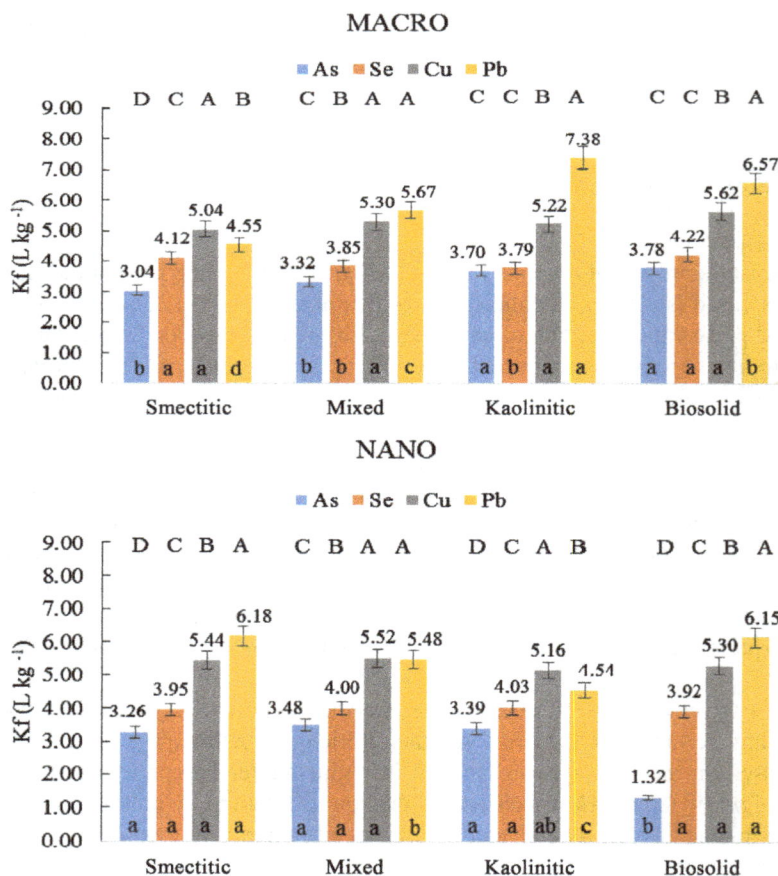

Figure 2. Multi-contaminant isotherm Kf values for As, Se, Cu, and Pb for macro- and nano-colloid fractions of different composition (capital letters portray trends across all four contaminants within each composition, while lower case letters display trends for each contaminant across all four compositions).

gle-contaminant isotherms. The alternating sorption affinities for Pb and Cu based on composition and size are consistent with conflicting sorption affinity exchanges reported for Cu and Pb in sediments, particularly in the presence of multiple contaminants [40]. Some studies suggest that preferential sorption of Pb over Cu occurs through inner sphere complexes due to its lower hydrolysis constant [22] [38], while others report preferential sorption of Cu over Pb due to its greater electronegativity and charge-to-radius ratio [41].

Conflicting preferential sorption trends for As and Se have also been reported in variable pH and anionic solution environments [18] [19] [28] [34]. Significant quantities of Cl^-, NO_3^-, PO_4^{3-}, and SO_4^{2-} found in the kaolinitic and bio-nanocolloid fractions, may have offered significant competition for sorption of anions, and may have inhibited some cation sorption in the multi-contaminant isotherms. Additionally, sorption trends may have been influenced by Fe/Al and organic coatings as well as bridging effects between solution cations (natural or added contaminant) and anionic contaminants or enhanced aggregation caused by sorbed cation contaminants with free positive charge attracting negatively charged colloid particles. The later may cause encapsulation of some contaminants and deflect available sorption surfaces for others [5] [13] [20].

3.3. Anionic vs. Cationic Contaminants

Average Kf values for the anion (As and Se) and cation (Cu and Pb) contaminants for the single- and multi-contaminant isotherms demonstrated greater sorption affinities for cation-over anion-contaminants (**Figure 3**). This trend is consistent with other findings [12] [23] [25], and is likely due to the greater attraction of cations to the negatively charged colloid surfaces, as well as due to greater charge-to-radius ratios of Cu and Pb compared

ANIONS

CATIONS

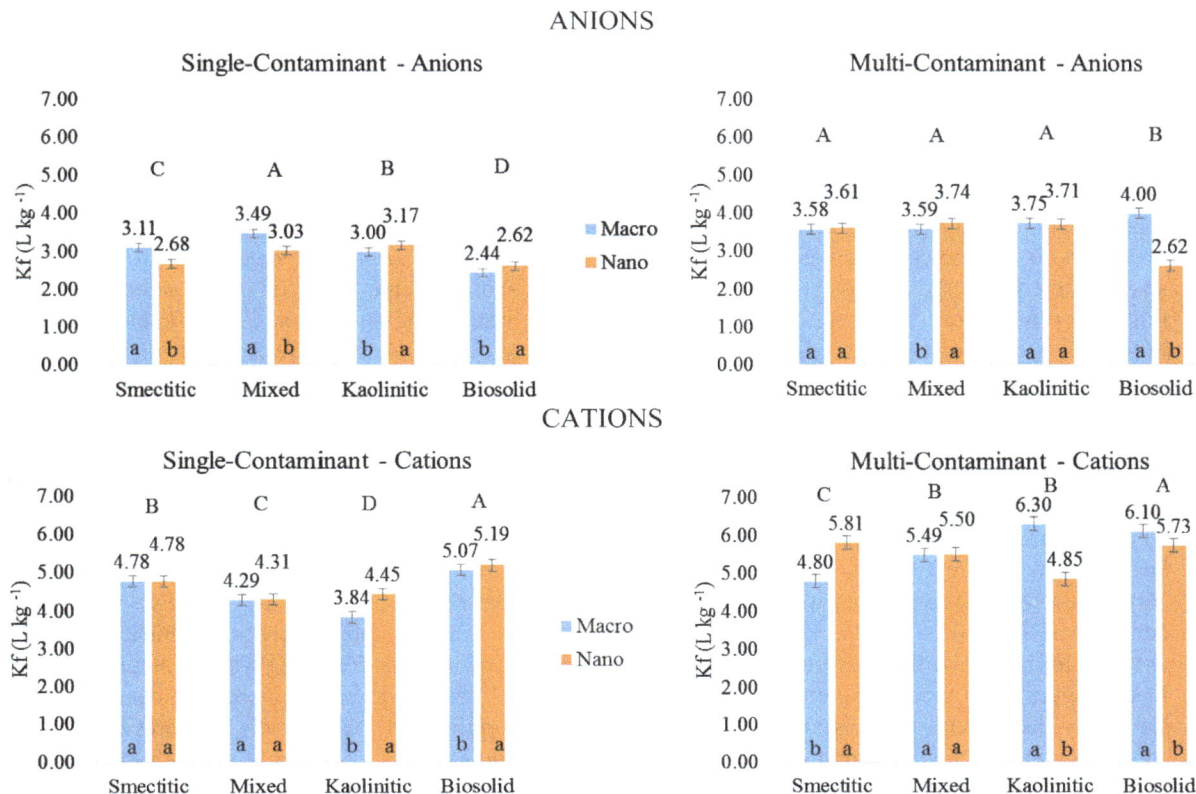

Figure 3. Average Kf values for anion-(As, Se) and cation-contaminants (Cu, Pb) determined from single- and multi-contaminant isotherms (capital letters portray trends based on composition, while lower case letters show significant differences based on size within each composition).

to As and Se [40]. Single-contaminant isotherms indicated the following sorption preference sequence based on composition for anionic contaminants: Mixed (A) > Kaolinitic (B) > Smectitic (C) > Biosolid (D). The multi-contaminant isotherms showed no significant sorption differences among the mineral colloids, but agreed with the single-contaminant isotherms by indicating greater overall affinities for anionic contaminants by mineral than bio-colloids (**Figure 3**). This is probably due to high concentrations of other anions in the bio-colloid fractions which may be out-competing the anionic contaminants for sorption sites [42]. Another likely explanation is the high negative/positive charge ratio of the bio-colloids inducing greater repulsion of the anionic contaminants [19] [43]. In contrast, both single- and multi-contaminant isotherms showed the bio-colloids having the strongest sorption affinity for both cationic contaminants (**Figure 3**). The sorption preference sequence for single-contaminant isotherms was: Biosolid (A) > Smectitic (B) > Mixed (C) > Kaolinitic (D), while for the multi-contaminant isotherms: Biosolid (A) > Mixed = Kaolinitic (B) > Smectitic (C). Sorption competition based on size showed mostly opposing trends between the single- and the multi-contaminant isotherms for both anion- and cation-contaminants (**Figure 3**). A comparison of average K_f values for anion- and cation-contaminant sorption regardless of colloid size and composition indicated significantly higher sorption affinities by multi- vs. single-contaminant isotherms. A similar comparison of average K_f values for all four contaminants suggested that smectitic and mixed nanocolloids can sorb significantly greater quantities of contaminants than corresponding macro-fractions while the opposite trend held true for the kaolinitic and biosolid colloids (**Figure 3**). Despite the fact that organic carbon and surface area have been shown in many studies to be key factors in understanding contaminant sorption patterns [12] [14] [32] [39], neither isotherm version (multi- or single-contaminant) gave consistent differences in size or composition-based trends when normalized to surface area and organic carbon content. Zeta potentials measured in the absence and presence of contaminants ($\alpha < 0.05$), ionic strength of contaminant added ($\alpha < 0.01$), contaminant type, OC, CEC and OC:SA were significant factors in the overall sorption model ($R^2 = 0.92$; $\alpha < 0.01$). The contradicting contaminant sorption trends for both cations and anions highlight the complexities associated with predicting the behavior of colloids of various size and composition in a multi-contaminant natural environment.

3.4. Surface Sorption Characteristics

Single point zeta potential measurements were taken in the absence of contaminants and with additions of 2 and 10 mg·L^{-1} mixed contaminants (**Table 2**) in order to decipher contaminant sorption mechanisms. In order to mimic natural conditions, the pH during the zeta potential measurements was not kept constant (**Table 2**), so that shifts in zeta potential are associated with pH and contaminant sorption changes through inner- or outer sphere attraction [14]. Zeta potentials without contaminants showed trends based on composition and size, with mineral nano-colloids exhibiting more negative values than corresponding macrocolloids (**Table 2**). Similarly, in the absence of contaminants all mineral colloid fractions showed more negative zeta potentials than the bio-colloid fractions (**Table 2**). The addition of contaminants caused significant shifts in zeta potential with opposing trends (positive vs. negative) for the mineral- and bio-colloid fractions, respectively. The large positive shift in zeta potential displayed by the smectitic macrocolloids and the mixed nano-colloids upon addition of contaminants indicates greater inner sphere cation contaminant attraction [14] (**Table 2**). Overall, the mineral colloid zeta potentials became more positive with the increased addition of contaminants, suggesting that the cations were out-competing the anions for inner sphere sorption (anions prefer outer sphere bonding and thus would have little to no effect on zeta potential values). Inner sphere bonding of Pb and Cu in the mineral colloid adsorption isotherm experiments was suggested through positive zeta potential shifts (**Table 2**) with increased contaminant loads ($\alpha = 0.05$) [14]. In contrast, outer sphere bonding was apparently the dominant sorption mechanism for the bio-colloids as indicated by negative zeta potential shifts (**Table 2**) with increased contaminant loads, suggesting a greater preference for anion contaminants. The pH drop with increasing contaminant loads shown by the mineral colloids also indicates inner sphere sorption [14]. Conversely, the prevalence of outer sphere bonding of the oxy-anion contaminants (As, Se) was indicated by increases in pH with increased contaminant loads. In single-contaminant isotherms the addition of anion contaminants increased the equilibrium solution pH by an average of 0.32 units over the initial pH, while the addition of cation contaminants caused a respective pH decrease of 0.31 units ($\alpha < 0.05$). The mixed macrocolloids showed the largest pH drop, while the bio-nanocolloids the largest increase with addition of contaminants, indicating prevalence of inner-sphere and outer-sphere bonding, respectively (**Table 2**).

4. Conclusions

The findings of this study demonstrate the complex characteristics of the competitive sorption behavior of natural colloid fractions of different sizes and compositions for As, Se, Cu, and Pb contaminants. It also emphasizes the potential shifting and even reversal of sorption trends predicted by different experimental approaches. Even though both single- and multi-contaminant isotherms suggested preferential sorption of cationic over anionic contaminants, sorption affinity estimates for each contaminant were greater by multi-contaminant isotherms and sorption sequences varied by composition and size. The varying and sometimes conflicting trends between mineralogical compositions and particle size were certainly affected by the presence of accessory minerals in the mixtures, surface OC and Fe/Al coatings, partial nano-aggregation phenomena, as well as the variable ionic composition of the solution environment. Zeta potential and pH changes after addition of different contaminant loads suggested mainly inner sphere bonding of cation contaminants to the nanocolloid surfaces and outer sphere bonding between oxy-anion contaminants and macro-colloid and bio-colloid surfaces.

These data also highlight the challenges posed by colloid heterogeneity (composition, size, surface characteristics) to predictions of contaminant interactions in natural environments and the importance of their comprehensive characterizations by water quality professionals and environmental consultants undertaking remediation tasks and by developers of engineered nanoparticles trying to model their environmental behavior.

References

[1]   McNaught, A.D. and Wilkinson, A. (1997) IUPAC Compendium of Chemical Terminology. 2nd Edition, Blackwell Science Publications, Oxford.

[2]   Christian, P., Von der Kammer, F., Baalousha, M. and Hofmann, T. (2008) Nanoparticles: Structure, Properties, Prep-

aration and Behaviour in Environmental Media. *Ecotoxicology*, **17**, 326-343. http://dx.doi.org/10.1007/s10646-008-0213-1

[3]   Maurice, P.A. and Hochella Jr., M.F. (2008) Nanoscale Particles and Processes: A New Dimension in Soil Science. *Advances in Agronomy*, **100**, 123-153. http://dx.doi.org/10.1016/S0065-2113(08)00605-6

[4]   Theng, B.K.G. and Yuan, G.D. (2008) Nanoparticles in the Soil Environment. *Elements*, **4**, 395-399. http://dx.doi.org/10.2113/gselements.4.6.395

[5]   Karathanasis, A.D. (2010) Composition and Transport Behavior of Soil Nanocolloids in Natural Porous Media. In: Frimmel, F.H. and NieBner, R., Eds., *Nanoparticles in the Water Cycle*, Chapter 4, Springer-Verlag, Berlin Heidelberg.

[6]   Tsao, T.M., Chen, Y.M. and Wang, M.K. (2011) Origin, Separation, and Identification of Environmental Nanoparticles: A Review. *Journal of Environmental Monitoring*, **13**, 1156-1163. http://dx.doi.org/10.1039/c1em10013k

[7]   Haering, K.C. and Evanylo, G.K. (2006) Mid-Atlantic Nutrient Management Handbook. CSREES Mid-Atlantic Regional Water Quality Program. MAWQP #06-02. http://www.mawaterquality.org/capacity_building/ma_nutrient_mgmt_handbook.html

[8]   Lowry, G.V., Majetich, S., Matyjaszewski, K., Sholl, D. and Tilton, R. (2006) Transport, Targeting, and Applications of Metallic Functional Nanoparticles for Degradation of DNAPL Chlorinated Organic Solvents. Technical Report, Carnegie Mellon University, Pittsburgh. http://dx.doi.org/10.2172/902659

[9]   Unrine, J.M., Bertsch, P.M. and Hunyadi, S. (2008) Bioavailability, Trophic Transfer and Toxicity of Manufactured Metal and Metal Oxide Nanoparticles in Terrestrial Environments. In: Grassian, V.H., Ed., *Nanoscience and Nanotechnology Environmental and Health Impacts*, Chapter 14, John Wiley and Sons, Hoboken, 345-360. http://dx.doi.org/10.1002/9780470396612.ch14

[10] De Momi, A. and Lead, J.R. (2008) Behaviour of Environmental Aquatic Nanocolloids When Separated by Split-Flow Thin-Cell Fractionation (SPLITT). *Science of the Total Environment*, **405**, 317-323. http://dx.doi.org/10.1016/j.scitotenv.2008.05.032

[11] Bolea, E., Laborda, F. and Castillo, J.R. (2010) Metal Associations to Microparticles, Nanocolloids and Macromolecules in Compost Leachates: Size Characterization by Assymetrical Flow Field-Flow Fractionation Coupled to ICP-MS. *Analytica Chimica Acta*, **661**, 206-214. http://dx.doi.org/10.1016/j.aca.2009.12.021

[12] Echeverría, J.C., Morera, M.T., Mazkiarán, C. and Garrido, J.J. (1998) Competitive Sorption of Heavy Metals by Soils. Isotherms and Fractional Factorial Experiments. *Environmental Pollution*, **101**, 275-284. http://dx.doi.org/10.1016/S0269-7491(98)00038-4

[13] Cruz-Guzmán, M., Celis, R., Hermosin, M.C., Leone, P., Nègre, M. and Cornejo, J. (2003) Sorption-Desorption of Lead (II) and Mercury (II) by Model Associations of Soil Colloids. *Soil Science Society of America Journal*, **67**, 1378-1387. http://dx.doi.org/10.2136/sssaj2003.1378

[14] Lair, G.J., Gerzabek, M.H., Haberhauer, G., Jakusch, M. and Kirchmann, H. (2006) Response of the Sorption Behavior of Cu, Cd, and Zn to Different Soil Management. *Journal of Plant Nutrition and Soil Science*, **169**, 60-68. http://dx.doi.org/10.1002/jpln.200521752

[15] Signes-Pastor, A., Burló, F., Mitra, K. and Carbonell-Barrachina, A.A. (2007) Arsenic Biogeochemistry as Affected by Phosphorus Fertilizer Addition, Redox Potential and pH in a West Bengal (India) Soil. *Geoderma*, **137**, 504-510. http://dx.doi.org/10.1016/j.geoderma.2006.10.012

[16] Su, C. and Suarez, D.L. (2000) Selenate and Selenite Sorption on Iron Oxides: An Infrared and Electrophoretic Study. *Soil Science Society of America Journal*, **64**, 101-111. http://dx.doi.org/10.2136/sssaj2000.641101x

[17] Covelo, E.F., Vega, F.A. and Andrade, M.L. (2007) Competitive Sorption and Desorption of Heavy Metals by Individual Soil Components. *Journal of Hazardous Materials*, **140**, 308-315. http://dx.doi.org/10.1016/j.jhazmat.2006.09.018

[18] Gao, X. (2008) Speciation and Geochemical Cycling of Lead, Arsenic, Chromium, and Cadmium in a Metal-Contaminated Histosol. Doctoral Dissertation, ProQuest Dissertations and Thesis, Accession Order No. 3294674.

[19] Violante, A. (2013) Chapter Three: Elucidating Mechanisms of Competitive Sorption at the Mineral/Water Interface. In: Donald, L.S., Ed., *Advances in Agronomy*, Academic Press, Waltham, 111-176.

[20] Redman, A.D., Macalady, D.L. and Ahmann, D. (2002) Natural Organic Matter Affects Arsenic Speciation and Sorption onto Hematite. *Environmental Science & Technology*, **36**, 2889-2896. http://dx.doi.org/10.1021/es0112801

[21] Saada, A.D., Breeze, D., Crouzet, C., Cornu, S. and Baranger, P. (2003) Adsorption of Arsenic (V) on Kaolinite and on Kaolinite-Humic Acid Complexes: Role of Humic Acid Nitrogen Groups. *Chemosphere*, **51**, 757-763. http://dx.doi.org/10.1016/S0045-6535(03)00219-4

[22] Heidmann, I., Christl, I., Leu, C. and Kretzschmar, R. (2005) Sorption of Cu and Pb to Kaolinite-Fulvic Acid Colloids:

Assessment of Sorbent Interactions. *Geochimica et Cosmochimica Acta*, **69**, 1675-1686. http://dx.doi.org/10.1016/j.gca.2004.10.002

[23] Karathanasis, A.D., Johnson, D.M.C. and Matocha, C.J. (2005) Biosolid Colloid-Mediated Transport of Copper, Zinc, and Lead in Waste-Amended Soils. *Journal of Environmental Quality*, **34**, 1153-1164. http://dx.doi.org/10.2134/jeq2004.0403

[24] Essington, M.E. (2004) Soil and Water Chemistry: An Integrative Approach. CRC Press LLC., Boca Raton.

[25] Smith, E., Naidu, R. and Alston, A.M. (2002) Chemistry of Inorganic Arsenic in Soils: II. Effect of Phosphorus, Sodium, and Calcium on Arsenic Sorption. *Journal of Environmental Quality*, **31**, 557-563. http://dx.doi.org/10.2134/jeq2002.0557

[26] Bowell, R.J. (1994) Sorption of Arsenic by Iron Oxides and Oxyhydroxides in Soils. *Applied Geochemistry*, **9**, 279-286. http://dx.doi.org/10.1016/0883-2927(94)90038-8

[27] Balasoiu, C.F., Zagury, G.J. and Deschênes, L. (2001) Partitioning and Speciation of Chromium, Copper, and Arsenic in CCA-Contaminated Soils: Influence of Soil Composition. *Science of the Total Environment*, **280**, 239-255. http://dx.doi.org/10.1016/S0048-9697(01)00833-6

[28] Lui, F., De Cristofaro, A. and Violante, A. (2001) Effect of pH, Phosphate and Oxalate on the Adsorption, Desorption of Arsenate on/from Goethite. *Soil Science*, **166**, 197-208. http://dx.doi.org/10.1097/00010694-200103000-00005

[29] Huang, P.M. (1975) Retention of Arsenic by Hydroxy-Aluminum on Surfaces of Micaceous Mineral Colloids. *Soil Science Society of America Journal*, **39**, 271-274. http://dx.doi.org/10.2136/sssaj1975.03615995003900020016x

[30] Bar-Yosef, B. and Meek, D. (1987) Selenium Sorption by Kaolinite and Montmorillonite. *Soil Science*, **144**, 11-19. http://dx.doi.org/10.1097/00010694-198707000-00003

[31] Goldberg, S. (2013) Modeling Selenite Adsorption Envelopes on Oxides, Clay Minerlas, and Soils Using the Triple Layer Model. *Soil Science Society of America Journal*, **77**, 64-71. http://dx.doi.org/10.2136/sssaj2012.0205

[32] Peak, D. and Sparks, D.L. (2002) Mechanisms of Selenate Adsorption on Iron Oxides and Hydroxides. *Environmental Science & Technology*, **36**, 1460-1466. http://dx.doi.org/10.1021/es0156643

[33] Waychunas, G.A., Kim, C.S. and Banfield, J.A. (2005) Nanoparticulate Iron Oxide Minerals in Soils and Sediments: Unique Properties and Contaminant Scavenging Mechanisms. *Journal of Nanoparticle Research*, **7**, 409-433. http://dx.doi.org/10.1007/s11051-005-6931-x

[34] Goh, K. and Lim, T. (2004) Geochemistry of Inorganic Arsenic and Selenium in a Tropical Soil: Effect of Reaction Time, pH, and Competitive Anions on Arsenic and Selenium Adsorption. *Chemosphere*, **55**, 849-859. http://dx.doi.org/10.1016/j.chemosphere.2003.11.041

[35] Jackson, B.P. and Miller, W.P. (1999) Soluble Arsenic and Selenium Species in Fly Ash/Organic Waste-Amended Soils Using Ion Chromatography-Inductively Coupled Plasma Mass Spectrometry. *Environmental Science & Technology*, **33**, 270-275. http://dx.doi.org/10.1021/es980409c

[36] Donner, E., Ryan, C.G., Howard, D.L., Zarcinas, B., Scheckel, K.G., McGrath, S.P., *et al.* (2012) A Multi-Technique Investigation of Copper and Zinc Distribution, Speciation and Potential Bioavailability in Biosolids. *Environmental Pollution*, **166**, 57-64. http://dx.doi.org/10.1016/j.envpol.2012.02.012

[37] Strawn, D.G., Palmer, N.E., Furnare, L.J., Goodell, C. and Amonette, J.E. (2004) Copper Sorption Mechanisms on Smectites. *Clays and Clay Minerals*, **52**, 321-333. http://dx.doi.org/10.1346/CCMN.2004.0520307

[38] Sipos, P., Németh, T., Kis, V.K. and Mohai, I. (2008) Sorption of Copper, Zinc and Lead on Soil Mineral Phases. *Chemosphere*, **73**, 461-469. http://dx.doi.org/10.1016/j.chemosphere.2008.06.046

[39] Morton, J.D., Semrau, J.D. and Hayes, K.F. (2001) An X-Ray Absorption Spectroscopy Study of the Structure and Reversibility of Copper Adsorbed to Montmorillonite Clay. *Geochimica et Cosmochimica Acta*, **65**, 2709-2722. http://dx.doi.org/10.1016/S0016-7037(01)00633-0

[40] Seo, D.C., Yu, K. and DeLaune, R.D. (2008) Comparison of Monometal and Multimetal Adsorption in Mississippi River Alluvial Wetland Sediment: Batch and Column Experiments. *Chemosphere*, **73**, 1757-1764. http://dx.doi.org/10.1016/j.chemosphere.2008.09.003

[41] Selim, H.M. (2012) Competitive Sorption and Transport of Heavy Metals in Soils and Geological Media. CRC Press, Boca Raton, 426. http://dx.doi.org/10.1201/b13041

[42] Goldberg, S. and Glaubig, R.A. (1987) Effect of Saturating Cation, pH, and Aluminum and Iron Oxides on the Flocculation of Kaolinite and Montmorillonite. *Clays and Clay Minerals*, **35**, 220-227. http://dx.doi.org/10.1346/CCMN.1987.0350308

[43] Shen, Y.H. (1999) Sorption of Humic Acid to Soil: The Role of Mineralogical Composition. *Chemosphere*, **38**, 2489-2499. http://dx.doi.org/10.1016/S0045-6535(98)00455-X

2

Reducing Land Degradation on the Highlands of Kilimanjaro Region: A Biogeographical Perspective

Christine Noe

Department of Geography, University of Dar es Salaam, Dar es Salaam, Tanzania
Email: cnpallangyo@gmail.com, tinanoe@yahoo.com

Abstract

In 2012, governments across the world adopted "The Future We Want" outcome document in Rio De Janeiro as a commitment to achieve a land-degradation-neutral world. This document reasserts the importance of sustainable land management in the top of the debates on sustainable development. This paper provides an overview of Tanzania's preparedness towards achieving these global objectives. The paper is based on a keynote address which was presented in the conference on reducing land degradation on the highlands of Kilimanjaro Region in Tanzania. Using a biogeographical perspective, the paper assesses challenges of adopting programmatic approach to sustainable land management in Tanzania. It also presents some opportunities that exist through Global Mechanism of the United Nations Convention to Combat Desertification, which promote actions leading to coordination, mobilization and channeling of financial resources to assist member countries to coordinate and sustain sustainable land management projects.

Keywords

Sustainable Land Management, Bioregional Planning, Kilimanjaro, Tanzania

1. Introduction

Recent analyses of global trend of land degradation reassert that there is a close relationship between land degradation and human welfare (poverty or development) [1]. A study by the Food and Agriculture Organization (FAO) (2011) demonstrates that 25% of global land is highly degraded and 40% of this degradation occurs in areas with high poverty rates. It is argued that poor people have the least access to land and water, they depend on small farms (most of which are in poor quality soils) and they have least access to modern farm technologies, which make them most vulnerable to land degradation and climatic uncertainties [1] [2]. Notably, Pingali *et al.* [1] demonstrate that across Sub-Saharan Africa, poverty coincides with marginal environments. This put Sustainable Land Management (SLM) at the top of the global sustainable development agenda. Indeed, in June 2012, governments adopted "The Future We Want" outcome document in Rio De Janeiro, which recognized that poverty eradication was an indispensable requirement for sustainable development (in paragraph 2) and the need for urgent action to reverse land degradation (in paragraph 206). Throughout this document, countries agreed to strive to achieve a land-degradation-neutral world in the context of sustainable development [3].

This paper was presented in April 2014 as a keynote address in a conference on reducing land degradation in the slopes of Mount Kilimanjaro, Tanzania. Being a human geographer, my approach to this task was guided by the bioregional planning model, which paid particular attention to human-nature interactions. Notably, the address focused on the relationship between land degradation and how nature-human landscapes were managed in Tanzania. Since the Rio Earth Summit of 1992, local responses to land degradation and related problems have varied from area-specific and nationwide to short-term and long-term projects and programs. Tanzania's involvement in these international development debates and the signing of various agreements attests to the country commitment to reducing land degradation and related socio-economic and environmental impacts. Among others, Tanzania agreed to the terms of the United Nations Convention to Combat Desertification (UNCCD), the Convention on Biological Diversity (CBD) and related protocols, the United Nations Framework Convention on Climate Change (UNFCCC), the Kyoto Protocol and the Convention on International Trade in Endangered Species of Wild Flora and Fauna (CITES), to mention a few.

These international conventions informed nationwide actions leading to policies and strategies to address environmental problems generally and land degradation in particular. The first National Environment Policy (NEP, 1997) and National Action Programme (NAP, 2004) formed the basis for the implementation of actions relating to sustainable resource management and combating degradation and desertification. There are several other policies and strategies that form a national framework for facilitating the achievement of these objectives including the Poverty Reduction Strategy Paper (PRSP) (2000), Tanzania Development Vision 2025 (2001), Rural Development Strategy (2001), the Agricultural Sector Development Strategy (ASDS) (2001), National Environmental Management Act (2004) as well as the National Strategy for Economic Growth and Reduction of Poverty (2005). Despite these national efforts, Tanzania still suffers from widespread land degradation [4]. Agricultural land in particular is continuously under pressure from factors associated with inappropriate land use practices such as poor crop cultivation practices, overgrazing, deforestation and bush fires, which have weakened land productive capacity. Specifically in the case of Kilimanjaro, I discuss the opportunities that exist for adopting programmatic approaches to SLM in Tanzania.

2. The Context of Kilimanjaro: The Bioregional Model and Sustainable Land Management

The Kilimanjaro Mountain is part of a large ecosystem which forms an important human-nature landscape that straddles local and international political borders. Precisely, the nature landscape encompasses the Kilimanjaro national park and forest reserve (in Tanzania) and Amboseli and Tsavo West national parks (in Kenya). There are wildlife corridors that establish ecological connectivity within and beyond the landscape including with the neighboring protected areas such as Arusha and Mkomazi national parks and Meru forest reserve. The Kilimanjaro ecosystem hosts about 2500 plant and 179 birds' species, some of which are endemic to the area. These characteristics, coupled with Kilimanjaro being the highest mountain in Africa, afforded the ecosystem the status of a Biosphere Reserve and World Heritage Site since 1989. In terms of human landscape, the Kilimanjaro administrative region is about 1.4% of the total land mass of Tanzania. According to the 2012 census, the region is a homeland of 4% of the total country population [5]. Combined, these statuses make Kilimanjaro an important bioregion, a description that I discuss in this section.

In the recent debates, the use of economic aspects of nature as a means for achieving both conservation and development objectives prompted the adoption of a bioregional planning model. The model is based on the UNESCO's Man and Biosphere Reserves concept of geographical zoning, which comprises clearly delineated and legally protected *core areas, buffer zones and cooperation areas* [6] [7]. While core areas are devoted to the protection of the environment and its biological diversity, each of them should be surrounded by a well defined *buffer zone* where only activities compatible with the conservation objectives may take place. To the extent possible, buffer zones provide cushion to core areas, provide connectivity in the landscape via ecological corridors and meet their function as stand-alone polygons.

Cooperation zones (also known as agro ecological zones/productive human landscapes) are important areas where new approaches to sustainable resource management initiatives and practices are encouraged, with the cooperation of the human population [6]. This is where multiple land uses such as forestry, agriculture, settlements and other human related activities are found. The zone is therefore a core of SLM practices and application of sustainable development principals and strategies.

As demonstrated in **Figure 1** (left), bioregional borders are metaphorical, which translate that bioregions are natural and therefore permeable. For this reason, different natural and human-induced forces have often redefined borders of bioregions. Often, these forces come from the cooperation/productive zone where actions are required to translate the use of resources into real development through control of soil erosion and enhancement of land productivity, water and biodiversity conservation and sustainable options for household food security and income. However, the proponents of bioregional model recognize the cooperation zone as an important tool for the harmonization of the often conflicting objectives of intensified economic and social development while maintaining and enhancing the ecological and global life support functions of land resources. It is against this background that, in my view, thinking in terms of the three components of a bioregional planning model would be a useful starting point for the evaluation of SLM projects that seek to improve land productivity and its subsequent impacts on development.

Different studies that have assessed the rate of land degradation in the Kilimanjaro bioregion support that the level of threats has progressively increased in the past two decades. Some of the major observable threats have been disconnection of core protected areas following the disappearance of ecological corridors. For example, with the exception of the Kilimanjaro-Amboseli migration route, all other ecological corridors that connected Kilimanjaro national park with the surrounding ecosystems in Tsavo West, Mkomazi and Arusha national parks have all been blocked by human activities [8]. Reports of different research projects support the view that the Kilimanjaro bioregion has increasingly been under threat from different forces. In particular, a collaborative research by the UNDP/UNEP, Kenya Wildlife Service (KWS), Wildlife Conservation Society of Tanzania (WCST) and the University of Bayreuth investigated threats to the forests of Kilimanjaro in 2002. The report from this research is the source of **Figure 1** (right), which suggest that the whole area of the forest belt of Mount Kilimanjaro is disturbed by human activities through illegal logging of indigenous trees, fire occurrences on the

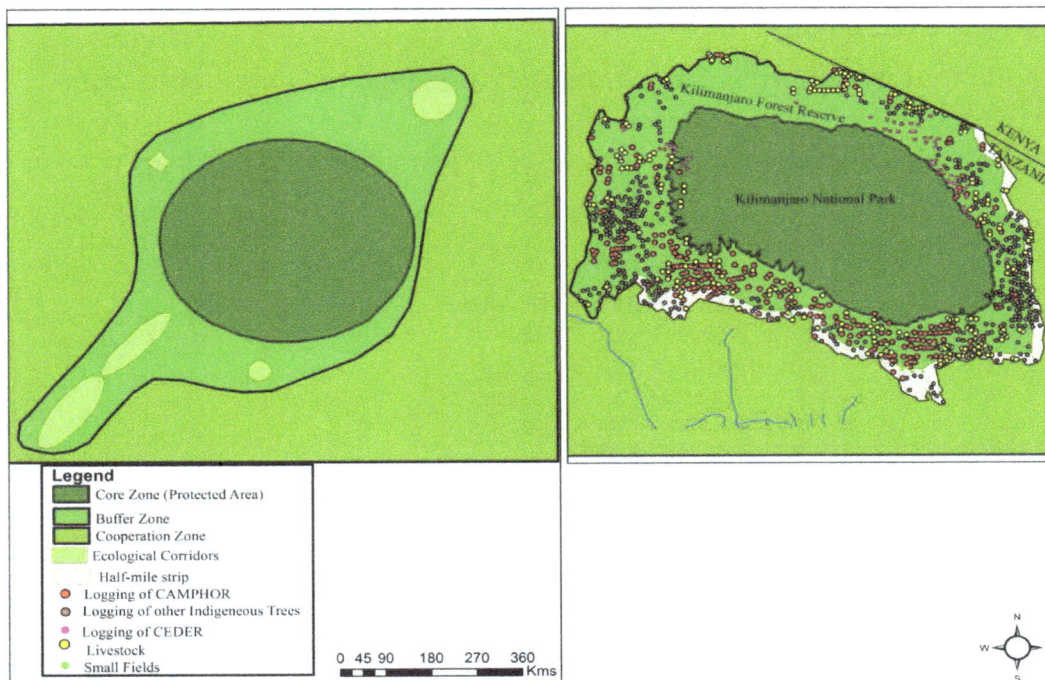

Figure 1. The Kilimanjaro bioregion. Source of the model: Adopted from [6] page 149.

south eastern slopes and the establishment of forest villages in the western and northern slopes [9].

Another project called Land Use Change, Impacts and Dynamics (LUCID) indicated that there has been a decreasing trend for pasture and fodder, native vegetation, accessibility to key resources in grazing areas, and wildlife movement corridors, diversity and distribution ranges of all ecosystems [10]. In the same time, the project recorded increasing trends for soil erosion, pastoral-farmer and human-wildlife conflicts, all of which were associated with water and pasture scarcity as well as wildlife habitat fragmentations [8] [11].

Other recent studies such as those by Soini [12] and Hemp [13] report declining agro-forestry system in Kilimanjaro following changes in Chagga home gardens which have long been a source of livelihoods, biodiversity habitat and refuge area for plants and animals. As in the previous, these studies associate the declining trend with land fragmentation particularly following changes in local land management and ownership (the *kihamba* system) and the management of commercial coffee plantations. The later has meant that, there is no need to retain a tree canopy layer anymore although demands for commercial and domestic wood continues to increase hence putting more pressure on both reserved and sacred forests. Indeed, Ylhäisi [14] demonstrates how indigenous forests in North Pare lost its cover by 95% during 1982-1997 while the cultivated land area increased by 68% [14]. Overall, changes in land cover and use have increasingly affected microclimate, land management and biodiversity conservation in the region [12] [13].

Although levels of threats to biodiversity have been growing, community participation in conservation that has been spearheaded by various actors has worked to protect nature from further encroachments. For example, the UNDP and the United Nations Foundation facilitated the Community Management of Protected Areas Conservation Project (COMPACT) around Mount Kilimanjaro since early 2000s. The project was significantly successful in bringing human and nature landscapes into the management practices of protected areas through local-based initiatives which have also proved to sustain local livelihoods [9] [15]. This project suggests therefore that although SLM projects in Kilimanjaro are implemented by different actors, they respond variously to the calls for sustaining nature and human landscape in this changing socio-economic and biophysical environment.

The government of Tanzania and its different development partners facilitate the implementation of the bioregional model through formulation and implementation of policies and project objectives that seek to achieve sustainable development by reconciling economic growth and conservation of resources. My assessment of these efforts reveals that since 2005 over 100 projects of bioregional nature have been implemented countrywide. Of these, five are on-going in Kilimanjaro Region (**Table 1**). This list is limited to those projects that are strictly operating on the ground hence excluding national wide programs and old projects. A more detailed list of these projects is provided by Mashauri [16].

3. Challenges of Adopting Programmatic Approach to Sustainable Land Management

The SLM project in Kilimanjaro commissioned various consultants who provided detailed technical assessments

Table 1. SLM/bioregional projects in Kilimanjaro Region.

S/N	Name	Project objectives	SLM component	Main actors
1	Sustainable land management	Reducing land degradation on the highlands of Kilimanjaro	All resources (agricultural land, forest, water, wildlife)	Vice President's Office/Kilimanjaro Regional Administration and UNDP/GEF
2	Maasai steppe heartland	Scaling up conservation and livelihood efforts in northern Tanzania (SCALE-TZ)	Biodiversity	African Wildlife Foundation/USAID
3	Wildlife management areas and environmental management policy program (SCAPES)	Implement environmental and natural resources policy programs through the application of wildlife management areas (WMAs) regulations and environmental management act	Wildlife and livelihoods	WWF/USAID
4	Women and organic agriculture	Improved livelihoods for women through organic agriculture	Agricultural development	Floresta/USAID
5	Participatory forest management (PFM)	Improved sustainable forests and woodland resources for sustainable livelihoods	Forest and livelihoods	DANIDA/Ministry of Natural Resources and Tourism

on specific thematic areas including on the issues of fuel efficient technologies for domestic, institutions and industrial use (by Camco Cleanenergy Ltd.); extension service delivery in SLM activities (by SUA); decision-making tools to facilitate village land-use planning (SUA-Bureau of Agricultural Consultancy and Advisory Service (BACAS); gender mainstreaming (by Moshi University of Cooperatives and Business Studies) and issues of policy review, harmonization, and traditional institutions (by Alpha and Omega Ltd.). These assessments contribute to this special issue. I should commend the project management for according due importance to these issues given their relevance in ensuring success of the project.

Meanwhile, the Global Mechanism of the United Nations Convention to Combat Desertification (UNCCD) commissioned different studies with the aim of assisting Tanzania to formulate an Integrated Investment Framework for SLM (IIF). These studies were conducted on different aspects including the economic valuation of land [17]; promotion, mobilization and innovation of financing mechanisms [18]; financial instruments associated with incentives and market-based mechanisms (IMBMs) [16] and on Aid for Trade (AfT) [19]. Among other things, these studies identify major constrains that hinder SLM progress in Tanzania generally and proposes a model for promoting SLM [4]. Important for the current section will be the challenges that these different studies have identified and ways that they reflect local circumstances in Kilimanjaro.

The various diagnostic studies indicate that although the country has implemented many programmes and projects to counteract degradation since colonial times, Tanzania still suffers from widespread land degradation. These studies suggest that the growing trend of degradation owes generally to systemic problems associated with the promotion of SLM in the country [4]. These have ranged from financial, organizational, technical and more so the innovations on how to upscale and mainstream SLM. Other causes of the problem are associated with lack of knowledge and reflection in national accounts or development policies and lack of understanding of the economic significance of land resources and the ecosystem services they produce. These constrains are summarized.

3.1. Lack of Finance to Implement SLM Related Programmes and Projects

Although different development partners have participated through facilitation of various SLM interventions, there are still many constrains to accessing resources. These constrains are caused by policy, fiscal, legal, institutional and human resource problems [16]. Consider, for example, that Tanzania is eligible to access funds from 15 sources of climate change finance for the implementation of SLM projects through the UNFCCC and other global climate change initiatives but it has only been able to access 0.14 percent of the available climate change financing [18]. The main reasons given relate to low capacity, particularly on the technical and scientific requirements for funding application. The analysis of district level institutional capacities and capabilities indicate, for example, that many local government agencies, NGOs and community organizations have resources limitations in terms of staff complements, skills and funds. Low technical capacity results in decision making without sufficient knowledge of the proposed interventions. The capacity deficiency has in particular led to knowledge and information gaps: the links between research and extension service remain weak and research findings largely un-communicated [16].

3.2. Lack of Prioritization

SLM is not tabled as a critical issue in government planning and thus tends to be subsumed in key Ministry of Agriculture without being singled out as a critical area of intervention in and of itself [4]. As a result, SLM activities receive only a fraction of the national budget. Even then, the budget is channeled through different sectors which are not coordinated leading to duplications of actions, lack of effective scaling-up and continued dependence on donor interventions.

3.3. Poor Sectoral Linkages

Different assessments of SLM projects suggest that there is little coordination between actors at all levels. Indeed, SLM projects are funded from different sources depending and their focus depend mainly on priority areas of donors, which does not always respond to specific local needs. Hence, SLM projects in the country have operated almost independent of each other making the assessment of impacts and scaling up of these projects difficult so as the promotion of investments in them. Taking an example of SLM projects in Kilimanjaro (as presented in **Table 1**), there is little to say about their linkages and sharing of experience among different actors.

Overall, poor coordination has proved difficult to take advantage of complementarities and opportunities for joint implementation. This sectoral approach has not only limited opportunities for sharing experiences among projects and programmes of different sponsors but has also constrained efforts to identify sustainable sources of funding.

3.4. Top-Down Approaches

Historically, SLM and environmental issues have been characterized by inappropriate policies designed from above with little regard for the unique features of livelihood systems in fragile ecosystems. This has led to two related problems; lack of knowledge about traditional land management practices and their integration in policy processes and contradictions between official and customary land management practices. These contradictions have threatened the political power base causing decision makers to refrain from implementing and enforcing policies.

3.5. Multiplicity of Institutional and Policy Frameworks

The multiplicity of policy and legal frameworks in the SLM sectors result in duplication of resources resulting in lack of a coherent, coordinated and integrated approach.

3.6. Lack of Financial Incentives for Adopting SLM

Alternative livelihood options require knowledge and financial incentives. Currently, over 80% of Tanzania's population depends on agriculture for their livelihood, and a similar percentage depends on local wood resources for energy and construction. Although several SLM options exist, their uptake is overly low because there is lack of awareness among beneficiaries about the benefits that can be accrued from sustainable natural resources utilization and management [16].

3.7. The Absence of a Comprehensive Monitoring System

Despite the formulation of national development strategies and plans, programme delivery by government in Tanzania is still largely sectoral. Some districts have been conducting isolated monitoring of environmental variables but there has not been coordination system at the regional/national level to assess degradation dynamics, leading to underestimation of the extent of problems and the actual costs of remedy.

4. Opportunities for Adopting Programmatic Approach to Sustainable Land Management

Combating land degradation and desertification in Tanzania, as it is in most sub-Sahara African countries, will require coordination and sustainable funding. Opportunities for improving coordination and funding that are necessary for adopting a programmatic approach to SLM are seen in the following ways.

1) The Global Mechanism (GM) of the UNCCD has taken the mandate to promote actions leading to coordination, mobilization and channeling of financial resources to assist developing member countries to coordinate and sustain SLM projects. The GM is seeking to formulate an Integrated Investment Framework (IIF) and Integrated Financing Strategy (IFS) for SLM in Tanzania. The IIF is a comprehensive and realistic roadmap of prioritized investments needed for the attainment of SLM in Tanzania. Notably, the IIF will enhance resource mobilization, build capacity and provide effective co-ordination of SLM programmes/projects on the basis of clearly identified priorities. As such IIF is based on five national level assessments, namely; economic valuation of land, promotion of mobilization of innovative financing mechanisms, including from climate change financial mechanisms, financial instruments associated with incentives and market-based mechanisms (IMBMs) and Aid for Trade (AfT) to finance SLM promotion [4].

2) Whereas IIF will provide a systematic framework to mobilize resources, integrated financing strategy (IFS) will focus on ensuring predictable and sustainable financing for SLM in Tanzania. The two will facilitate the innovative blending and application of various financing instruments from different sources, and channeling these to investments in SLM [4]. Some work has already been done to identify available financial resources at the country level that could be allocated to finance investments in SLM. However, the achievement of these ob-

jectives will require institutional and policy reforms. It is proposed that the National SLM Agency be established. The National SLM Agency will play the catalytic role in aligning SLM activities to the proposed IIF/IFS. It is envisaged that the proposed Agency will be empowered by law. At the national level, some work has been done to take stock of relevant projects and programmes (planned and on-going) with a view to proposing an entry point for the Agency and specific SLM budget allocations [16].

3) Up-scaling of current SLM adaptation activities and creating synergies across the agro-ecological zones in relevant SLM sectors (Wildlife, Agriculture, Water and Forestry). Proposals for up-scaling SLM projects have already been prepared as summarized in **Table 2**.

4) While local financial contributions to SLM have remained overly low, it is confirmed that Tanzania has not utilized the available external sources. For example, there is almost no any support drawn from the available Special Climate Change Fund (SCCF), Strategic Priority for Adaptation (SPA), Least Developed Countries Fund (LDCF), etc. There are also wide range of funding opportunities that have been identified including innovative sources such as the trade and market access and carbon trading that offers prospects for investments in SLM [18]. With the proposed IIF and IFS, Tanzania stands a chance to mobilize resources and coordinate SLM activities in ways that will address current institutional and financial problems.

5. Conclusions

This paper provides a situational analysis of environmental resources of the Kilimanjaro ecosystem, which suggests that land degradation has increasingly manifested in different ways: high trends for soil erosion, pastoral-farmer and human-wildlife conflicts; low trend for pasture and fodder, native vegetation, accessibility to key resources in grazing areas and wildlife corridors as well as the diversity and distribution ranges of all the surrounding ecosystems. It is also suggested that there is a declining agro-forestry system particularly following land fragmentations. These conditions sustain land degradation, which in turn, supports the growth of poor population. Coupled with challenges of climate change and the increase of different kinds of livelihood insecurities in rural areas, poverty continues to be a challenge in global efforts to attain SLM and sustainable development.

The bioregional model as used in this paper facilitates reflections on human-nature relations and how threats to land resources in Kilimanjaro are a function of such relations. Coordinated efforts towards improved

Table 2. Proposed strategies for up-scaling SLM activities across the agro-ecological zones in Tanzania.

Objectives	Proposed activities
To mainstream SLM issues in the national agenda and relevant sectors of the national budgeting framework	1) Harmonize SLM policies/entry points in national budgeting process in the budget preparation cycles of local government authorities (LGAs) and sectoral ministries. 2) Support policy dialogue on SLM-related initiatives in national and regional initiatives on CAADP and trade. 3) Develop a legislative/regulatory framework for SLM-related activities. 4) Facilitate technical assistance, EIF national implementation. 5) Harmonize SLM policies/entry points in national budgeting process in the budget preparation cycles of local government authorities (LGAs) and sectoral ministries. 6) Support policy dialogue on SLM-related initiatives in national and regional initiatives on CAADP and trade. 7) Develop a legislative/regulatory framework for SLM-related activities. 8) Facilitate technical assistance, EIF national implementation.
To promote alternative sources of energy for both industrial and domestic use and adoption of clean technologies in industry	1) Upscale existing miombo project to develop the national co-generation framework for the tobacco and mining industries (finance incentive structure). 2) Develop a strategic utilization framework for the biofuel industry, codes and standards, promotion of alternative sources of energy and co-financing of clean technology practices. 3) Assist stakeholders to access CDM funds by providing technical assistance to mobilize technical and financial resources. 4) Promote green businesses.
To enhance incomes of farmers by addressing structural market inefficiencies in the agriculture, forestry and wildlife sectors	1) Enhance extension services in rangeland management and agriculture sectors (credit facility for livestock owners, bulking centres and inputs). 2) Support national certification schemes. 3) Promote on-farm dairy farming.

Source: [4]: page 34.

livelihoods through SLM remain an important tool for achieving sustainable development in the region. It is also noted that although the trend of land degradation is still high at the national level, the approach taken by the Global Mechanism of the United Nations Convention to Combat Desertification (UNCCD) offers an opportunity for improvements in SLM through financial mobilization and coordination of activities. Yet, specific-site assessments of different thematic areas of SLM will be required to arrive at the desired goal of coordinating, financing and mainstreaming SLM activities. These assessments will be useful for correct interpretations of existing challenges and opportunities for adopting pragmatic approaches to SLM.

Acknowledgements

I would like to extend my appreciation to the Kilimanjaro Regional Office and the UNDP/GEF Kilimanjaro project office for inviting me to give a keynote address in the conference. Special appreciation goes to Dr. Fransis Mkanda, the project technical adviser, for encouraging me to provide the note and the preparation of this paper. Thanks to my anonymous reviewers for their useful comments.

References

[1]   Pingali, P., Schneider, K. and Zurek, M. (2014) Poverty, Agriculture and the Environment: The Case of Sub-Saharan Africa. Marginality. Springer, Berlin.

[2]   FAO (2011) The State of the World's Land and Water Resources for Food and Agriculture (SOLAW)—Managing Systems at Risk. Earthscan, Rome.

[3]   United Nations (2012) The Future We Want. Washington DC.

[4]   Awere-Gyekye, K. (2014) Tanzania: Integrated Investment Framework (IIF) and Integrated Financing Strategy (IFS) for Sustainable Land Management. Draft Report for Global Mechanism, Rome.

[5]   United Republic of Tanzania (URT) (2012) Tanzania in Figures, Dar es Salaam, Government Printers.

[6]   Ajathi, H. and Krumme, K. (2002) Ecosystem-Based Conservation Strategies for Protected Areas in Savanna: With Special Reference to East Africa. University of Essen, Essen.

[7]   Brunckhorst, D. (2000) Bioregionalism Planning: Resource Management beyond the New Millennium. Routledge, London.

[8]   Noe, C. (2003) The Dynamics of Land Use Changes and Their Impacts on the Wildlife Corridor between Mt. Kilimanjaro and Amboseli National Park, Tanzania, Nairobi. LUCID Project, International Livestock Research Institute.

[9]   Lambrechts, C., Woodley, B., Hemp, A., Hemp, C. and Nnyiti, P. (2002) Aerial Survey of the Threats to Mt. Kilimanjaro Forests. UNDP, Dar es Salaam.

[10] Misana, S., Majule, A. and Lyaruu, H. (2003) Linkages between Changes in Land Use, Biodiversity and Land Degradation on the Slopes of Mount Kilimanjaro, Tanzania, Nairobi. LUCID Project, International Livestock Research Institute.

[11] Maitima, J. and Olson, J. (2002) Biodiversity in Agricultural Productive Systems. Land Use Impacts and Dynamics, Working Paper 3.

[12] Soini, E. (2005) Changing Livelihoods on the Slopes of Mt. Kilimanjaro, Tanzania: Challenges and Opportunities in the Chagga Homegarden System. *Agroforestry Systems*, **64**, 157-167. http://dx.doi.org/10.1007/s10457-004-1023-y

[13] Hemp, A. (2006) The Banana Forests of Kilimanjaro: Biodiversity and Conservation of the Chagga Homegardens. *Forest Diversity and Management*, **2**, 133-155.

[14] Ylhäisi, J. (2004) Indigenous Forests Fragmentation and the Significance of Ethnic Forests for Conservation in the North Pare, the Eastern Arc Mountains, Tanzania. *Fennia*, **182**, 109-132.

[15] Murusuri, N. and Nderumaki, V. (2013) Bringing Communities into the Management of Protected Areas: Experience from COMPACT Mt. Kilimanjaro. In: Brown, J. and Hay-Edie, T., Eds., *COMPACT Engaging Local Communities in Stewardship of World Heritage*, United Nations Development Programme, New York, 68-78.

[16] Mashauri, S. (2014) Integrated Investment Framework for SLM in Tanzania: Financial Diagnosis and Stocktaking. Volume 3. Global Mechanism, United Nations Convention to Combat Desertification, Rome.

[17] Soussan, J. (2014) Integrated Investment Framework for SLM in Tanzania: The Value of Land in Tabora Region, Tanzania. The Global Mechanism, United Nations Convention to Combat Desertification, Rome.

[18] Yanda, P. (2010) Sustainable Land Management and Climate Change Finance in Tanzania: National Level Assessment

Report. Vol. 5. Global Mechanism, the United Nations Convention to Combat Desertification, Rome.

[19] Canigiani, E. and Bingi, S. (2014) Integrated Investment Framework for SLM in Tanzania. Mobilizing Aid for Trade for Agriculture: Context Analysis. Vol. 6. The Global Mechanism, United Nations Convention to Combat Desertification, Rome.

Ecophysiological Effects of Nitrogen on Soybean [*Glycine max* (L.) Merr.]

Ifeanyichukwu O. Onor[1], Gabriel I. Onor Junior[2], Murty S. Kambhampati[3]

[1]College of Pharmacy, Xavier University of Louisiana, New Orleans, USA
[2]Department of Stem Cell and Regenerative Biology, Harvard University, Cambridge, USA
[3]Department of Natural Sciences, Southern University at New Orleans, New Orleans, USA
Email: ionor@xula.edu

Abstract

Soybean [*Glycine max* (L.) Merr.] is a leguminous plant with high nutritional and medicinal value. The goal of this research was to determine the optimal concentration of nitrogen, using Hoagland nutrient solution, which will enhance the productivity of soybeans. The specific objective of the study was to assess the effect of variation of nitrogen concentration on soybean growth and leaf chlorophyll concentrations. Soybeans were grown under three soil nitrogen amendments: low, medium, and high concentration of Hoagland nutrient solution and a control group. Soybeans were grown under controlled environmental conditions in the Biotronette® environmental chamber. Temperature of the environmental chamber was regulated at 27°C and the photoperiod was set to 10 L: 14D. Soybeans grown in the low treatment group had the highest growth rate (1.03 ± 0.03 cm/day) compared to the control, medium, and high treatment groups. During the first chlorophyll analyses, the control group had the highest total chlorophyll concentration (216.25 ± 4.09 μg/mL/g). During the second chlorophyll analyses, the low treatment group had the highest total chlorophyll concentration (102.81 ± 14.54 μg/mL/g). Although no finding was statistically significant between groups, the low nitrogen treatment conditions had a trend towards producing more favorable physiological outcomes on soybeans.

Keywords

Soybeans, Nitrogen, Growth, Chlorophyll

1. Introduction

The soybean [*Glycine max* (L.) Merr.] plant is a member of the Leguminoceae family which includes other legumes such as peas, beans, lentils, peanuts, and other podded plants [1] [2]. Soybean has a history as a domesticated plant with earliest accounts of its use dating as far back as the eleventh century BC in China [1]. Today, soybean is a major crop of global importance due to its nutritional and medicinal value [3]. As of 2007, annual global production of soybeans reached 206.4 million tons compared to 27 million tons produced annually across the globe during the 1960s [1].

Soybeans, as with all legumes, are well recognized as excellent sources of dietary protein [2]. Soybean protein is considered a complete protein because it contains ample amounts of the essential amino acids that are found in animal protein [4]. Isolated protein from soybean has a protein digestibility-corrected amino acid (PDCAA) score of 1.0, which is similar to casein and egg protein [4] [5]. This high quality protein content from soybean has contributed to its dietary popularity and has led to the proliferation of soy-based food products such as soy flour, soy tofu, soy protein concentrate, soy milk, soy-based medical nutrition products, and soy-meat products such as hotdogs and sausages [6] [7].

Soybeans are a unique source of the isoflavones, genistein and diadzein [5]. These isoflavones are naturally occurring phytoestrogens similar to mammalian estrogens and selectively bind to and activate estrogen receptor beta more than estrogen receptor alpha, and thus may have similar action as selective estrogen receptor modulators (SERMs) with beneficial effects on bone and heart without detrimental effects on breast and tissues [8]. Soybean and soy foods have potentially multifaceted health promoting effects including cholesterol reduction, improved vascular health, preserved bone mineral density and reduction of menopausal symptoms [5]. A study by Ho et al. indicated that soy protein and soy isoflavones independently had a modest raise in hip bone mineral density as well as total bone mineral concentration in women who have had menopause for 4 or more years [9]. Soybean oil has also gained clinical attention and was found to show no different effect than olive oil on infectious and noninfectious complications, glycemic control, inflammatory and oxidative stress markers, and immune function in critically ill patients [10]. There is enduring scientific interest on the health benefits of soybeans in both in vitro and in vivo studies.

Nitrogen is an essential mineral macronutrient required in the greatest amounts by plant [11]. Nitrogen is a major limiting factor for plant development, growth and overall crop yield [11]-[13]. Soybeans, as well as most leguminous plants, possess the ability to acquire nitrogen for growth through nitrogen fixation and inorganic nitrogen uptake from the soil [14] [15]. Soybeans perform nitrogen fixation through its symbiotic association with Bradyrhizobioum japonicum bacteria [16] [17]. Bradyrhizobioum japonicum form nodules in soybean roots and facilitate the process of nitrogen fixation [16]. Nitrogen fixation is a biological phenomenon characterized by conversion of atmospheric molecular dinitrogen (N_2) to ammonium ion (NH_4^+) which is further assimilated by cytosolic plant enzymes [17]. In addition, soybeans can also absorb and assimilate inorganic nitrogen, in the form of nitrates, from the soil [18]. The combined process of nitrogen fixation from atmospheric nitrogen and nitrogen assimilation from the soil ensure that soybean meet its large nitrogen demand for the production of protein rich seeds [18].

Previous studies comparing sources of nitrogen for soybeans suggest that nitrogen fixation, facilitated by Bradyrhizobioum japonicum, is ideal for soybean growth and productivity [14] [19]. Two studies by de Veau et al. and Kaschuk et al. have reported that Bradyrhizobium nodulated soybeans, in the absence of soil nitrogen supplementation, have higher photosynthetic rates compared to soybeans subjected to soil nitrogen supplementation [14] [19]. Other studies have also shown that soil nitrogen supplementation, albeit at low nitrogen concentration, may be beneficial for the growth and productivity of soybeans. More specifically, a study by Sabaratnam et al. noted an improved relative growth rate and higher net photosynthetic rate in soybeans supplemented with low concentration of nitrogen solution [20] [21]. Invariably, the study observed a decline in both relative growth rate and net photosynthetic rate in soybeans supplemented with higher concentration of nitrogen solution [20] [21].

The goal of this research was to determine the optimal concentration of nitrogen, using Hoagland nutrient solution, which will enhance the productivity of soybeans. The specific objective of the study was to assess the effect of variation of nitrogen concentration on soybean growth and leaf chlorophyll concentrations. The findings from this study will contribute vital information on the ideal concentration of nitrogen required for optimum soybean productivity.

2. Materials and Methods

2.1. Experimental Design

This is an experimental study with three treatment groups (low, medium, and high) and a control group. The three treatment groups' designation was based on variation of nitrogen concentrations using Hoagland nutrient solution. Four replicates, containing two plants per replicate, were assigned to each group: control group, and low, medium, and high treatment groups to maintain the statistical validity of the data.

2.2. Soil Preparation

Soil was prepared by using the ratio of 5:2:8:1. Five pots of potting soil, two pots of garden soil, eight pots of top soil, and one pot of river sand were mixed thoroughly for homogeneity and to give the soil a loamy quality. The soil mixture was then placed into 16 plant pots (~2 lbs capacity), with four pot replicates for each of the three treatment groups and the control group.

2.3. Sowing of Soybean Seeds

Soybean seeds were washed with five percent Clorox® bleach for approximately five minutes to eliminate contaminants. Five seeds were sown in each pot and 100 mL of distilled water was added to each pot and left under room conditions (26°C). As seedlings started sprouting, the plants were thinned to two plants per pot and were ready for the experiment.

2.4. Hoagland Nutrient Solution Preparation

The Hoagland nutrient solution [22] was used to make the desired concentration of nitrogen for the three experimental treatment groups. The compounds and concentration used for the preparation are presented on **Table 1**. The final volume of each of the three treatment group solution and the control group was 500 mL.

The concentration of the compounds for the low treatment group is the standard concentration for the Hoagland Nutrient Solution [22]. The concentrations of the compounds used for the medium and high treatment group are the modified concentrations of the Hoagland Nutrient Solution. It is worth noting that the concentration of $Ca(NO_3)_2$ and KNO_3 were increased exponentially in the treatment group pots because of their contribution in modifying the nitrogen concentration of the Hoagland nutrient solution. The concentration of the non-nitrogen contributing compounds was not altered as depicted on **Table 1**.

2.5. Application of Modified Hoagland Nutrient Solution to the Soil

The nutrient solution was applied to the soil on two occasions separated by a 7-day interval. The first nutrient application occurred 7 days after seeds were sown. In the first nutrient application, 50 mL of distilled water was added to the control and 50 mL of the low Hoagland nutrient solution, 50 mL of the medium Hoagland nutrient

Table 1. Concentration of compounds used for treatment groups based on the Hoagland Nutrient Solution.

Compounds	Control (ppm)	Low (ppm) (Standard) [22]	Medium (ppm) (Modified)	High (ppm) (Modified)
$Ca(NO_3)_2$	0	1653	3310	6611
KNO_3	0	506	1011	2022
KH_2PO_4	0	326	326	326
$MgSO_4$	0	493	493	493
Trace Elements Constituents: H_3BO_3—2.8 gm/L $MnCl_2 \cdot 4H_2O$—1.8 gm/L $ZnSO_4 \cdot 7H_2O$—0.2 gm/L $CuSO_4 \cdot 5H_2O$—0.1 gm/L $NaMoO_4$—0.025 gm/L	0	4.93	4.93	4.93
FeEDTA Constituents: $EDTA \cdot 2Na$—10.4 gm/L $FeSO_4 \cdot 7H_2O$—7.8 gm/L KOH—56.1 gm/L	0	74.3	74.3	74.3

solution, and 50 mL of the high concentrated Hoagland nutrient solution was added to each pot in the low, medium, and high treatment groups, respectively. A week later, a second 50 mL of the nutrient doses were applied to each pot similar to the first nutrient application. On the same day of the second application, 50 mL of distilled water was added to each of the 16 pots to ensure the soil remain moist.

2.6. Environmental Regulation for Plant Growth

Following the first application of the nutrient solutions to the soil, plant pots were transferred into the Biotronette® Environmental chamber for photoperiod and temperature regulation. Temperature was regulated at about 27°C and photoperiod was regulated at 10 light (L): 14 dark (D) hours because soybeans are short day plants [23].

2.7. Measurement of Plant Growth

Three plant growth measurements were taken on days 7, 17, and 31 after sowing of the soybean seeds. Plant growth rate per day was calculated by dividing the average plant growth per treatment group by the number of days until the last plant growth measurement (31 days after seeds were sown).

2.8. Leaf Chlorophyll Analysis

Soybean leaves were collected twice for the chlorophyll analysis using the method outlined by Einhellig and Rasmussen [24]. The chlorophyll analysis was conducted in two sets (17th and 31st days of growth). After leaf collection, leaves were weighed using an analytical balance and transferred into a 50 mL test tube with 10 mL of 95% ethanol. The test tubes were labeled according to their respective experimental unit and wrapped in aluminum foil to avoid light. The test tubes were kept in the dark at room temperature. After 48 hours in ethanol solution, chlorophyll extract was decanted into a cuvette for absorbance (A) measurement at both 649 and 665 nanometer (nm) wavelengths using the Spectronic 20D+ spectrophotometer. Chlorophyll a and chlorophyll b concentrations were calculated using the following equations:

[µg Chlorophyll a/mL solution = (13.70) (Absorbance [A] at 665nm) – (5.76) (Absorbance [A] at 649nm)
[µg Chlorophyll b/mL solution = (25.80) (Absorbance [A] at 649nm) – (7.60) (Absorbance [A] at 665nm)

Due to variation in the weight of the soybean leaves used for chlorophyll analysis, chlorophyll concentration of the soybean leaves was calculated per gram using the formula:

Chlorophyll a concentration = [µg Chlorophyll a/mL ÷ weight (g)]
Chlorophyll b concentration = [µg Chlorophyll b/mL ÷ weight (g)]

2.9. Statistics

Data analysis was performed using IBM SPSS Statistics for Windows, Version 19.0. Armonk, NY: IBM Corporation. Analysis of variance (ANOVA) was used to compare the differences between the means in the three treatment groups and the control group. Data are expressed as mean values ± standard deviation (SD).

3. Results and Discussion

3.1. Results

3.1.1. Growth Measurement

Soybean growth was measured on three occasions as described in the experimental section. The data on soybean growth rate and 31-day growth are presented on **Table 2**. Soybeans grown under the low treatment group had the highest growth rate (1.03 ± 0.03 cm/day) compared to the control (0.935 ± 0.05 cm/day), medium (0.959 ± 0.07 cm/day), and high (0.928 ± 0.11 cm/day) treatment groups. The 31-day growth findings were similar to the growth rate results as shown in **Table 2**. There was no statistically significant difference between the groups when evaluating soybean growth rate and 31-day growth (P = 0.147 for both). There was, however, a trend towards better growth in the low treatment group.

3.1.2. Chlorophyll Analysis

Chlorophyll a, chlorophyll b, and total chlorophyll were assessed twice as described in the experimental section.

There was no statistically significant difference between chlorophyll a, chlorophyll b, and total chlorophyll in either the first or second set of chlorophyll analysis as depicted in **Tables 3** and **4**. There was a general decline in the leaf chlorophyll concentration in the second chlorophyll analysis when compared to the first chlorophyll analysis as depicted in **Tables 3** and **4** and **Figures 1** and **2**.

Table 2. Growth measurement result.

	Control	**Low**	**Medium**	**High**	**P-Value**
Growth Rate (cm/day)	0.935 ± 0.05	1.038 ± 0.03	0.959 ± 0.07	0.928 ± 0.11	0.147
31-Day Growth (cm)	28.975 ± 1.40	32.175 ± 0.96	29.725 ± 2.07	28.775 ± 3.31	0.147

Table 3. First chlorophyll analysis result.

	Control	**Low**	**Medium**	**High**	**P-Value**
Chlorophyll a Concentration [μg/mL/weight (g)]	82.62 ± 2.93	81.36 ± 2.38	83.71 ± 8.12	77.75 ± 12.63	0.723
Chlorophyll b Concentration [μg/mL/weight (g)]	133.63 ± 2.27	131.51 ± 7.97	116.65 ± 33.02	125.531 ± 20.25	0.633
Total Chlorophyll [μg/mL/weight (g)]	216.25 ± 4.09	212.87 ± 8.67	200.36 ± 38.23	203.28 ± 32.85	0.791

Table 4. Second chlorophyll analysis result.

	Control	**Low**	**Medium**	**High**	**P-Value**
Chlorophyll a Concentration [μg/mL/weight (g)]	34.80 ± 2.59	38.36 ± 5.53	36.09 ± 3.86	36.40 ± 8.75	0.847
Chlorophyll b Concentration [μg/mL/weight (g)]	55.28 ± 4.55	64.45 ± 9.04	61.56 ± 7.39	65.14 ± 12.43	0.411
Total Chlorophyll [μg/mL/weight (g)]	90.08 ± 7.03	102.81 ± 14.54	97.65 ± 11.23	101.54 ± 21.14	0.610

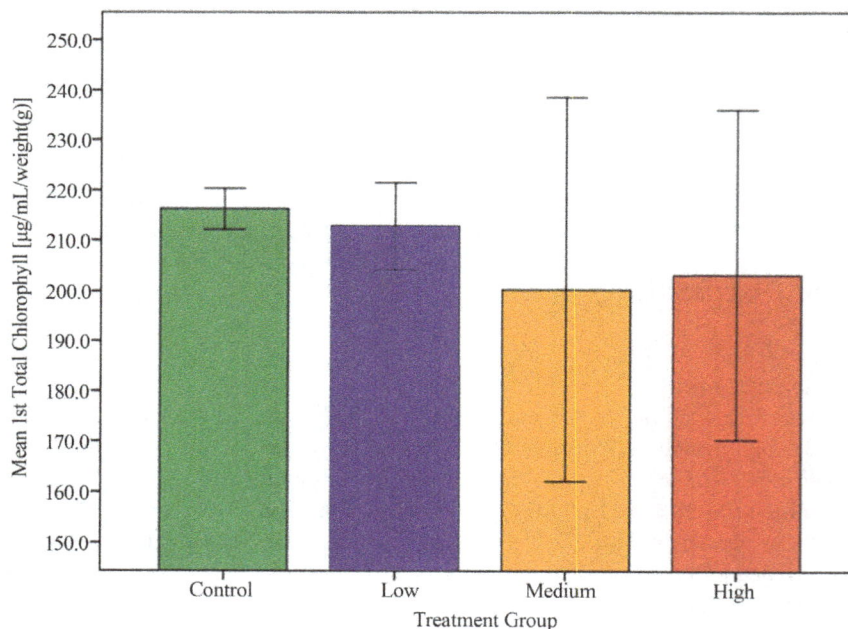

Figure 1. Mean total chlorophyll concentration [μg/mL/g] from the first chlorophyll analysis. All values are mean ± SD. Vertical bars indicate the SD within each treatment group.

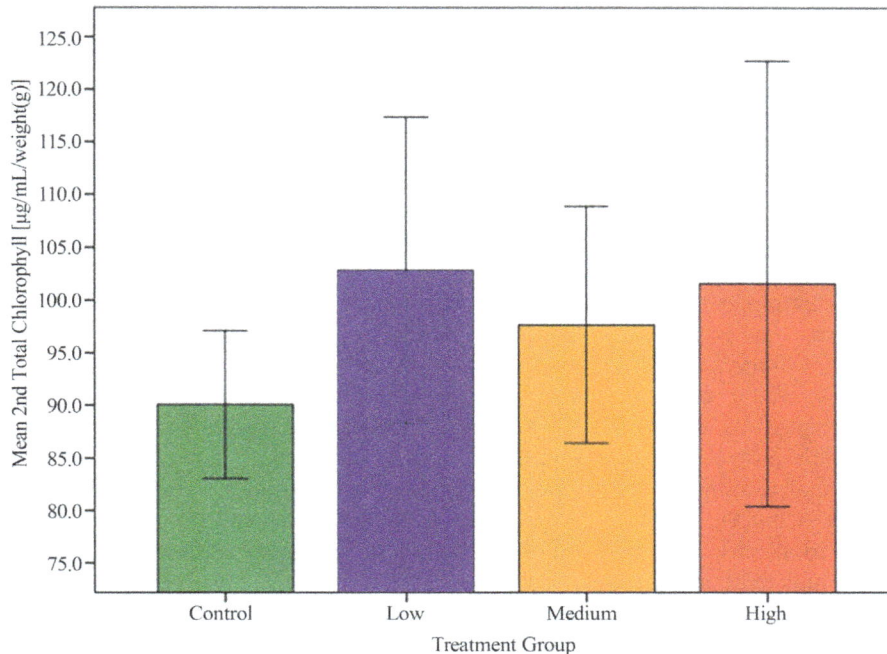

Figure 2. Mean total Chlorophyll concentration [μg/mL/g] from the second chlorophyll analysis. All values are mean ± SD. Vertical bars indicate the SD within each treatment group.

The results of chlorophyll a, chlorophyll b, and total chlorophyll from the first set of chlorophyll analysis are presented in **Table 3** with **Figure 1** showing the total chlorophyll concentration from the first chlorophyll analysis. The soybeans grown under the control group had the highest total chlorophyll concentration (216.25 ± 4.09 μg/mL/g) compared to soybeans grown under the low (212.87 ± 8.67 μg/mL/g), medium (200.36 ± 38.23 μg/mL/g) and high (203.28 ± 32.85 μg/mL/g) treatment groups. The soybeans treated in the low treatment group had the second highest total chlorophyll concentration. Although there was no statistically significant difference between the groups, there was a trend suggesting higher total chlorophyll concentration in the control and low treatment groups compared to the medium and high treatment groups. The soybean in the medium treatment group had the lowest total chlorophyll concentration in the first chlorophyll analysis.

The results of chlorophyll a, chlorophyll b, and total chlorophyll from the second set of chlorophyll analysis are presented in **Table 4** with **Figure 2** showing the total chlorophyll concentration from the second chlorophyll analysis. In general, there was a decline in the leaf chlorophyll concentration in the second chlorophyll analysis compared to the first chlorophyll analysis. In the second chlorophyll analysis, the soybeans in the low treatment group had the highest total chlorophyll concentration (102.81 ± 14.54 μg/mL/g) compared to the control (90.08 ± 7.03 μg/mL/g), medium (97.65 ± 11.23 μg/mL/g), and high (101.54 ± 21.14 μg/mL/g) treatment groups. Although there was no statistically significant difference between the groups, there was a higher total chlorophyll concentration in the low treatment groups compared to the control, medium and high treatment groups. The soybeans in the control group had the lowest total chlorophyll concentration in the second chlorophyll analysis.

3.2. Discussion

In our study, there was no statistically significant difference observed between the treatment groups when evaluating growth rate and 31-day growth. There was however, a trend to higher growth rate in the low treatment group. A study by Sabaratnam and colleagues evaluated growth characteristics of soybean subjected to different concentration of nitrogen as relative growth rate (RGR) [21]. The study revealed a statistically significant reduction in RGR by 47% in soybean treated with higher concentration of NO_2 (0.5 ppm) compared to soybean treated with lower concentrations of NO_2 (0.1, 0.2, and 0.3 ppm) and the control group [21]. The finding from this study supports our finding because the soybeans grown under the high concentration of nitrogen in our study had the lowest growth rate and 31-day growth compared to the soybeans under the control, low, and medium treatment groups. Our finding provides supporting evidence that supplementation of soybean with low concentration of nitrogen may improve the growth of soybeans.

Another study by de Veau et al. assessed leaf area of soybean grown under three different nitrogen regimens [14]. The three regimens include: a) Nod+/+ group which was inoculated with *Bradyrhizobium japonicum* and received a nutrient solution containing 6 millimolar NH_4NO_3; b) Nod+/− group which was inoculated with *Bradyrhizobium japonicum* and did not receive a nutrient solution containing nitrogen; c) Nod− group was not inoculated with *Bradyrhizobium japonicum* but received a nutrient solution containing 6 millimolar NH_4NO_3 [14]. This study found that the Nod+/− group which did not receive a nutrient solution containing nitrogen had a statistically significant smaller leaf area which is an indication that the growth of the Nod+/− group was nitrogen limited [14]. The finding from this study suggests that supplementation of soybeans with lower concentration of nitrogen may enhance the growth characteristic of soybeans.

In our study, we observed no statistically significant differences between the treatment groups in the first and second chlorophyll analyses. In the first chlorophyll analysis, the control and low treatment groups had the highest total chlorophyll concentration compared to the medium and high treatment groups although this finding was not significantly different between the groups. In the second chlorophyll analysis, the low treatment group had the highest total chlorophyll concentration compared to the control, low, and medium treatment groups. Overall, when assessing the cumulative concentration of leaf chlorophyll from the two chlorophyll analyses, there was a trend to increase in leaf chlorophyll concentration in the low treatment groups compared to the control, medium and high treatment groups.

Sabaratnam et al. in their study observed a significant reduction in chlorophyll a and total chlorophyll content in soybeans treated with higher concentration of NO_2 (0.5 ppm) compared to soybeans treated with lower concentrations of NO_2 (0.1, 0.2, and 0.3 ppm) and in the control group [20] [21]. The reduction in chlorophyll a and total chlorophyll were 45% and 47%, respectively [20] [21]. This finding is similar with our finding and suggests that with higher concentrations of nitrogen there is a decline in leaf chlorophyll concentration. Sabaratnam et al. also reported decline in net photosynthetic rate in soybean treated with 0.5 ppm NO_2 compared to soybean treated with lower concentrations of NO_2 (0.1, 0.2, and 0.3 ppm) and controls [20] [21]. This finding supports prior studies reporting photosynthetic rate decline with lower chlorophyll content [25] [26].

deVeau et al. in their study assessed the leaf chlorophyll content of the soybean grown under three different nitrogen regimens [14]. The three regimens have been described earlier and include: a) Nod+/+, b) Nod+/− and c) Nod− [14]. The study found that the Nod+/− group which did not receive a nutrient solution containing nitrogen had a statistically significant lower leaf chlorophyll content compared to soybean grown with nutrient solution, which is an indication that the leaf chlorophyll content of the Nod+/− group was nitrogen limited relative to Nod+/+ and Nod− plants [14]. This study, however, did find a statistically significant higher photosynthetic rate in the Nod+/− soybeans suggesting that these plants may be more efficient in utilizing their chlorophyll for photosynthetic CO_2 uptake [14].

Certain limitations are applicable to our study. First, we did not evaluate the effect of variation of nitrogen on soybean photosynthetic rate. Second, this study is susceptible to a type II error and may not have adequate power to detect a statistically significant difference between the three treatment groups and the control group. Our future research will focus on the photosynthetic rates of soybean under different concentrations of nitrogen and its impact on overall yield.

4. Conclusion

This study evaluated the effects of variation of nitrogen concentration, using Hoagland nutrient solution, on soybean growth and leaf chlorophyll concentrations. There was no statistically significant difference between groups when assessing growth rate, 31-day growth, and leaf chlorophyll concentration. This study, however, provides some positive evidence to support that supplementation of soybean with low concentration of nitrogen (based on Hoagland nutrient solution) may produce relatively ideal physiological outcomes for soybeans.

Acknowledgements

Thanks to the National Science Foundation for sponsoring this research project and financial support (HRD-0102620). Special thanks to Dr. Murty Kambhampati, Professor of Biology, for providing mentorship and resources for this experiment.

References

[1]   Barnes, S. (2010) The Biochemistry, Chemistry and Physiology of the Isoflavones in Soybeans and Their Food Products. *Lymphatic Research and Biology*, **8**, 89-98. http://dx.doi.org/10.1089/lrb.2009.0030

[2]   Messina, M.J. (1999) Legumes and Soybeans: Overview of Their Nutritional Profiles and Health Effects. *The American Journal of Clinical Nutrition*, **70**, 439S-450S.

[3]   Kim, M.Y., Van, K., Kang, Y.J., Kim, K.H. and Lee, S.H. (2012) Tracing Soybean Domestication History: From Nucleotide to Genome. *Breeding Science*, **61**, 445-452. http://dx.doi.org/10.1270/jsbbs.61.445

[4]   Velasquez, M.T. and Bhathena, S.J. (2007) Role of Dietary Soy Protein in Obesity. *International Journal of Medical Sciences*, **4**, 72-82. http://dx.doi.org/10.7150/ijms.4.72

[5]   Anderson, J.W., Smith, B.M. and Washnock, C.S. (1999) Cardiovascular and Renal Benefits of Dry Bean and Soybean Intake. *The American Journal of Clinical Nutrition*, **70**, 464S-474S.

[6]   Slavin, J. (1991) Nutritional Benefits of Soy Protein and Soy Fiber. *Journal of the American Dietetic Association*, **91**, 816-819.

[7]   Friedman, M. and Brandon, D.L. (2001) Nutritional and Health Benefits of Soy Proteins. *Journal of Agricultural and Food Chemistry*, **49**, 1069-1086. http://dx.doi.org/10.1021/jf0009246

[8]   Kenny, A.M., Mangano, K.M., Abourizk, R.H., Bruno, R., Anamani, D.E., Kleppinger, A., Walsh, S.J., Prestwood, K.M. and Kerstetter, J.E. (2009) Soy Proteins and Isoflavones Affect Bone Mineral Density in Older Women: A Randomized Controlled Trial. *The American Journal of Clinical Nutrition*, **90**, 234-242. http://dx.doi.org/10.3945/ajcn.2009.27600

[9]   Ho, S.C., Woo, J., Lam, S., Chen, Y., Sham, A. and Lau, J. (2003) Soy Protein Consumption and Bone Mass in Early Postmenopausal Chinese Women. *Osteoporosis International*, **14**, 835-842. http://dx.doi.org/10.1007/s00198-003-1453-9

[10]   Umpierrez, G.E., Spiegelman, R., Zhao, V., Smiley, D.D., Pinzon, I., Griffith, D.P., Peng, L., Morris, T., Luo, M., Garcia, H., Thomas, C., Newton, C.A. and Ziegler, T.R. (2012) A Double-Blind, Randomized Clinical Trial Comparing Soybean Oil-Based versus Olive Oil-Based Lipid Emulsions in Adult Medical-Surgical Intensive Care Unit Patients Requiring Parenteral Nutrition. *Critical Care Medicine*, **40**, 1792-1798. http://dx.doi.org/10.1097/CCM.0b013e3182474bf9

[11]   Kraiser, T., Gras, D.E., Gutiérrez, A.G., González, B. and Gutierrez, R.A. (2011) A Holistic View of Nitrogen Acquisition in Plants. *Journal of Experimental Botany*, **62**, 1455-1466. http://dx.doi.org/10.1093/jxb/erq425

[12]   Gutiérrez, R.A. (2012) Systems Biology for Enhanced Plant Nitrogen Nutrition. *Science*, **336**, 1673-1675. http://dx.doi.org/10.1126/science.1217620

[13]   White, P.J. and Brown, P.H. (2010) Plant Nutrition for Sustainable Development and Global Health. *Annals of Botany*, **105**, 1073-1080. http://dx.doi.org/10.1093/aob/mcq085

[14]   de Veau, E.J., Robinson, J.M., Warmbrodt, R.D. and van Berkum, P. (1990) Photosynthesis and Photosynthate Partitioning in N_2-Fixing Soybeans. *Plant Physiology*, **94**, 259-267. http://dx.doi.org/10.1104/pp.94.1.259

[15]   Albrecht, S.L., Maier, R.J., Hanus, F.J., Russell, S.A., Emerich, D.W. and Evans, H.J. (1979) Hydrogenase in *Rhizobium japonicum* Increases Nitrogen Fixation by Nodulated Soybeans. *Science*, **203**, 1255-1257. http://dx.doi.org/10.1126/science.203.4386.1255

[16]   Abu-Shakra, S.S., Phillips, D.A. and Huffaker, R.C. (1978) Nitrogen Fixation and Delayed Leaf Senescence in Soybeans. *Science*, **199**, 973-975. http://dx.doi.org/10.1126/science.199.4332.973

[17]   Hennecke, H. (1990) Nitrogen Fixation Genes Involved in the *Bradyrhizobium japonicum*-Soybean Symbiosis. *FEBS Letters*, **268**, 422-426.

[18]   Wych, R.D. and Rains, D.W. (1978) Simultaneous Measurement of Nitrogen Fixation Estimated by Acetylene-Ethylene Assay and Nitrate Absorption by Soybeans. *Plant Physiology*, **62**, 443-448. http://dx.doi.org/10.1104/pp.62.3.443

[19]   Kaschuk, G., Hungria, M., Leffelaar, P. A., Giller, K.E. and Kuyper, T.W. (2010) Differences in Photosynthetic Behaviour and Leaf Senescence of Soybean (*Glycine max* [L.] Merrill) Dependent on N_2 Fixation or Nitrate Supply. *Plant Biology*, **12**, 60-69. http://dx.doi.org/10.1111/j.1438-8677.2009.00211.x

[20]   Sabaratnam, S., Gupta, G. and Mulchi, C. (1988) Effects of Nitrogen Dioxide on Leaf Chlorophyll and Nitrogen Content of Soybean. *Environmental Pollution*, **51**, 113-120. http://dx.doi.org/10.1016/0269-7491(88)90200-X

[21]   Sabaratnam, S. and Gupta, G. (1988) Effects of Nitrogen Dioxide on Biochemical and Physiological Characteristics of Soybean. *Environmental Pollution*, **55**, 149-158. http://dx.doi.org/10.1016/0269-7491(88)90125-X

[22]   Hoagland, D.R. and Arnon, D.I. (1950) The Water-Culture Method for Growing Plants without Soil. *Circular & Cali-*

fornia Agricultural Experiment Station, **347**, 32.

[23] Devlin, P.F. and Kay, S.A. (2000) Flower Arranging in *Arabidopsis*. *Science*, **288**, 1600-1602.
      http://dx.doi.org/10.1126/science.288.5471.1600

[24] Einhellig, F.A. and Rasmussen, J.A. (1979) Effects of Three Phenolic Acids on Chlorophyll Content and Growth of
      Soybean and Grain Sorghum Seedlings. *Journal of Chemical Ecology*, **5**, 815-824.
      http://dx.doi.org/10.1007/BF00986566

[25] Ryle, G. and Hesketh, J. (1969) Carbon Dioxide Uptake in Nitrogen-Deficient Plants. *Crop Science*, **9**, 451-454.
      http://dx.doi.org/10.2135/cropsci1969.0011183X000900040019x

[26] Falkowski, P.G., Sukenik, A. and Herzig, R. (1989) Nitrogen Limitation in *Isochrysis galbana* (Haptophyceae). II. Re-
      lative Abundance of Chloroplast Proteins. *Journal of Phycology*, **25**, 471-478.
      http://dx.doi.org/10.1111/j.1529-8817.1989.tb00252.x

4

Fluoride Uptake and Net Primary Productivity of Selected Crops

P. C. Mishra*, S. K. Sahu, A. K. Bhoi, S. C. Mohapatra

Department of Environmental Sciences, Sambalpur University, Jyoti Vihar, Sambalpur, Odisha, India
Email: *profpcmishra@gmail.com

Abstract

Crop field soil collected from Sambalpur University campus of Odisha and treated with various fluoride concentrations was used to raise selected local crops. Background concentration of total and leachable fluoride content in soil was 95.19 and 8.89 ppm respectively. At the time of harvest of the crops, the total fluoride content was found to decrease and leachable fluoride content was found to increase both in control and experimental sets. This might be due to the addition of fluoride to soil in the experimental set up as well as availability of background fluoride content in soil and the irrigated water (*i.e.* 0.5 ppm). The fluoride accumulation in plant tissue increased with increase in the fluoride content in soil. Net Primary Productivity (NPP) of fluoride treated plants decreased in Brinjal by 6.64% - 56.72%, Tomato by 14.46% - 62.24% and Mung by 10.27% - 53.61%, all in 20 - 100 ppm fluoride range. However, NPP of Mustard, Ladies finger and Chili decreased by 15.58% - 61.21%, 12.28% - 52.78% and 40.8% - 90.65% in 10 - 50 ppm fluoride treated sets respectively in 10 - 50 ppm fluoride range. Maize NPP decreased by 12.17% - 61.20% in 20 - 100 ppm fluoride range as Rice NPP decreased by 6.64% - 56.72% in 20 - 100 ppm fluoride range. Pod formation was inhibited at 100 ppm fluoride amended soil in case of Mung, and 50 ppm in Ladies finger, 40 - 100 ppm in Maize and 30 - 50 ppm fluoride amended soil in case of Chilli. Thus, Maize and Chilli are more sensitive to fluoride contamination than other crops. In all the crops NPP decreased with increase in fluoride content in soil with significant decrease in highest concentration of fluoride.

Keywords

Fluoride, Uptake, Crops, Net Primary Productivity

---

*Corresponding author.

1. Introduction

Fluorine (F) is an element of the halogen family and Fluoride (F^-) is the anion the reduced form of fluorine. Both organic and inorganic fluorine compounds are sometimes called fluorides. Fluoride, like other halides, is a monovalent ion (−1 charge). Its compounds often have properties that are distinctly relative to other halides. Structurally, and to some extent chemically, the fluoride ion resembles the hydroxides ion. Fluoride-containing compounds range from potent toxins such as Sarin to life-saving pharmaceuticals such as Efavirenz and from refractory materials such as calcium fluoride to the highly reactive sulfur tetrafluoride. The range of fluorine-containing compounds is considerable as fluorine is capable of forming compounds with all the elements except for helium and neon [1]. Fluorine in the environment is therefore found as fluorides which together represent about 0.06 - 0.09 percent of the earth's crust. The average crustal abundance is 300 mg·kg^{-1} [2]. Fluorides are found at significant levels in a wide variety of minerals, including fluorspar, rock phosphate, cryolite, apatite, mica, hornblende and others [3]. Fluorite (CaF_2) is a common fluoride mineral of low solubility occurring in both igneous and sedimentary rocks. Fluoride is commonly associated with volcanic activity and fumarolic gases. Thermal waters, especially those of high pH, are also rich in fluoride [4]. Minerals of commercial importance include cryolite and rock phosphates. The fluoride salt cryolite is used for the production of aluminium [3] and as a pesticide [5]. Rock phosphates are converted into phosphate fertilizers by the removal up to 4.2 percent fluoride. The purified fluoride (as fluorosilicates) is added to drinking-water in some countries in order to protect against dental caries [6] and [7]. It forms inorganic and organic compounds called fluorides. Living organisms are mainly exposed to inorganic fluorides through food and water. Based on quantities released and concentrations presented naturally in the environment as well as the effects on living organisms, the important inorganic fluorides are hydrogen fluoride (HF), calcium fluoride (CaF_2), sodium fluoride (NaF), Sulphur hexafluoride (SF_6) and Silico fluorides. Fluoride in the form of HF or SiF_4 is one of the most important and damaging air pollutants affecting forests, crops and natural vegetation [8]. Fluoride occurs naturally in plants, but its presence has attracted attention primarily in certain areas where concentrations are elevated above normal by accumulation from the atmosphere.

Human activities releasing fluorides into the environment are mainly the mining and processing of phosphate rock and its use as agricultural fertilizer, as well as the manufacture of aluminums. Other fluoride sources include the combustion of coal (containing fluoride impurities) and other manufacturing processes (steel, copper, nickel, glass, brick, ceramic, glues and adhesives). In addition, the use of fluoride-containing pesticides in agriculture and fluoride in drinking water supplies also contribute to the release of fluorides into the environment. However, the greatest concentrations are found near anthropogenic point sources. In air, because of its extensive industrial use, hydrogen fluoride is probably the greatest single atmospheric fluoride contaminant [9]. Fluorides can be present as gases or particulates. They can be transported by wind over large distances before depositing on the earth's surface or dissolving in water. In general, fluoride compounds do not remain in the troposphere for long periods, nor do they move up to the stratosphere. In areas where fluoride-containing coal is burned or phosphate fertilizers are produced or used, the fluoride concentration in air is elevated leading to increased exposure by inhalation and absorption routes. High levels of atmospheric fluoride occur in areas of Morocco and China [10] and [11]. In some provinces of China, fluoride concentrations in indoor air ranging from 16 to 46 µg/m^3 owing to indoor combustion of high-fluoride coal for cooking, or drying and curing food [12]. Indeed, more than 10 million people in China are reported to suffer from fluorosis, related in part to the burning of high fluoride coal [13].

Fluoride is a component of most types of soil, with total concentrations ranging from 20 to 1000 µg/g in areas without natural phosphate or fluoride deposits and up to several thousand mg/g in mineral soils with deposits of fluoride [14]. Airborne gaseous and particulate fluorides tend to accumulate within the surface layer of soils, but they may be displaced throughout the root zone, even in calcareous soils [15]. Calcium fluoride is the most common in alkaline soils, and fluoroaluminate complexes are the most common in acidic soils. Thus, exposure to hydrofluoric acid will occur at a hazardous waste site only if someone comes in to contact with material leaking from a storage container or contaminated air before it is dispersed. Once in a stable form, fluoride persists in the environment for a relatively long time unless transforming to another compound or decomposed by radiation. The clay and organic carbon content as well as the pH of soil is primarily responsible for the origin and/or retention of fluoride in soils. It has also been reported that in saline soils the bioavailability of fluoride to plants is related to the water-soluble component of the fluoride present [14]. In living organisms, the quantity of fluoride accumulation depends on the route of exposure, on how well the particular fluorides are absorbed by the body and on how quickly they are taken up and excreted. Soluble fluorides are bioaccumulated by some aquatic and terrestrial biota. However, information concerning the biomagnification of fluoride in aquatic or terrestrial food-chains is scanty [16] and [17]. Inorganic fluorides tend to accumulate preferentially in the skeletal and dental hard tissues of vertebrates, exoskeletons of invertebrates and cell walls of plants [14] [16] [18]-[20]. Bio-concentration factors greater than 10 (expressed on a wet weight basis) were reported in both aquatic plants and animals following exposure to solutions up to 50 mg/l of fluoride [21]. Moeri [22] has reported effect of fluoride emission on enzyme activity in metabolism of plants.

The undivided Sambalpur District of Odisha is an industrial belt where two aluminium industries—Vedanta Aluminum Company at Jharsuguda and Hindalco Industries Limited at Hirakud Town have been established. The former started operating from 1965 while Vedanta, a major industry, is operating since last 8 years. Both the industries have their coal-based captive thermal power plants. The local people in Hirakud complain every year of crop damage during growing season due to emissions from aluminium and power industries in Hirakud. There are research reports on fluoride accumulation in soil, plants and animals to the extent of 40 - 80 ppm in Hirakud, which may be because of emissions from industrial activities [23] [24]. Effects of such accumulations on various environmental segments in Hirakud have also been studied [25] [26]. A survey of field soil around Vedanta Aluminium Limited shows a total and leachable fluoride content ranging from 94.01 - 467.7 mg/kg and 10.60 - 104.86 mg/kg respectively [27]. Therefore, the present study was undertaken to assess the fluoride uptake by selected crops and its effect on NPP.

2. Materials and Methods

Sodium fluoride (NaF) was used to prepare fluoride (F) solutions in various concentrations, *i.e.* 20, 40, 60, 80, and 100 ppm with distilled water for treatment. Tap water was used as control. Culture experiments were set up to monitor the growth of the plants in various concentrations of F^- amended soil. Plant seeds were collected from a Government authorized seed store located at Goshala, Sambalpur and the seedlings were collected from OUAT Chipilima, Sambalpur for pot culture. For pot culture experiment, 21 days old healthy seedlings were collected from OUAT, Chiplima and transplanted in to the pots containing F^- treated crop field soil. After harvest total yield (NPP) was calculated. Following local crops were assessed:

Winter crops—Brinjal (*Solanum melongena* L.), Tomato (*Lycopersicon esculentum*), Mustard (*Brassica compestris*) and Mung (*Vigna radiata*);

Summer crops—Ladies finger (*Abelmoschus esculentus*) and Maize (*Zea mays* L.);

Rainy crops—Paddy (*Oryza sativa*) and Chilli (*Capsicum annuum*).

NPP is usually estimated by harvest technique, in which above ground plant biomass (AGP) is harvested from all the pots and below ground Plant Biomass (BGP) were washed from soil cores. The short term harvest method [28] was employed for biomass estimation at 15 days intervals. The plant parts were oven dried at 80°C for 24 hours. After harvest total yield was calculated. The F^- content in soil and plants was estimated by Ion Selective Analyzer.

3. Results

3.1. Brinjal

The total and leachable F^- content in soil at the beginning of the experiment was found to be 95.19 and 8.89 ppm respectively. The F^- content in plant sample after harvest was found to be 1.45, 1.89, 2.12, 2.85, 3.24 and 3.83 mg/kg in 0, 20, 40, 60, 80 and 100 ppm of F treated soils (**Table 1**). NPP on the day of harvest (*i.e.* 75th day) was 117.4, 109.6, 89.6, 69.1, 60.3 and 50.8 g dry wt/plant in 0, 20, 40, 60, 80 and 100 ppm F^- amended soils respectively (**Table 2**). Compared to control, NPP was decreased by 6.64% and 56.72% in 20 and 100 ppm of F^- concentrations respectively. The yield (pod weight) decreased by 78% at 100 ppm F^- application compared to control.

3.2. Tomato

The total and leachable F^- content in soil at the beginning was 95.19 and 8.89 ppm respectively. At the time of harvest *i.e.* on 75th day, the total F^- content showed decreasing trend whereas leachable F^- showed an increasing trend in both control and treated soil (**Table 3**). The F^- content in plant sample after harvest was found to be

Table 1. Fluoride content in soil and Brinjal plant.

| Conc. | At Start in Soil (mg/Kg) | | | At Harvest (mg/Kg) | |
	Total Fluoride	Leachable Fluoride	Total Fluoride in Soil	Leachable Fluoride in Soil	Total F⁻ in Plant Sample
Control	95.19	8.89	86.40	18.56	1.45
20 ppm	95.19 + 20	8.89	92.40	21.58	1.89
40 ppm	95.19 + 40	8.89	110.20	25.63	2.12
60 ppm	95.19 + 60	8.89	124.60	29.76	2.85
80 ppm	95.19 + 80	8.89	138.50	35.80	3.24
100 ppm	95.19 + 100	8.89	158.40	38.14	3.83

Table 2. NPP (g dry wt/plant) in Brinjal in fluoride treated soil.

| Days | Concentration | BGP Root | AGP | | | NPP (BGP + AGP) | % Decrease in NPP over control |
			Leaf	Stem	Pod		
	60 Days Control	7.1	16.6	32.2	61.5	117.4	00
	20 ppm	6.8	15.8	29.7	57.3	109.6	6.64
	40 ppm	6.2	15.1	27.2	41.1	89.6	23.67
	60 ppm	5.9	13.7	24.2	25.3	69.1	41.14
	80 ppm	5.1	12.1	23.5	19.6	60.3	48.63
	100 ppm	4.2	11.5	21.6	13.5	50.8	56.72

Table 3. Fluoride content in soils of Tomato crop.

| Soil Conc. | At the Start (mg/Kg) | | | During Harvest (mg/Kg) | |
	Total Fluoride	Leachable Fluoride	Total Fluoride in Soil	Leachable Fluoride in Soil	Total F⁻ in Plant Sample
Control	95.19	8.89	86.40	22.87	1.05
20 ppm	95.19 + 20	8.89	92.40	23.70	1.66
40 ppm	95.19 + 40	8.89	110.20	27.52	2.08
60 ppm	95.19 + 60	8.89	124.60	31.47	2.68
80 ppm	95.19 + 80	8.89	138.50	33.30	3.17
100 ppm	95.19 + 100	8.89	158.40	37.28	3.49

1.05, 1.66, 2.08, 2.68, 3.17 and 3.49 mg/kg in 0, 20, 40, 60, 80 and 100 ppm of F⁻ amended soils respectively.

NPP on the day of harvest (*i.e.* 60th day) was 116.0, 99.2, 69.3, 65.2, 54.4 and 43.8 g dry wt/plant in 0, 20, 40, 60, 80 and 100 ppm F⁻ amended soils respectively (**Table 4**). NPP decreased by 14.48% - 62.24% in 20 - 100 ppm F⁻ amended soils compared to control set. The pod weight decreased by 83% in 1000 ppm F⁻ treated soils over control.

3.3. Mustard

The total and leachable F⁻ content in soil in the beginning was 95.19 and 8.89 ppm respectively. At the time of harvest *i.e.* on 60th day, the total F⁻ content decreased whereas leachable F⁻ increased both in control and treated soils (**Table 5**). The F⁻ content in plant sample on harvest was found to be 1.13, 1.46, 1.76, 1.95, 2.18 and 2.37 mg/kg in control, 10, 20, 30, 40 and 50 ppm of F⁻ treated soils respectively.

On 60th day, the highest NPP (0.263 g dry wt/plant) was recorded in case of control set and least (0.102 g dry wt/plant) in 50 ppm F⁻ treated soil. The percentage decrease in NPP on 60th day was found to be 15.58, 27.37, 49.04, 58.93 and 61.21% in 10, 20, 30, 40 and 50 ppm F⁻ treated soils respectively (**Table 6**) whereas Mustard yield got reduced by 62% in 50 ppm F⁻ treated soil.

Table 4. NPP (g dry wt/plant) of Tomato in fluoride treated soils.

Days	Concentration	BGP Root	AGP			NPP (BGP + AGP)	% Decrease in NPP over Control
			Leaf	Stem	Pod		
	Control	6.3	20.0	31.3	58.4	116.0	00
	20 ppm	5.0	18.6	28.0	47.6	99.2	14.48
60 Days	40 ppm	4.8	16.4	26.2	39.9	69.3	40.25
	60 ppm	4.3	14.3	24.1	22.5	65.2	43.79
	80 ppm	3.1	13.7	22.8	14.8	54.4	53.10
	100 ppm	2.2	11.5	20.4	9.7	43.8	62.24

Table 5. Fluoride content in soils and Mustard plant.

Soil	At the Start (mg/Kg)		During Harvest (mg/Kg)		
Conc.	Total Fluoride	Leachable Fluoride	Total Fluoride in Soil	Leachable Fluoride in Soil	Total F^- in Plant Sample
Control	95.19	8.89	81.38	16.37	1.13
10 ppm	95.19 + 10	8.89	87.70	18.95	1.46
20 ppm	95.19 + 20	8.89	93.37	22.61	1.76
30 ppm	95.19 + 30	8.89	99.61	26.57	1.95
40 ppm	95.19 + 40	8.89	106.72	30.59	2.18
50 ppm	95.19 + 50	8.89	112.58	32.18	2.37

Table 6. NPP (g dry wt/plant) of Mustard in fluoride treated soils.

Days	Concentration	BGP Root	AGP			NPP (BGP + AGP)	% Decrease in NPP over Control
			Leaf	Stem	Pod		
	Control	0.034	0.046	0.091	0.092	0.263	0
	10 ppm	0.033	0.043	0.07	0.076	0.222	15.58
	20 ppm	0.03	0.035	0.063	0.063	0.191	27.37
60 Days	30 ppm	0.024	0.031	0.054	0.025	0.134	49.04
	40 ppm	0.022	0.026	0.043	0.017	0.108	58.93
	50 ppm	0.021	0.021	0.036	0.014	0.102	61.21

3.4. Mung

The total and leachable F^- content in soil in the beginning was 95.19 and 8.89 ppm respectively. At the time of harvest *i.e.* on 60th day, the total F^- content decreased whereas leachable F^- increased both in control and treated soils (**Table 7**). The F^- content in plant sample after harvest was found to be 1.23, 1.72, 2.18, 2.76, 3.35 and 3.63 mg/kg in control, 10, 20, 30, 40 and 50 ppm of F^- treated soils respectively.

NPP on the harvest day was 36, 32.3, 28.8, 26.9, 25.0 & 16.7 g of dry wt/plant in control, 20, 40, 60, 80 and 100 ppm F^- treated soils respectively (**Table 8**). In comparison with control, NPP decreased by 10.27%, 20%, 25.27%, 30.55% and 53.61% in 20, 40, 60, 80 and 100 ppm F^- amended sets respectively. Mung yield decreased by 14% at 80 ppm F^- treatment and no pod formation took place at 100 ppm F^- concentration.

3.5. Ladies Finger

On harvest day *i.e.* 60th day, the total F^- content decreased and leachable F^- increased both in control and treated soils (**Table 9**). The total and leachable F^- content in soil in the beginning of the experiment was 95.19 and 8.89 ppm respectively. The F^- content in ladies finger plants during harvest was found to be 1.44, 1.55, 1.86, 2.03, 2.35 and 2.72 mg/kg in control, 10, 20, 30, 40 and 50 ppm of F^- treated soils respectively.

On the day of harvest (*i.e.* 60th day), NPP was found to be 87.9, 77.1, 69.9, 64.6, 60.8 and 41.5 g of dry

Table 7. Fluoride content in soils and Mung plants.

Soil Conc.	At the Start (mg/Kg)		During Harvest (mg/Kg)		
	Total Fluoride	Leachable Fluoride	Total Fluoride in Soil	Leachable Fluoride in Soil	Total F⁻ in Plant Sample
Control	95.19	8.89	86.40	22.93	1.23
20 ppm	95.19 + 20	8.89	92.40	24.52	1.72
40 ppm	95.19 + 40	8.89	110.20	29.25	2.18
60 ppm	95.19 + 60	8.89	124.60	33.07	2.76
80 ppm	95.19 + 80	8.89	138.50	36.76	3.35
100 ppm	95.19 + 100	8.89	158.40	42.04	3.63

Table 8. NPP (g dry wt/plant) of Mung in fluoride treated soils.

Days	Concentration	BGP Root	AGP			NPP (BGP + AGP)	% Decrease in NPP over control
			Leaf	Stem	Pod		
	Control	2.5	8.6	14.7	10.2	36.0	00
	20 ppm	2.2	7.4	13.6	10.2	32.3	10.27
	40 ppm	1.9	7.0	10.0	9.9	28.8	20
60 Days	60 ppm	1.6	6.4	10.0	8.9	26.9	25.27
	80 ppm	1.1	6.0	10.1	8.8	25.0	30.55
	100 ppm	1.0	5.8	9.9	---	16.7	53.61

Table 9. Fluoride content in soils and ladies finger plants.

Soil Conc.	At the Start (mg/Kg)		During Harvest (mg/Kg)		
	Total Fluoride	Leachable Fluoride	Total Fluoride in Soil	Leachable Fluoride in Soil	Total F⁻ in Plant Sample
Control	95.19	8.89	81.38	17.89	1.44
10 ppm	95.19 + 10	8.89	87.70	19.24	1.55
20 ppm	95.19 + 20	8.89	93.37	23.56	1.86
30 ppm	95.19 + 30	8.89	99.61	28.67	2.03
40 ppm	95.19 + 40	8.89	106.72	32.46	2.35
50 ppm	95.19 + 50	8.89	112.58	35.04	2.72

wt/plant in control, 10, 20, 30, 40 and 50 ppm F⁻ treated soils respectively (**Table 10**). NPP decreased by 12.28% - 52.78% in 10 - 50 ppm F⁻ treated soils. There was no pod found at 50 ppm F⁻ treatment.

3.6. Maize

On the harvest day (75th day), the total F⁻ content decreased and leachable F⁻ increased both in control and treated soils (**Table 11**). The total and leachable F⁻ content in soils in the beginning of the experiment was 95.19 and 8.89 ppm respectively. In maize samples F⁻ was found to be 1.98, 2.46, 2.65, 3.22, 3.54 and 4.02 mg/kg in control, 20, 40, 60, 80 and 100 ppm of F⁻ treated soils respectively on the harvest day.

NPP on 75th day was 11.42 g of dry wt/plant in control and 4.43 g in 100 ppm F⁻ amended soils. The decrease in NPP was found to be 12.17%, 36.66%, 40.45%, 52.62% and 61.20% in 20, 40, 60, 80 and 100 ppm F⁻ treated sets respectively (**Table 12**). No pod formation was found in sets beyond 20 ppm treatment.

3.7. Paddy

On the harvest day, the total and leachable F⁻ content in soil at the start of the experiment was 95.19 and 8.89 ppm respectively. The F⁻ content in plant sample on harvest was found to be 2.16, 2.63, 2.95, 3.62, 3.84 and 4.37 mg/kg in control, 20, 40, 60, 80 and 100 ppm of F⁻ treated soils respectively (**Table 13**).

Table 10. NPP (g dry wt/plant) of Ladies finger in fluoride treated soils.

Days	Concentration	BGP Root	AGP			NPP (BGP + AGP)	% Decrease in NPP over Control
			Leaf	Stem	Pod		
60 Days	Control	7.8	10.0	39.3	30.1	87.9	00
	10 ppm	7.7	9.6	23.0	29.8	77.1	12.28
	20 ppm	6.3	9.4	25.1	29.1	69.9	20.47
	30 ppm	5.9	9.2	20.8	28.7	64.6	26.50
	40 ppm	4.3	8.7	20.8	27.0	60.8	30.83
	50 ppm	4.1	8.5	28.9	---	41.5	52.78

Table 11. Fluoride content in soils and Maize plants.

Soil Conc.	At the Start (mg/Kg)		During Harvest (mg/Kg)		
	Total Fluoride	Leachable Fluoride	Total Fluoride in Soil	Leachable Fluoride in Soil	Total F⁻ in Plant Sample
Control	95.19	8.89	86.40	20.90	1.98
20 ppm	95.19 + 20	8.89	92.40	22.52	2.46
40 ppm	95.19 + 40	8.89	110.20	26.85	2.65
60 ppm	95.19 + 60	8.89	124.60	29.82	3.22
80 ppm	95.19 + 80	8.89	138.50	35.57	3.54
100 ppm	95.19 + 100	8.89	158.40	41.02	4.02

Table 12. NPP (g dry wt/plant) of Maize in F⁻ treated soils.

Days	Concentration	BGP Root	AGP			NPP (BGP + AGP)	% Decrease in NPP over Control
			Leaf	Stem	Pod		
75 days	Control	2.63	0.78	5.73	2.28	11.42	0
	20 ppm	2.46	0.74	5.33	1.5	10.03	12.17
	40 ppm	1.95	0.663	4.62	xxx	7.233	36.66
	60 ppm	1.89	0.62	4.29	xxx	6.8	40.45
	80 ppm	1.81	0.6	3.0	xxx	5.41	52.62
	100 ppm	1.4	0.55	2.48	xxx	4.43	61.20

Table 13. Fluoride content in soils and Paddy samples.

Soil Conc.	At the Start (mg/Kg)		During Harvest (mg/Kg)		
	Total Fluoride	Leachable Fluoride	Total Fluoride in Soil	Leachable Fluoride in Soil	Total F⁻ in Plant Sample
Control	95.19	8.89	86.40	21.87	2.16
20 ppm	95.19 + 20	8.89	92.40	23.45	2.63
40 ppm	95.19 + 40	8.89	110.20	29.40	2.95
60 ppm	95.19 + 60	8.89	124.60	32.52	3.62
80 ppm	95.19 + 80	8.89	138.50	35.73	3.84
100 ppm	95.19 + 100	8.89	158.40	41.13	4.37

On the day of harvest (*i.e.* 75th day), the NPP of Paddy was 105.9, 97.2, 84.9, 72.1, 65.8 and 56.6 g dry wt/plant in 0, 20, 40, 60, 80 and 100 ppm F⁻ treated soil (**Table 14**). Compared with control, the NPP decreased by 8.2% - 46.5% and yield by 9.5 - 55.5 in 20 - 100 ppm F⁻ treated soils respectively.

3.8. Chilli

The total and leachable F⁻ content in soils in the beginning was 95.19 and 8.89 ppm respectively. On the harvest day *i.e.* on 75th day, the total F⁻ content decreased and leachable F⁻ increased both in control and experimental soils (**Table 15**).

F⁻ content in plant samples on harvest was found to be 1.56, 1.76, 1.98, 2.35, 2.58 and 2.94 mg/kg in control, 10, 20, 30, 40 and 50 ppm of fluoride treated soils respectively. NPP on the day of harvest (*i.e.* 75th day) was 7.56, 4.47, 2.77, 1.74, 1.06 and 0.71 g of dry wt/plant in 0, 10, 20, 30, 40 and 50 ppm F⁻ treated soils (**Table 16**). Thus NPP decreased by 40.8% - 90.65% in 20 - 100 ppm in F⁻ treated sets. No pod formation was seen beyond 20 ppm treatment.

Table 17 presents a comparative figure on plant uptake of fluoride and respective NPP and yield in the crops tested. Mung, Ladies finger, Maize and Chilli had no pod yield at the highest concentration of F⁻, indicating that they are very sensitive plants compared to other four crops. **Figure 1** shows a comparative picture of percent reduction of NPP and yield.

Figure 1 presents a comparative figure on the rate of NPP and yield reduction at the highest concentration of F tested. All such crops are sensitive to F with Mung, Ladies finger, Maize and Chilli being far more sensitive than other four.

Table 14. NPP (g dry wt/plant) of Paddy in F⁻ treated soils.

Days	Concentration	BGP Root	AGP			NPP (BGP + AGP)	% Decrease in NPP over Control
			Leaf	Stem	Pod		
	60 Days Control	7.9	21.0	25.3	51.7	105.9	00
	20 ppm	6.8	20.6	23.0	46.8	97.2	6.64
	40 ppm	6.1	18.4	21.1	39.3	84.9	23.67
	60 ppm	5.7	16.1	18.7	31.6	72.1	41.14
	80 ppm	5.1	14.2	16.8	29.7	65.8	48.63
	100 ppm	4.7	13.5	14.9	23.5	56.6	56.72

Table 15. Fluoride content in soils and Chilli plants.

Soil Conc.	At the Start (mg/Kg)		During Harvest (mg/Kg)		
	Total Fluoride	Leachable Fluoride	Total Fluoride in Soil	Leachable Fluoride in Soil	Total F⁻ in Plant Sample
Control	95.19	8.89	81.38	12.50	1.56
10 ppm	95.19 + 10	8.89	87.70	16.72	1.76
20 ppm	95.19 + 20	8.89	93.37	21.47	1.98
30 ppm	95.19 + 30	8.89	99.61	25.45	2.35
40 ppm	95.19 + 40	8.89	106.72	28.56	2.58
50 ppm	95.19 + 50	8.89	112.58	33.70	2.94

Table 16. NPP (g dry wt/plant) of Chilli in F⁻ treated soils.

Days	Concentration	BGP Root	AGP			NPP (BGP + AGP)	% Decrease in NPP over Control
			Leaf	Stem	Pod		
	Control	1.22	0.89	2.95	2.5	7.56	0
	10 ppm	1.03	0.8	2.06	0.56	4.47	40.8
	20 ppm	0.87	0.8	1.68	0.42	2.773	63.32
75 days	30 ppm	0.42	0.38	0.93	xxx	1.74	76.9
	40 ppm	0.30	0.37	0.39	xxx	1.06	85.9
	50 ppm	0.26	0.24	0.21	xxx	0.71	90.65

Table 17. Leachable Fluoride (ppm) content, NPP and yield (g dry wt/plant) of Crops in highest concentration of F^-.

Crop	F Conc.	Soil F	Plant F	NPP	Yield
Brinjal	100	38.14	3.83	50.8	13.5
Tomato	100	37.28	3.49	43.8	9.7
Mustard	50	32.18	2.37	0.102	0.014
Mung	100	42.04	3.63	16.7	Nil
Ladies Finger	50	35.04	2.72	41.5	Nil
Maize	100	41.02	4.02	4.43	Nil
Paddy	100	41.13	4.37	56.6	23.5
Chilli	50	33.7	2.94	0.71	Nil

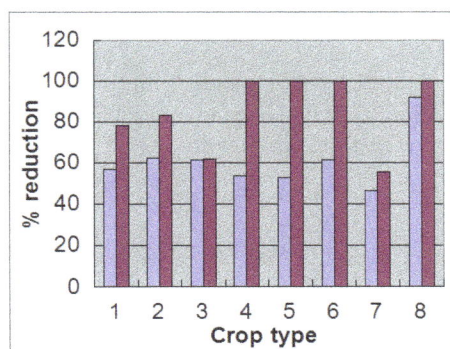

(1: Brinjal; 2: Tomato; 3: Mustard; 4: Mung; 5: Ladies finger; 6: Maize; 7: Paddy; 8: Chilli.)

Figure 1. Percent reduction in NPP and yield in crops.

4. Discussion

Terrestrial plants may accumulate inorganic fluorides following airborne deposition and uptake from the soil [14]. Sloof *et al.* [21] reported that the main route of uptake of fluoride by plants was from aerial deposition on the plant surface. Plant uptake from soil is generally low (except for accumulators) unless the fluoride has been added suddenly, such as following amendment with sludge or phosphate fertilizer. The availability of plants tends to decrease with time following application of fluoride. The degree of accumulation depends on several factors, including soil type and most prominently pH. At acidic pH (below pH 5.5), fluoride becomes more phyto-available through complexation with soluble aluminium fluoride species, which are themselves taken up by plants or increase the potential for the fluoride ion to be taken up by the plant [29] (Stevens *et al.* (1997)). Plant uptake of fluoride from solution culture is dependent on plant species and positively related to the ionic strength of the growth solution. Once a threshold fluoride ion activity in nutrient solutions is reached, fluoride concentrations in plants increase rapidly [30]. Ample evidences are also available relating to fluoride accumulation in different parts of the plants [31]. When fluoride contaminated plant tissue is ingested by the animals including human beings, fluoride associated problems are encountered. In the present study the same thing is also observed. Maximum fluoride accumulation in plant tissue was found in higher concentration of fluoride treated soils. Among all soils, it is the soluble fluoride content that is biologically important to plants and animals. It has also been reported that in saline soils, the bioavailability of fluoride of plants is related to the water-soluble component of the fluoride present [14]. Hocking *et al.* [32] indicated that although there was some accumulation of inorganic fluoride in marine vegetation, the rate of accumulation was far lower than that of airborne inorganic fluoride by terrestrial plants. Fluoride (F^-) contamination of soil, water and vegetation has been a continuing problem in the world. Various sources of F^- and their impact on the biology of plants and animals have been well documented. The detrimental effects of F^- on plants and animals have been known for more than hundred years [8]. Agricultural soils high in fluoride are common due to long term accumulation of fluoride from multisources. It may be extensive application of phosphate fertilizers, leaching from fluoride deposited rocks or industrial activities. Overall, the present study shows varied tolerance towards fluoride in the plant species tested and provides information about how fluoride can affect their germination, biochemistry and growth. Such knowledge is potentially useful for farmers to help them avoid excessive application of F⁻ containing fertilizers and selection of crops. As the fluoride endemicity is a great problem in recent days, much more studies on fluoride toxicity are needed not only to explore some viable remedial measures but also to save the present and future generations form fluoride related hazards. The study reveals that among the vegetable crops tested Chilli is more sensitive to fluoride contamination whereas Brinjal shows less sensitivity. Similarly, in case of grain crops (*i.e.* Maize and Rice) Maize shows more sensitivity to fluoride contamination as compared with Paddy.

The present work also indicates that the biomass of plants (both Mustard and Maize) grown in fluoride amended soil is less as compared with the control soil. This may be due to changes in above biochemical parameters which in consequence retard the growth and biomass of plants. Fluoride concentration more than 28 ppm significantly decreases dry weight of shoots [30], thus significantly reducing leaf surface area and weight in mature and immature leaves resulting in the inhibition of growth. This may be the reason for decrease in RGI with increase in fluoride concentration. Black [33] has reported that Fluoride in mesophyll cells disturbs mineral metabolism, reduces chlorophyll pigments and alters other morphological and physiological parameters such as height, number of leaves, biomass productivity, fruiting and yield of the plant, while higher fluoride accumulation causes leaf damage [31] and thus may retard growth.

References

[1]   Greenwood, Norman, N. and Earnshaw, A. (1997) Chemistry of the Elements. 2nd Edition, Butterworth-Heinemann, Oxford, 1340 p.

[2]   Tebutt, T.H.Y. (1983) Relationship between Natural Water Quality and Health. United Nations Educational, Scientific and Cultural Organization, Paris.

[3]   Murray, J.J. (1986) Appropriate Use of Fluorides for Human Health. World Health Organization, Geneva.

[4]   Edmunds, W.M. and Smedley, P.L. (1996) Groundwater Geochemistry and Health: An Overview. In: Appleton, Fuge and McCall, Eds., *Environmental Geochemistry and Health*, Vol. 113, Geological Society Special Publication, London, 91-105.

[5]   USEPA (1996) R.E.D. FACTS, Cryolite, EPA-738-F-96-016. United States Environmental Protection Agency.

[6]   Reeves, T.G. (1986) Water Fluoridation: A Manual for Engineers and Technicians. United States Department of Health and Human Services, Centres for Disease Control and Prevention, 138 p.

[7]   Reeves, T.G. (1994) Water Fluoridation. A Manual for Water Plant Operators. United States Department of Health and Human Services, Center for Disease Control and Prevention, 99 p.

[8]   Weinstein, L.H. and Davison, A.W. (2004) Fluorides in the Environment. CABI Publishing, Wallingford.

[9]   WHO (1984) Guidelines for Drinking Water Equality. World Health Organisation, Geneva, 2-249.

[10] Haikel, Y. (1986) Fluoride Content of Water, Dust, Soils and Cereals in the Endemic Dental Fluorosis Area of Khouribga (Morocco). *Archives of Oral Biology*, **31**, 279-286. http://dx.doi.org/10.1016/0003-9969(86)90041-5

[11] Haikel, Y. (1989) The Effects of Airborne Fluorides on Oral Conditions in Morocco. *Journal of Dental Research*, **68**, 1238-1241.

[12] WHO (1996) Volume 2: Health Criteria and Other Supporting Information. In: *Guidelines for Drinking-Water Quality*, 2nd Edition, World Health Organization, Geneva.

[13] Gu, S.L., Rongli, J. and Shouren, C. (1990) The Physical and Chemical Characteristics of Particles in Indoor Air Where High Fluoride Coal Burning Takes Place. *Biomedical and Environmental Sciences*, **3**, 384-390.

[14] Davison, A. (1983) Uptake, Transport and Accumulation of Soil and Airborne Fluorides by Vegetation. In: Shupe, J., Peterson, H. and Leone, N., Eds., *Fluorides: Effects on Vegetation, Animals and Humans*, Paragon Press, Salt Lake City, 61-82.

[15] Polomski, J., Fluhler, H. and Blaser, P. (1982) Accumulation of Airborne Fluoride in Soils. *Journal of Environmental Quality*, **11**, 457-461.

[16] Hemens, J. and Warwick, R.J. (1972) The Effects of Fluoride on Estuarine Organisms. *Water Research*, **6**, 1301-1308.

[17] ATSDR (1993) Toxicological Profile for Fluorides, Hydrogen Fluoride, and Fluorine. US Department of Health and Human Services, Agency for Toxic Substances and Disease Registry (TP-91/17), Atlanta.

[18] Michel, J., Suttie, J. and Sunde, M. (1984) Fluorine Deposition in Bone as Related to Physiological State. *Poultry*

Science, **63**, 1407-1411.

[19] Kierdorf, H. and Kierdorf, U. (1997) Disturbances of the Secretory Stage of Amelogenesis in Fluorosed Deer Teeth: A Scanning Electron-Microscopic Study. *Cell and Tissue Research*, **289**, 125-135.

[20] Sands, M., Nicol, S. and McMinn, A. (1998) Fluoride in Antarctic Marine Crustaceans. *Marine Biology*, **132**, 591-598.

[21] Sloof, W., Eerens, H., Janus, J. and Ros, J. (1989) Integrated Criteria Document: Fluorides. Bilthoven, National Institute of Public Health and Environmental Protection (Report No. 758474010).

[22] Moeri, P.B. (1980) Effect of Fluoride Emission in Enzyme Activity in Metabolism of Agric Plants. *Fluoride*, **13**, 122-128.

[23] Rath, S.P., Sarangi, P.K. and Mishra, P.C. (1998) Bioaccumulation and Bioconcentration of Fluoride in Environmental Segments of Hirakud, India. *Indian Journal of Environmental Protection*, **18**, 199-202.

[24] Mishra, P.C., Pradhan, K., Meher, K. and Bhosagar, D. (2009) Fluoride in the Environmental Segments at Hirakud of Western Orissa, India. *African Journal of Environmental Science and Technology*, **3**, 260-264.

[25] Mishra, P.C. and Mohapatra, A.K. (1998) Haematological Characteristics and Bone Fluoride Content in *Bufo melanostictus* from an Aluminium Industrial Site. *Environmental Pollution*, **99**, 421-423.

[26] Mishra, P.C. and Pradhan, K. (2006) Prevalence of Fluorosis among School Children and Cattle Population in Hirakud Town of Orissa. *Bioscan*, **2**, 31-36.

[27] Mishra, P.C. and Sahu, S.K. (2013) Effect of Fluoride on Locally Available Crops. Project Report Submitted to Vedanta Aluminium Company Ltd., Jharsuguda, 124 p.

[28] Odum, E.P. (1960) Organic Production and Turnover in Old Field Succession. *Ecology*, **41**, 34-49.

[29] Stevens, D.P., McLaughlin, M.J. and Alston, A.M. (1997) Phytotoxicity of Aluminium-Fluoride Complexes Culture by *Avena sativa* and *Lycopersicon esclentum*. *Plant and Soil*, **192**, 81-93.

[30] Stevens, D.P., McLaughlin, M.J. and Alston, A.M. (1998) Phytotoxicity of Hydrogen Fluoride and Fluoroborate and Their Uptake from Solution Culture by *Lycopersicon esculentum* and *Avena sativa*. *Plant and Soil*, **200**, 175-184.

[31] Klumpp, A., Klump, G., Domingos, M. and Silva, M.D.D. (1996) Fluoride Impact on Native Tree Species of the Atlantic Forest near Cubatao, Brazil. *Water, Air and Soil Pollution*, **87**, 57-71. http://dx.doi.org/10.1007/BF00696829

[32] Hocking, M.B., Hocking, D. and Smyth, T.A. (1980) Fluoride Distribution and Dispersion Processes about an Industrial Point Source in a Forested Coastal Zone. *Water, Air, & Soil Pollution*, **14**, 133-157.

[33] Black, C.A. (1968) Nitrogen, Phosphorus and Potassium. In: *Soil Plant Relationships*, 2nd Edition, John Wiley and Sons, New York, 405-773.

5

Adsorption and Transport of Ciprofloxacin in Quartz Sand at Different pH and Ionic Strength

Xiujiao Xu*, Jianglong He, Yu Li, Zhaoxi Fang, Shaohui Xu#

Department of Environment Science, Qingdao University, Qingdao, China
Email: *xjxuqd@sina.com, #shhxu@qdu.edu.cn

Abstract

Effects of pH and ionic strength on ciprofloxacin adsorption in quartz sand were studied through a batch equilibrium adsorption experiment in this paper. The experimental data were fitted by empirical formulas from Langmuir and Freundlich adsorption isothermal curves, and the transport experiments in quartz sand at different pH and ionic strength were conducted to investigate the transport characteristics of ciprofloxacin. It was found that with the increase of pH value or ionic strength, adsorption capacity of ciprofloxacin decreased, so that it could move easier. The results indicated that low pH or ionic strength was conductive to the adsorption of ciprofloxacin in quartz sand. Meanwhile, a higher initial concentration or stronger ionic strength could result in a smaller linear distribution coefficient of ciprofloxacin, which meant a low adsorption capacity. According to the fitting results, the adsorption of ciprofloxacin in quartz sand could be described well by both Langmuir and Freundlich equations, of which Freundlich equation had a better efficacy.

Keywords

Ciprofloxacin, Quartz Sand, pH, Ionic Strength, Isotherm Adsorption

1. Introduction

In large-scale livestock farming, the increasing consumption of various antibiotics including quinolones, tetracyclines, sulfonamides, etc. has become an irreversible trend around the world [1] [2]. However, most antibiotics cannot be completely assimilated into the animal body, and 40% to 90% of the antibiotic dose will be ex-

---

*The first author: Xiujiao Xu (1988), Female, Hebei, Master, is mainly engaged in the numerical simulation of underground water flow and solute transport in the environment.
#Corresponding author.

creted with the feces and urine in the form of precursor or metabolite [3]. As a result, the final antibiotics in animal faeces may reach tens of mg/kg or more [4] [5], which finally go into the soil environment through the application of agricultural manure. Relevant studies indicate that the amount of antibiotics entering into the soil is comparable to the pesticide application rate [6]. Antibiotics not only affect soil fertility and the safety of agricultural products, but also pollute water by surface runoff and leaching, ultimately threatening animal and human health. Nowadays, the antibiotics directly entering into the soil environment through the application of agriculture manure have become a potential risk of environmental pollution [7]-[9]. Therefore, the antibiotic pollution has become an important environmental organic pollutant and meanwhile a hotspot in environmental science.

Ciprofloxacin (CIP) is a quinolone antibiotic widely used in livestock and poultry breeding industry because of its broad-spectrum antibacterial and efficiency. Along with its popularization, various environmental problems also appear, and have attracted more and more attentions from domestic and foreign scholars. However, current research on CIP mainly focuses on the clinical application of animal diseases, animal in vivo pharmacokinetics, drug residues in animal products, etc. [10]-[13], but seldom concerns the transport behaviors of CIP after it enters into the environment from the perspective of ecological security. There are less research on CIP adsorption and transport in the soil, and the situation is even worse in China.

Prior to this study, the CIP adsorption characteristics in the soil at different pH and ionic strength have been studied using the measurement methods including HPLC, MS, GC, etc. [14]. However, due to the restriction of current laboratory conditions, the samples were analyzed by UV spectrophotometer, and the experimental data were not ideal. There were probably two reasons: Firstly, the soil contained a lot of organic matter that interfered with the analytical measurement, so that the peaks were not obvious at 273 nm wavelength under UV detection conditions; secondly, the mineral composition in the soil, introduced by inappropriate centrifugal filtration, led to the filtrate turbidity, which resulted in inaccurate measurements. At the beginning, it was suspected that the filtration operation was not good enough, but the experimental data were not improved even with qualitative and quantitative filter paper. Further, after EDTA-2Na salt was added in the filtrate as reagent, the results of UV spectral scanning were not improved yet. Considering UV could not accurately measure the CIP in the soil, ICP was tried to measure the CIP in the soil indirectly through measuring the zinc content by creating CIP—zinc complexes in the sample. However, the complex concentration was too low to produce obvious precipitation, and no meaningful experimental data were obtained.

Therefore, considering the complexity of the soil system, in order to analyze the adsorption characteristics of ciprofloxacin under a single factor, the quartz sand of low reactivity was taken as a medium to explore the CIP absorption in quartz sand at different pH and ionic strength, and the mixed displacement experiments were carried out indoors to demonstrate the vertical transport characteristics of CIP in the soil profile. This study can provide a scientific basis for further revealing adsorption mechanism in the soil and the ecological risk assessment and management, and meanwhile plays an important role in the rational use of ciprofloxacin in animal husbandry.

2. Materials and Methods

2.1. Instruments and Reagents

Instruments: Oscillator THZ-92A, Haibo Motion Industries Limited Medical Equipment Factory; Centrifuge TDL-4A, Heidfeld Pradesh Analytical Instruments Co., Ltd.; UV spectrophotometer, Beijing Lebo Tektronix Instrument Co., Ltd.; pH meter PHS-3C, Shanghai Jing Branch Instrument Co., Ltd.; Soil column with a height of 10 cm and an inner diameter of 5 cm, own product made of plexiglass; Peristaltic pump, BT100-1F, Baoding Lange cross-flow pump, Ltd.; Automatic part collection, BSZ-100, Shanghai Huxi Analytical Instrument Factory.

Reagents: CIP reference (purity 98.0%, Tokyo Kasei Kogyo Co., Ltd.); other chemical reagents (analytical grade); test water, crystal-D ultrapure water.

2.2. Test Materials

Quartz sand is a kind of hard and wear-resistant quartzte mineral with stable chemical properties. Its main mineral component is SiO_2 (99.5 wt%), along with tiny iron oxide (<0.02 wt%), clay, mica and organic impurities.

Quartz sand is milky white or colorless translucent solid with the relative density of 2.65. It is insoluble in acid but slightly soluble in KOH solution. The quartz crystals own porous structure, which can absorb various particles or molecules. In general, the adsorption and chemical properties of SiO_2 surface are mainly determined by the surface hydroxyl groups, and CIP is just adsorbed on the surface of the medium by forming hydrogen bonds with hydroxyl groups [15]. Before each experiment, the quartz sand was sieved over 20 - 30 mesh sieve, and then soaked in 0.01 mol/L NaOH for 24 h and 0.01 mol/L HCl for 24 h successively to remove the metal oxide on the surface. After soaked in NaOH or HCl solution, the quartz sand was washed with distilled water until the pH was close to neutral. After dried at 105°C, the spare test materials were obtained [16].

2.3. Experimental Methods

2.3.1. Adsorption Isothermal Experiment

In this experiment, eighteen 50 mL centrifuge tubes were divided into three groups, and 1 g sample of quartz sand was added into each tube respectively. Then, CIP-KCl solutions of different concentrations were prepared using 0.01 mol/L KCl solution as supporting electrolyte, and the CIP concentrations in the same group were 0, 10, 20, 30, 40, 50 mg/L respectively. pH value of the three groups was adjusted to 5, 6 and 7 respectively, and 20 mL CIP solution with different pH was added into each centrifuge tube respectively according to grouping requirements. The solution was sealed and shaken at 25°C for 24 h, and then centrifuged. The supernatant was removed by filtration and the filtrate was diluted by 5 times. The CIP concentration of the supernatant was measured by UV spectrophotometer, and then the adsorption amount was calculated. Three replicates were done for the above operations, and the blank (CIP solution without soil) was set. In order to avoid oscillations during the photo degradation of antibiotics, the whole process was carried out in the dark.

2.3.2. Effects of Ionic Strength on CIP Adsorption

Monovalent metal ions such as Na^+ and K^+ have an impact on the adsorption by competing adsorption sites with antibiotics of cationic state or zero valence state [17]. With the same experimental method described above, the pH of CIP was keep as 6 and the ionic strength were 0.01, 0.05 and 0.1 mol/L, respectively. Then, the CIP concentration in solution was measured and the adsorption capacity was calculated. The measurements above were repeated three times.

2.3.3. Soil Column Outflow Experiments

To analyze the effects of pH and ionic strength on CIP transport in quartz sand, the soil column outflow experiments were carried out. The experiments were conducted indoors in one-dimensional saturated soil column with a height of 10 cm and a diameter of 5 cm. The washed and dried quartz sand was filled into the soil column in 5 times, and the soil column was compacted by plastic compactor each time to ensure uniform distribution of quartz sand particles. Meanwhile, some cotton wool was added to the higher and lower ends of each interface in the soil column in order to prevent quartz sand particles from clogging the water hole. At the each end there was a layer of filter paper which ensured that the solution could be uniformly penetrated into the quartz sand [18]. Taking a beaker and a peristaltic pump as the water supply devices, the peristaltic pump at a fixed speed pushed the background solution (KCL solution concentrations of 0.01 mol/L, 0.05 mol/L and 0.1 mol/L) with adjusted pH (5, 6 and 7) from the bottom slowly into the soil column in order to drain the air in soil column. When the soil column was saturated and the steady flow field was formed, with the pulse input method, 1.0 pv CIP solution of 50 mg/L was input from the upper soil column, and then the soil column was rinsed with KCl background solution. The auto collector collected the flow fluid until the observed CIP concentration tended to zero in the flow fluid (two replicates for each experiment). By the way, $pv = vt/l$ is a dimensionless time, where v represents the pore water velocity (cm/h), t represents time (h), and l represents the soil column length (cm).

2.4. Data Processing

2.4.1. Adsorption Amount

$$S = \frac{(C - C_0)V}{m} \tag{1}$$

where S is the CIP equilibrium adsorption on quartz sand (mg/kg), C_0 is the CIP concentration in the added solution (mg/L), C is the CIP concentration in the balanced solution (mg/L), V is the equilibrium liquid volume (L), and m is the mass of added quartz sand (kg).

2.4.2. Fitting Equation

Empirical formulas of Langmuir and Freundlich isothermal adsorption curves can well fit the adsorption experimental data. By fitting equations, each coefficient and certainty coefficient can be determined, and thus the CIP adsorption capacity on the quartz sand can be analyzed [19].

Langmuir equation:

$$S = \frac{K_L S_m C}{1 + K_L C} \tag{2}$$

Freundlich equation:

$$S = K_F C^n \tag{3}$$

In the above equations, S_m represents the maximum CIP adsorption amount on the quartz sand (mg/kg), K_L represents the affinity of the quartz sand to CIP adsorption, K_F is a constant, and n is the adsorption reaction order (usually less than 1).

3. Results and Discussion

3.1. Static Adsorption

3.1.1. pH Influence on the Adsorption of CIP on Quartz Sand

As **Figure 1** shows, in the pH range of 5.0 to 7.0, the higher pH leads to the lower CIP adsorption amount. pH generally affects the adsorption behaviors of adsorbent in water environment by changing the state of the soil surface and the morphology of solute molecules in solution. The adsorption on quartz sand mainly depends on its surface properties, and the surface reactivity mainly concerns about the surface hydroxyl groups. The CIP contains $-NH_3$ and $-COOH$, which can combine with the H^+ and OH^- in solution respectively, so that the CIP can exist in solution in three forms including cation, zwitterion and anion. According to the pKa value of CIP ($pKa_1 = 6.10$, $pKa_2 = 8.70$) [20], the proportion of CIP in different forms can be calculated as shown in **Figure 2**. When pH is less than pKa1, $-NH_3$ in CIP can form $CIPH^+$ by combining with H^+ in solution, which benefits the CIP adsorption on the negatively-charged surface of quartz sand. In the pH range of 4 to 6, most of CIP is cationic. With pH increasing, the CIP cations in the solution decrease, while neutral ions gradually increase. When pH = 7.5, the CIP almost exists in electrically neutral zwitterionic form (+ - 0) [21], and the CIP cationic groups can combine with negatively charged cations on the CIP surface by cation exchange. With the proportion decline of positive charge in CIP, the proportion of zwitterionic molecules gradually increases, so that the adsorp-

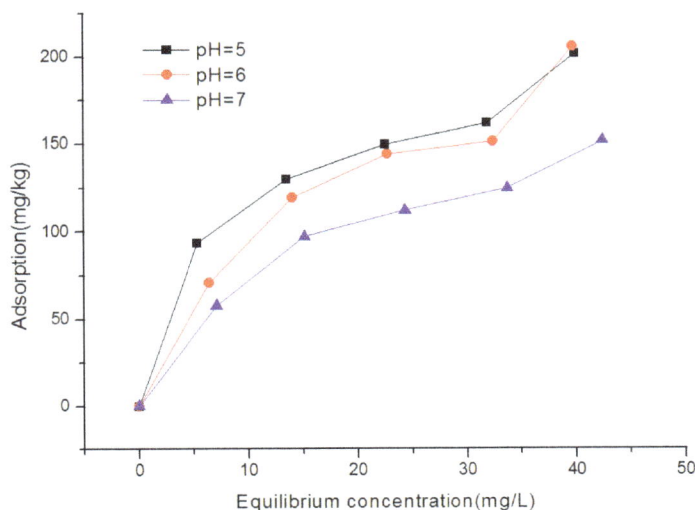

Figure 1. The adsorption isothermal of CIP at different pH in quartz sand.

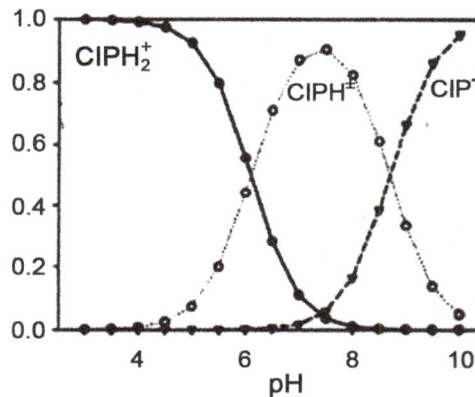

Figure 2. Speciation of CIP at different pH [20].

tion capacity of CIP on the quartz sand gradually weakens. Thus, the absorption of CIP on the quartz sand proceeds mainly through cation exchange adsorption [22]. Meanwhile, the adsorption of CIP on quartz sand also relates to its own pH to some extent. Under acidic conditions, cation exchange effect is helpful to adsorb CIP on quartz sand, the CIP cations can combine with negatively charged quartz sand by cation exchange action; while under neutral conditions, the electrically neutral CIP molecules are absorbed on the quartz sand mainly by Van der Waals force.

3.1.2. Influence of Ionic Strength on the Adsorption of CIP on Quartz Sand

Figure 3 shows the comparative adsorption of CIP at different ionic strength (0.01 mol/L, 0.05 mol/L, 0.1 mol/L) when pH = 6. As can be seen, the bigger the ionic strength is, the lower the adsorption amount of CIP on quartz sand is. Probably, the increase of ionic strength makes the concentration of the competitive cation K^+ on quartz sand increase, so that a large amount of negative charge points are occupied by K^+ on quartz sand surface. Consequently, ions adsorbed by adsorption point tend to be saturated, which reduces the electrical adsorption ability of the quartz sand and thus the CIP adsorption amount on the quartz sand.

Table 1 shows the K_d value distribution coefficient of CIP in quartz sand at different ionic strength when pH = 6 (K_d is the ratio of CIP absorption amount per unit mass of soil to the CIP concentration in equilibrium solution, $L \cdot kg^{-1}$). K_d value can be used to describe the distribution of different cations between solid phase and liquid phase in soil. Low K_d value means that most CIP exists in the soil solution with potential activity and transport ability, while high K_d value indicates that there is a strong affinity between CIP and the surface of soil particles so that CIP can be easily adsorbed by the surface of soil particles.

As shown in **Table 1**, K_d value on the quartz sand gradually decreases with the increase of CIP concentration in equilibration solution. Meanwhile, as ionic strength of the solution increases, K_d value of CIP on quartz sand decreases. Through the analysis above, the adsorption of CIP on quartz sand proceeds mainly by the ion exchange between the negatively charged quartz sand surface and CIP cationic groups. With the increase of cation concentration in quartz sand solution, they compete with CIP cation active groups for active adsorption sites, so that the adsorption amount of CIP decreases. With the increase of initial CIP concentration, K_d value of CIP on quartz sand decreases, indicating the decrease of absorption rate in balanced solution. In low equilibrium concentration, specific adsorption with high energy absorption sites dominates the adsorption process. With the increase of CIP concentration in balanced solution, specific adsorption sites are gradually occupied and non-specific adsorption increases. As a result, the adsorption capacity of quartz sand relatively reduces, and K_d value of CIP decreases [23]-[24].

3.1.3. Data Fitting for the CIP Adsorption on Quartz Sand

Figure 4 shows the fitting results of the CIP adsorption on the quartz sand by the Langmuir and Freundlich adsorption isothermal models. Obviously, the CIP adsorption characteristics on the quartz sand are nonlinear, and the obtained adsorption parameters are shown in **Table 2**. In the fitting process by Langmuir equation, the correlation coefficient R^2 is larger than 0.806, while the correlation coefficient R^2 is larger than 0.886 by Freundlich equation. Meanwhile, n is between 0.3 and 0.6, which shows that the adsorption isotherms are S-type adsorption

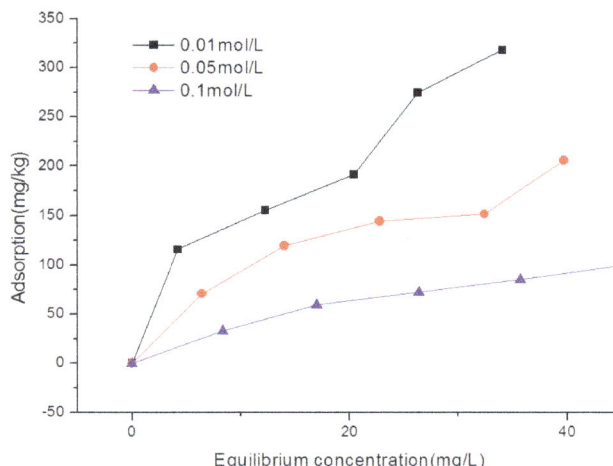

Figure 3. The adsorption isothermal of ciprofloxacin at different ionic strength in quartz sand.

Table 1. The distribution coefficient of CIP at different ionic strength in quartz sand when pH = 6.

Ionic strength (mol/L)	Concentration (mg/L)				
	10	20	30	40	50
0.01	15.7	14.5	8.7	10.5	9.3
0.05	11.0	8.5	6.3	4.7	5.2
0.1	2.9	3.4	4.4	3.7	3.9

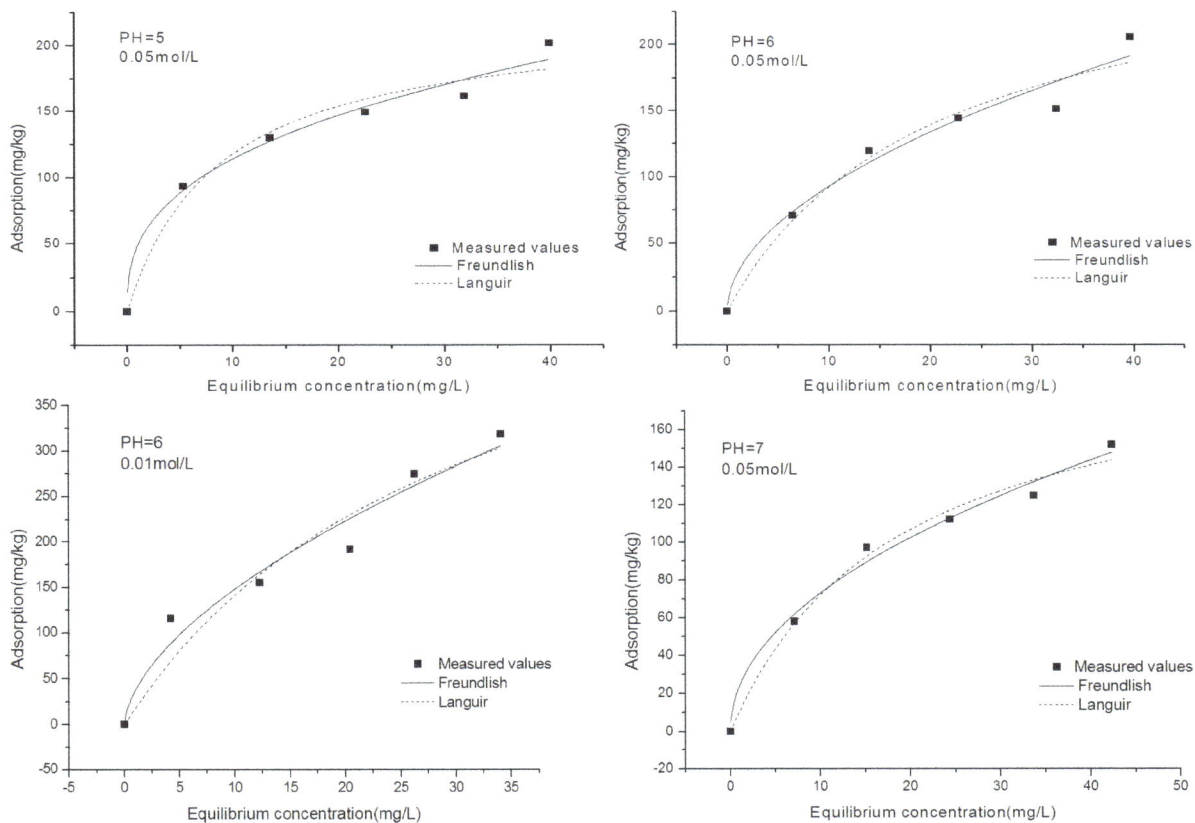

Figure 4. Adsorption curves of ciprofloxacin at different pH and ionic strength.

Table 2. The adsorption isotherm parameters of CIP at different pH and ionic strength.

pH	Ionic strength (mol/L)	Freundlich			Langmuir		
		K_F	n	R^2	S_m	K_L	R^2
5	0.05	48.8	0.368	0.929	223.3	0.111	0.842
	0.01	37.9	0.592	0.886	579.7	0.032	0.806
6	0.05	27.5	0.527	0.901	286.2	0.047	0.876
	0.1	10.0	0.603	0.985	173.8	0.028	0.989
7	0.05	23.8	0.488	0.960	207.4	0.053	0.950

Note: R^2 is the certainty coefficient, which represents the fitness of fitting values and measured values. The closer to 1 means the better fitness.

isotherms and that the adsorption type is nonlinear adsorption. Generally, this adsorption type can be attributed to the competition between water molecules and CIP molecules for adsorption sites on the soil surface. At low CIP concentrations, the affinity of CIP to the soil is weaker than that to the water phase; while with the increase of CIP concentration, its affinity to the soil increases, so that the adsorption amount increases. In contrast, if "$n > 1$", the adsorption isotherm belongs to L-type adsorption isotherms. At low concentrations, the affinity of CIP to the soil is stronger than that to the water phase; while with the increase of CIP concentration, its affinity to the soil decreases, so that the adsorption weakens [25]. K_F and K_L can be used to evaluate the adsorption affinity of CIP on the quartz sand. As can be seen from **Table 2**, at a constant pH, the higher the ion intensity is, the smaller the affinity of CIP to the soil is, and thus the greater the fitted maximum adsorption amount is. The result is consistent with the experimental result. Under the same ionic strength, for Freundlich model, the higher the pH is, the smaller the affinity of CIP to the soil is, which is consistent with experimental result; while the adsorption isotherm fitting result by Langmuir model is not consistent with the experimental result. That is to say, Langmuir equation has certain limitations to fit the adsorption isotherm of competitive ions.

Adsorption reaction is the most important process during the transport and transformation of CIP in quartz sand, which can reflect the degree of interaction between CIP and the quartz sand and predict the stability of CIP in the quartz sand. Generally, the CIP is very stable if it can be strongly adsorbed by quartz sand, so that it is easier to accumulate in quartz sand instead of migration. In contrast, if the CIP cannot be absorbed and fixed easily on quartz sand, it is easy to be transported into surface water or groundwater under eluviations effect. The above results show that low pH and low ionic strength are conductive to the fixing of CIP in the soil, which eliminate its migration with the water.

3.2. Soil Column Transport Experiments

3.2.1. The Breakthrough Curves of CIP Transport in Quartz Sand at Different pH

Figure 5 shows the breakthrough curves of CIP in quartz sand at different pH (5, 6, 7) when the ionic strength is 0.05 mol/L. The CIP solution of pH = 7 starts to outflow when pv = 0.47, then the CIP solution of pH = 6 starts to outflow when pv = 0.51, and finally the CIP solution of pH = 5 starts to outflow when pv = 0.54. The greater the pH is, the sooner the outflow is. The reason is that the greater pH usually weakens the CIP adsorption capacity on quartz sand, so that the CIP can outflow very fast, with a stronger migration. Prior to 1.0 pv, it is the adsorption stage, followed by the desorption stage. When pv = 1.0, the backwash stage (adsorption-desorption phase) starts, and then the CIP concentration gradually increases. When pv = 1.63, the solution of pH = 5 reaches the maximum concentration, with a relative concentration C/C_0 of 0.88; when pv = 1.56, the solution of pH = 6 reaches the maximum concentration, with a relative concentration C/C_0 of 0.96; when pv = 1.65, the solution of pH = 7 reaches the maximum concentration, with a relative concentration C/C_0 of 0.89. After that, the CIP concentration decreases rapidly. During this process, the solution of pH = 7 desorbs most rapidly, followed by the solution of pH = 6 and the solution of pH = 5. The higher the pH is, the smaller the adsorption amount is, and thus the greater the desorption amount is. In the transport process from solution outflow to the concentration being zero, the volume of the solution of pH = 5 is 2.50 pv, that of pH = 6 is 2.48 pv, and that of pH = 7 is 2.36 pv. The CIP starts to outflow earliest when pH = 7, followed by pH = 6 and pH = 5. This difference may be related to the different adsorption capacity of CIP at different pH. The lower the pH is, the stronger the adsorption capacity is, and thus the longer the outflow time is. On the contrary, the greater the pH is, the shorter the outflow

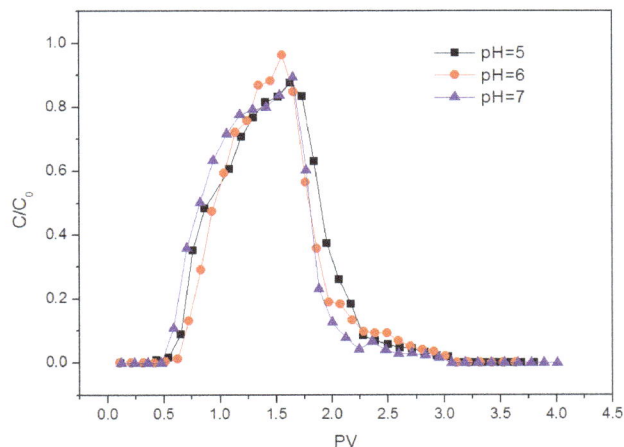

Figure 5. The breakthrough curves at different pH when the ionic strength is 0.05 mol/L.

time is. When pH = 5, the desorption phase has a long tail due to the slow desorption caused by large adsorption amount. Hence, if the soil is in alkaline environment, the CIP concentration in the soil will increase, and the CIP is easier to migrate in the soil, thus increasing the risk of groundwater contamination.

3.2.2. The Breakthrough Curves of CIP Transport in the Quartz Sand at Different Ionic Strength

Figure 6 shows the breakthrough curves of CIP in quartz sand at different ionic strength (0.01, 0.05, 0.1 mol/L) when pH = 6. In the adsorption phase, the CIP adsorbs fast and outflows slowly at the ionic strength of 0.01 mol/L. The greater the ionic strength is, the faster the outflow is. The reason may be that a large number of K^+ ions adsorbed on the quartz solid surface compete with CIP for the adsorption sites, resulting in the decrease of CIP adsorption amount on quartz sand and further delaying the outflow. In the desorption stage, at the ionic strength of 0.01 mol/L, CIP desorbs slowly, and there presents a tail. This process is a continuous and slow desorption process. With the ionic strength increasing, the desorption becomes fast, which promotes the transport of CIP in quartz sand. At the same time, when the ionic strength is large, a large number of K^+ ions occupy the negatively charged adsorption sites on the surface of medium, and the negatively charged adsorption sites tend to be saturated, which reduces the adsorption of CIP. In addition, the increase of ionic strength will enhance the ionic interaction in systems and reduce the ion activity coefficient, which decreases the effective concentration of CIP and accelerates its transport in the soil column. Hence, if the soil is contaminated by a high-salt wastewater, it will accelerate the desorption of CIP and promote its transport in the soil, thus polluting the groundwater.

4. Conclusions

In summary, the CIP adsorption and transport characteristics in quartz sand were investigated at different pH and ionic strength conditions, and some conclusions can be drawn as follows:

1) The adsorption of CIP on the quartz sand belonged to cation exchange adsorption, and the variations of pH and ion morphology had a significant influence on the adsorption capacity.

2) When pH was in the range from 5 to 7, the higher the pH was, the lower the adsorption capacity of CIP on quartz sand was; the greater the ionic strength was, the lower the CIP adsorption amount on quartz sand was. That was to say, the low pH and low ionic strength were in favor of CIP adsorption on quartz sand. With the increase of CIP concentration in balanced solution, the K_d value of CIP in quartz sand decreased; meanwhile, with the increase of ionic strength in the solution, the K_d value of CIP in quartz sand decreased.

3) The CIP adsorption on quartz sand could be described by the Langmuir and Freundlich equations, and the Freundlich equation fitted the adsorption isotherms better.

4) When pH was in the range from 5 to 7, the higher the pH was, the faster the transport of the CIP on quartz sand was, and thus the earlier the outflow was. That was to say, the higher the pH was, the stronger the adsorption capacity was, and thus the shorter the outflow time was. Meanwhile, the increase of ionic strength would

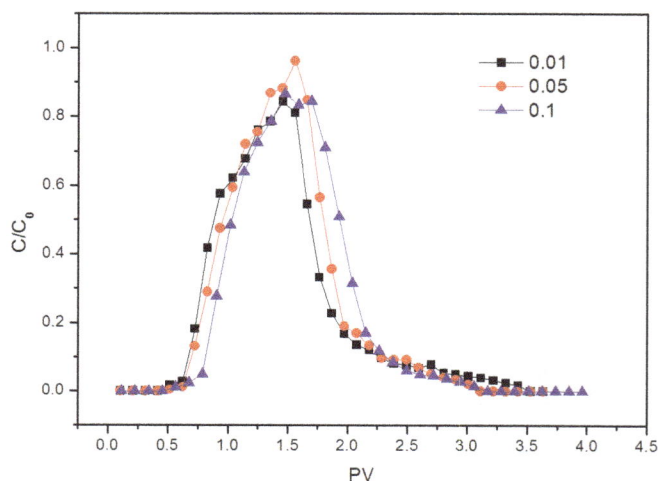

Figure 6. The breakthrough curves at different ionic strength when pH = 6.

reduce the absorption of CIP on quartz sand, which led to the faster transport and the earlier outflow. As a large number of K^+ ions occupied the negatively charged adsorption sites, the medium surface tended to be saturated, so that the adsorption capacity of CIP on quartz sand was reduced.

Acknowledgements

This study was accomplished with participation of the undergraduates innovative experimental group members Jiang-Long He, Yu Li, Zhao-Xi Fang, Xin Li and Ya-Kun Hu.

Foundation Item

National Natural Science Foundation of China (40771095), Qingdao people's livelihood project (13-1-3-132-nsh).

References

[1]   Kools, S.A.E., Moltmann, J.F. and Knacker, T. (2008) Estimating the Use of Veterinary Medicines in the European Union. *Regulatory Toxicology and Pharmacology*, **50**, 59-65. http://dx.doi.org/10.1016/j.yrtph.2007.06.003

[2]   Hou, F.L. (2003) Feed Additive Application Da Quan. China Agriculture Press, Beijing, 1-45.

[3]   Cui, H. and Wang, S.P. (2012) Adsorption Characteristics of Ciprofloxacin in Ustic Cambosols. *Chinese Journal of Environmental Science*, **33**, 2895-2900.

[4]   Hu, X.G., Luo, Y., Zhou, Q.X., *et al.* (2008) Determination of Thirteen Antibiotics Residues in Manure by Solid Phase Extraction and High Performance Liquid Chromatography. *Chinese Journal of Analytical Chemistry*, **36**, 1162-1166. http://dx.doi.org/10.1016/S1872-2040(08)60063-8

[5]   Liu, X.C., Dong, Y.H. and Wang, H. (2008) Residues of Tetracyclines in Animal Manure from Intensive Farm in Jiangsu Provilace. *Journal of Agro-Environment Science*, **27**, 1177-1182.

[6]   Haller, M.Y., Müller, S.R., McArdell, C.S., *et al.* (2002) Quantification of Veterinary Antibiotics (Sulfonamides and Trimethoprim) in Animal Manure by Liquid Chromatography-Mass Spectrometry. *Journal of Chromatography A*, **952**, 111-120. http://dx.doi.org/10.1016/S0021-9673(02)00083-3

[7]   Halling-Sorensen, B., Nors Nielsen, S., Lanzky, P.F., *et al.* (1998) Occurrence, Fate and Effects of Pharmaceutical Substances in the Environment—A Review. *Chemosphere*, **36**, 357-393. http://dx.doi.org/10.1016/S0045-6535(97)00354-8

[8]   Jemba, P.K. (2002) The Potential Impact of Veterinary and Human Therapeutic Agents in Manure and Biosoilds on Plants Grown on Arable Land: A Review. *Agriculture, Ecosystems and Environment*, **93**, 267-278.

[9]   Kumar, K., Gupta, S.C., Chander, Y., *et al.* (2005) Antibiotic Use in Agriculture and Its Impact on the Terrestrial Environment. *Advances in Agronomy*, **87**, 1-54. http://dx.doi.org/10.1016/S0065-2113(05)87001-4

[10] Andon, A., Martinez-Larranage, M.R., Diaz, M.J., *et al.* (1995) Pharmacokinetics and Residues of Enrofloxacin in

Chickens. *American Journal of Veterinary Research*, **56**, 502-506.

[11] Manceau, J., Gicque, M., Laurentie, M., *et al.* (1999) Simultaneous Determination of Enrofloxacin and Ciprofloxacin in Animal Biological Fluids by High-Performance Liquid Chromatography Application in Pharmacokinetic Studies in Pig and Rabbit. *Journal of Chromatography B*, **726**, 175-184.

[12] Dosogne, H., Meyer, E., Sturk, A., *et al.* (2002) Effect of Enrofloxacin Treatment on Plasma Endotoxin during Bovine *Escherichia coli* Mastitis. *Inflammation Research*, **51**, 201-205. http://dx.doi.org/10.1007/PL00000293

[13] Vancutsem, P.M., Babish, J.G. and Schwark, W.S. (1990) The Fluoroquinolone Antimicrobials: Structure, Antimicrobial Activity, Pharmacokinetics, Clinical Use in Domestic Animals and Toxicity. *The Cornell Veterinarian*, **80**, 173-186.

[14] Ye, X.Q., Liu, D.H. and Chen, J.C. (2005) Advances in the Rapid Detection of Antibiotics Residues in Dairy Products. *Transactions of the CSAE*, **21**, 181-185.

[15] Yang, J.W., Xu, L.Z., Jiang, X., *et al.* (2005) Study on the Adsorption of Aluminum Ions on Silica by X-Ray Photoelectron Spectroscopy. *Chinese Journal of Spectroscopy Laboratory*, **22**, 666-668.

[16] Chu, L.Y., Si, Y.B. and Zhou, D.M. (2011) Study on the Migration of Nano-Hydroxyapatite and Carrying Heavy Metals-Cu in Quartz sand Column. Anhui Agricultural University, Hefei.

[17] Qi, H.M., Lu, L. and Qiao, X.L. (2009) Progress in Sorption of Antibiotics to Soils. *Soils*, **41**, 703-708.

[18] Sun, Y.Y. and Xu, S.H. (2013) Characteristic of $Zn^{2+}/Cd^{2+}/NH^{4+}$ Transport in Soils with Different pH Value and Ionic Strength. *Transactions of the Chinese Society of Agricultural Engineering*, **29**, 218-227.

[19] Wang, K.L., Xu, S.H., Yang, Y.L., *et al.* (2011) Study on Zn and Cd Colloid-Affected Adsorption in Three Different Soils. *Soils*, **43**, 239-246.

[20] Wu, T.X. (2012) Adsorption of Ciprofloxacin on Three Different Soils. *Northern Environmental*, **25**, 54-56.

[21] Gu, C. and Karthikeyan, K.G. (2005) Sorption of the Antimicrobial Ciprofloxacin to Aluminum and Iron Hydrous Oxides. *Environmental Science & Technology*, **39**, 9166-9173. http://dx.doi.org/10.1021/es051109f

[22] Wu, T.X. (2009) Studies on the Adsorption of Antibiotics in Soil. Northwest Normal University, Lanzhou.

[23] Basta, N.T. and Tabatai, M.A. (1992) Effect of Cropping Systems on Adsorption of Metals by Soils: III. Competitive Adsorption. *Soil Science*, **153**, 331-337. http://dx.doi.org/10.1097/00010694-199204000-00010

[24] Yu, S., He, Z.L., Huang, C.Y., Chen, G.C. and Calvert, D.V. (2002) Adsorption-Desorption Behavior of Copper at Contaminated Levels in Red Soils from China. *Journal of Environmental Quality*, **31**, 1129-1139. http://dx.doi.org/10.2134/jeq2002.1129

[25] Zhang, W., Wang, W.Q., Wang, J.J. and Li, G. (2010) *Journal of Sichuan Normal University*, **33**, 366-371.

6

Denitrification in a Soil under Wheat Crop in the Humid Pampas of Argentina

Liliana Inés Picone[1*], Cecilia Videla[1], Calypso Lisa Picaud[2], Fernando Oscar García[3], Roberto Héctor Rizzalli[1]

[1]Facultad Ciencias Agrarias, Unidad Integrada Balcarce Instituto Nacional Tecnología Agropecuaria (INTA) UNMdP, Balcarce, Argentina
[2]AgroParisTech, Institut des sciences et industries du vivant et de l'environnement, 16 rue Claude Bernard, Paris, France
[3]International Plant Nutrition Institute (IPNI), Programa Cono Sur de Latinoamérica, Buenos Aires, Argentina
Email: [*]picone.liliana@inta.gob.ar

Abstract

The need to accurately estimate gaseous nitrogen losses from soils is required to have a better understanding of the processes involved as well as soil and environmental conditions, and management practices contributing to these emissions. The objective was to quantify the denitrification rate using undisturbed cores with acetylene, as related to nitrogen (N) fertilization rate in a spring wheat crop (*Triticum aestivum L.*) under conventional tillage. Soil denitrification losses remained low throughout most of the growing season, when water-filled pore space (WFPS) was below 60%, ranging from 0.79 to 447.3 g N_2O-N $ha^{-1} \cdot day^{-1}$ in the fertilized plot and was less than 47.3 g N_2O-N $ha^{-1} \cdot day^{-1}$ in the control. Denitrification rates were the highest when N fertilizer was applied after frequent and intensive rain. A good correlation was found between the logarithm of the daily denitrification rate and WFPS (r = 0.67, n = 90); however the NO_3-N concentration was not a good indicator (r = 0.21, n = 90). Cumulative N_2O-N losses by denitrification averaged 3.5 and 0.9 kg N_2O-N ha^{-1} in the fertilized and unfertilized treatment, respectively, during a period of 4 months this difference was not significant. Most N_2O-N losses occurred early in the spring; therefore sampling schedules need to focus on this period.

Keywords

Nitrogen, Urea-N, Losses, Water Content, Soluble Organic Carbon

---

[*]Corresponding author.

1. Introduction

The intensive use of natural resources by man, especially non-renewable natural resources has led to a gradual degradation of environmental quality, compromising the sustainability of ecosystems. One consequence of this intensification is global warming, caused by increasing concentration of atmospheric greenhouse gases such as nitrous oxide (N_2O). Comparatively, N_2O has a global warming potential 298 times higher than CO_2 over a 100-year time horizon [1]. This gas accounts for about 8% of the global annual emissions of anthropogenic greenhouse gases, and its concentration has increased considerably over the past few decades and continues to increase at an annual rate of 0.25% [1].

Nitrous oxide is produced in soils by microbial transformations during nitrification and denitrification processes [2]. Nitrification is the dominant process contributing to N_2O emissions at water-filled pore space (WFPS) between 35% and 60% [3] while denitrification increases with increasing WFPS above 60% [4]. Therefore, denitrification is considered as the predominant pathway responsible for N_2O production, particularly under humid climates, and in very wet and waterlogged soils [5] [6]. Soil water content is commonly identified as the most important regulator of soil denitrification [7], but this process is also controlled by nitrate (NO_3-N) concentration, available carbon (C) and other soil properties such as temperature and pH [8]. Due to the complex interactions among these factors, large temporal and spatial variations of N_2O emissions are usually observed in cropland soils.

The soil and crop management practices can also regulate the N_2O emissions through its effect on the mentioned soil properties. One agricultural management practice is the application of nitrogen (N) fertilizers that provides substrate for denitrification. Several studies have shown that the N_2O emissions from agricultural soils increase with N application [9]-[12]. Worldwide, IPCC [1] has estimated that the emission factor is 1% of applied N. Zhang *et al.* [13] reported N losses by denitrification that increase with the fertilization rate, ranging from 0.28% to 0.49% of the applied N fertilizer in a winter wheat crop. In contrast, a study that evaluated the effects of water table management on denitrification during the corn growing season, showed no consistent differences between 120 and 200 kg N ha^{-1} rates [14]. The synchrony between the supply and demand of N is an important factor in determining the availability of soil N and release of N_2O from agricultural soils. Fertilizer N application prior to crop planting results in increased soil N with no N uptake by plant and greater potential of N_2O emissions. However, if the N fertilizer is used efficiently by the crop, for example by adjusting N applications to crop needs, less N_2O should be generated and released to the atmosphere. In fact, in study conducted on maize crop under no-tillage, denitrification losses were greater when urea was applied at planting than those when fertilizer was applied at six leaf stage (V6) [10].

The Pampas region is the main producer of maize, wheat, sunflower and soybeans in Argentina and is one of the most important agricultural areas in the world [15], with climatic and soil conditions suitable for grain crop production. This region includes several sub-regions, one of which is the southeastern area of Buenos Aires province, with a cropped area of 0.7 million ha and a wheat production of 2.5 million Mg [16]. Field experiments conducted in that area have shown that the wheat crop yield has increased an average rate of 38 kg·ha^{-1}·yr^{-1}, as a result of increments in applied N; however the N recovery efficiency in grain plant was low, ranging on average from 32% to 41% [17], depending on the timing of N fertilizer application and weather conditions. Then, it is probably that some mechanism of N loss is occurring in the soil and may help to explain the low N use efficiency by plants. In fact, the southeast wheat belt is characterized by a high probability of rainfall during the fallow and the early stages of the wheat crop periods, along with low evaporative demand [18]. This situation combined with availability of NO_3-N from mineralization of organic N during the fallow period or from applied N fertilizer at sowing, likely results in favorable conditions for denitrification and associated N_2O emissions.

In Argentina, the total greenhouse gas emissions were estimated at 238702.9 Gg of CO_2 equivalents in 2000. Of this amount, N_2O emissions from agricultural activities accounted for approximately 43% [19]. The estimation of Argentina's greenhouse gas emissions is based on the IPCC methodology, which includes gas emission factors based on international gas emission studies. Research on N_2O emissions in Argentina is crucial in order to set up an appropriate national inventory of greenhouse gases and to calculate the emission factors based on specific experimental measurements that could then be included in the IPCC methodology [20]. Therefore, more field investigation on N_2O fluxes, about regulating factors and processes that contribute to N emissions, is needed to assess their contribution to the N cycling in order to adopt management practices leading to reduced emissions from agricultural soils.

Therefore, the aims of this study were: 1) to quantify the denitrification rate related to N fertilization rate and 2) to identify the soil factors, such as water content and NO_3-N concentration that control this process, in a wheat crop under conventional tillage.

2. Methods and Materials

2.1. Site Description

The denitrification measurements and the field experiment were conducted at the Balcarce-Experimental Station of the National Institute of Agricultural Technology (INTA), located in the southeastern area of Buenos Aires province, Argentina (37°45'S lat. 58°18'W long.; 130 m above sea level). The soil is a complex of a fine, mixed, thermic Typic Argiudoll and a fine, illitic, thermic Petrocalcic Paleudoll (USDA Soil Clasification). It has a loam texture, with an organic matter content of 58.4 $g \cdot kg^{-1}$, 23.7 $cmolc \cdot kg^{-1}$ of cation-exchange capacity and a pH of 5.8 (ratio soil:water, 1:2.5) in the top soil (0 - 20 cm depth).

The region has a humid-subhumid mesothermal climate with maximum and minimum monthly mean temperatures of 27.4°C and 3.1°C in January and July, respectively. The 40-yr average annual precipitation is 922.4 mm, 45% of which occurs during the growing season of wheat. Air temperature as well as precipitation during the experiment is shown in **Figure 1**.

2.2. Experimental Design

The overall field experiment was designed to compare the effects of two tillage systems: no tillage (NT) and conventional tillage (CT), each with two rates of N fertilization: 0 and 120 kg N ha^{-1}. The experimental design was a split-plot arrangement set as a randomized complete block with three replications. Tillage systems were applied to the main plots while the two fertility treatments were applied to the subplots. However, measurements of denitrification fluxes were only made on the CT treatment. The CT consisted of disking to mix crop residues into the soil, one moldboard plowing to the depth of 20 cm followed by one to three disking to the depth of 8 to 10 cm, before wheat planting date. Plots were seeded with wheat (*Triticum aestivum L.*) in the second week of July and it was harvested early in December.

Nitrogen fertilizer was applied as urea at a rate of 120 kg N ha^{-1} and it was broadcasted on the soil surface, three days before wheat planting. At sowing, the experiments were fertilized with P as triple superphosphate at a rate of 31 kg P ha^{-1}.

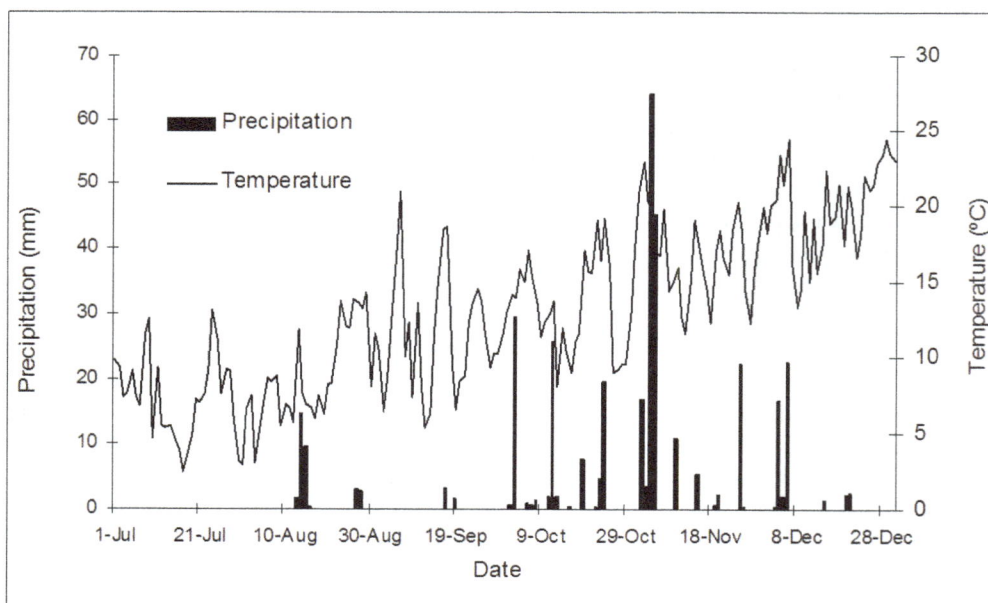

Figure 1. Rainfall and daily mean air temperatures during the growing season of spring wheat (July-December).

2.3. Denitrification Measurements

Denitrification rate measurements were made approximately weekly, during the wheat growing season, from early August (16 August) to mid-December (15 December) by the acetylene inhibition method [21]. Eight intact soil cores (4.2 cm in diameter by 15 cm long) were randomly taken from between rows in each plot using poly-vinyl chloride cylinders (PVC) of 20-cm length. The cylinders were immediately brought to the lab, and both ends were capped with rubber stoppers, the upper stopper has a rubber septum for gas sampling. Approximately 10% of the headspace volume in the cylinder was removed using syringe and then an equivalent volume of ace-tylene (generated from calcium carbide and distilled water) was injected to the headspace. The cores were then incubated for 24 hours, outside in the shade. Gas samples were removed from each cylinder after 0 and 24 hours of incubation and subsequently stored in evacuated vials. The N_2O concentration in a 1 mL gas sample was de-termined using a 5890 series-II Hewlett Packard (Palo Alto, CA) gas chromatograph equipped with a Porapak Q column at 35°C and a [63]Ni-electron capture detector (ECD). The injector was set at 50°C and the ECD at 300°C, and the carrier gas was N_2 flowing at a rate of 15 mL·min^{-1}.

Daily denitrification rates were calculated as the change in N_2O concentration over the time using linear re-gression analysis. Denitrification rates were corrected for the N_2O dissolved in the liquid phase using the Bunsen absorption coefficient [22]. Cumulative N_2O losses over 4 months were calculated by interpolating linearly the daily denitrification rates between consecutives sampling dates and integrating the area.

2.4. Soil Measurements

At each date of denitrification rate measurement, the surface soil (0 - 15 cm depth) was sampled by collecting soil cores beside each cylinder for determination of NO_3-N and gravimetric water content. Soil NO_3-N was ex-tracted by shaking 20 g of field moist soil with 80 mL of 0.5 M K_2SO_4 solution (extractant:soil ratio of 4:1, w:w) for 1 hour at 200 rpm, and then filtering through filter paper (Whatman N°42). Filtrates were kept frozen until analyzed by steam distillation [23]. Gravimetric soil moisture was determined after oven drying of fresh sub-samples at 105°C for 24 hours.

Soil samples (0 - 15 cm depth) for water-extractable organic C (WEOC) were taken monthly during the wheat growing season. The method of WEOC determination was adapted from Mebius [24]. Briefly, 10 g of fresh soil sample was shaken with 20 mL of distilled water for 30 min (140 rpm) and centrifuged at 19,500 g for 5 min. The supernatant was filtered through a 0.22 um membrane filter (Millipore Corp.) and analyzed for total organic C by dichromate oxidation in the presence of H_2SO_4, involving boiling and refluxing conditions during 30 min at 150°C. The excess of dichromate was titrated with ferrous ammonium sulfate.

Soil bulk density was determined by the cylinder method [25]. Undisturbed soil cores (5 cm in diameter by 5 cm long) were taken from 0 to 15 cm depth. On average, bulk density was 1.20 Mg·m^{-3} in the CT plots.

Gravimetric water content was converted to WFPS using the equation:

$$WFPS = (\theta * \delta)/f$$

where: θ = gravimetric water content; δ = soil bulk density; f = total soil porosity. Soil porosity was calculated from the average bulk density, assuming a particle density of 2.65 Mg·m^{-3}.

2.5. Statistical Analysis

The denitrification rate values were checked for normality using the Shapiro-Wilk test [26], and they were log$_{10}$ transformed prior to statistical analysis due to the skewed distribution and unequal variance. The effect of ferti-lization rate on daily denitrification fluxes, WFPS, and NO_3-N and WEOC contents was evaluated for each sampling date using the General Linear Model (GLM) procedure of SAS statistical program [27]. The PROC CORR function of SAS was used to determine the Pearson correlation coefficients between denitrification rates and NO_3-N content using the individual data of each sampling date. A probability level of p = 0.05 was used to indicate significant differences.

3. Results and Discussion

3.1. Air Temperature and Precipitation

Over the 4-month evaluation period, the mean daily air temperature was 14.0°C ± 5.0°C, and ranged from 3°C (3

August) to 24.4°C (6 and 29 December). The coldest month was August with air temperatures below 10°C, and the warmest months were November and December with average temperatures equal or above 16.8°C (**Figure 1**). During the evaluation period, 38 days received precipitation, approximately 28% of the total period. Precipitations were more frequent and intensive in November and December, and less frequent in September. The lowest total precipitation occurred in September with only 4.9 mm of rainfall while the highest total precipitation was in October and November with 94.1 and 171.5 mm, respectively (**Figure 1**).

3.2. Soil Mineral Nitrogen, Water-Filled Pore Space and Water Extractable Organic Carbon

There was a significant effect of the fertilized treatment on NO_3-N concentration in the 0 to 15 cm depth. Application of N fertilizer resulted in higher NO_3-N concentrations ($p < 0.05$) in August, September, October and 17 November compared with the unfertilized treatment. Nitrate concentration was more than twice in the fertilized soil in comparison with the control soil. However, on 3, 8 and 22 November and during December, there were no differences ($p > 0.05$) in NO_3-N level between both treatments (**Figure 2**).

The maximum value of soil NO_3-N concentration of the control was 22.10 mg N kg^{-1} at the beginning of the growing season, and gradually decreased to low values in December, below 5.88 mg N kg^{-1} (**Figure 2**). Previous study conducted in these soils also showed relatively low inorganic N contents early in the growing season of wheat related to the fertilized treatment [28]. This initial concentration could be attributed to mineralization and subsequent nitrification of organic N from previous crop residues. For this site, the mineralized organic N estimated during a wheat season under CT was 65 kg N ha^{-1} [29].

The highest content of NO_3-N (82.22 mg N kg^{-1}) was found shortly after 1 month fertilizer application (**Figure 2**) and represents 108 kg N ha^{-1} (subtracted the control), about 90% of the applied N. This result indicates that urea hydrolysis and NH_4 oxidation was almost completed during this time. Ferrari [30] reported that soils from the study area exhibit high urease activity, 43.8 mg N $kg^{-1} \cdot h^{-1}$, and thus high potential for urea hydrolysis. On the other hand, Navarro *et al.* [31] measuring nitrification rates under laboratory conditions, found that there was no delay in the nitrification process even with the highest concentration of ammonium, suggesting a high density of nitrifying microorganisms. This peak in NO_3-N content was followed by a sharp decrease, probably caused by a combination of immobilization, plant uptake and gas emission by denitrification/ nitrification. A second fall in inorganic N in mid-October coincided with the early stem elongation to anthesis wheat stages, the period during which the crop can accumulate up to 75% of total N in above-ground biomass at maturity [32] [33]. At wheat physiological maturity, the soil N content reached values similar to those measured in the control.

The WFPS was not affected ($p > 0.05$) by N application but fluctuated with time. It decreased from August to September, increased rapidly in response to rainfall events in October and November, and again decreased thereafter. The WFPS peaked on 16 August, 13 October and 8 November, reaching values higher than 50% (**Figure 3**). These values of water content correspond to aeration conditions that encourage N_2O production by both

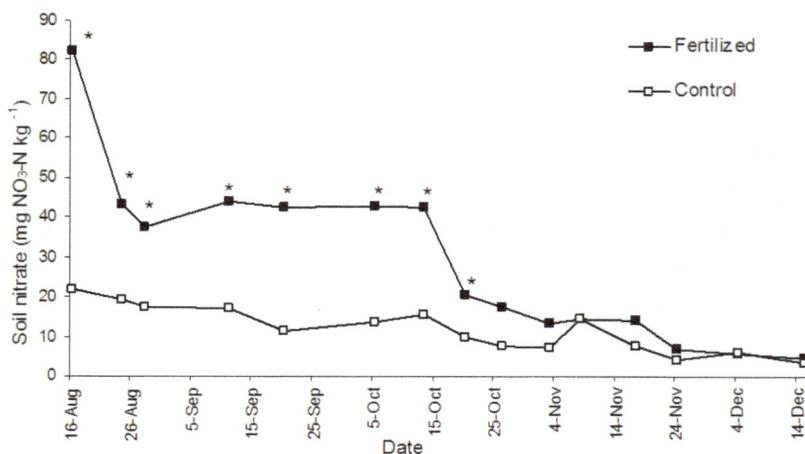

Figure 2. Soil nitrate (NO_3-N) content in the 0 to 15 cm depth for the control (without N) and fertilized treatments during the growing season of spring wheat. Asterisks indicate significant differences between both treatments ($p < 0.05$).

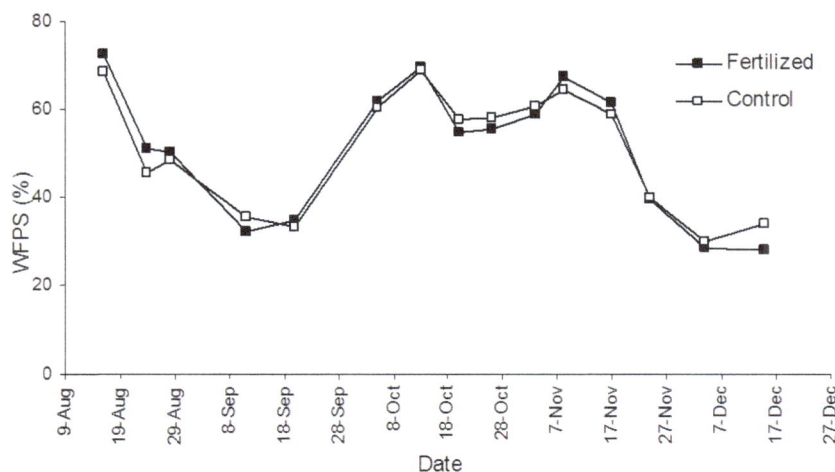

Figure 3. Distribution of soil water-filled pore space (WFPS) in the 0 to 15 cm depth for the control and fertilized treatments throughout the wheat growing season.

nitrification and denitrification [34]. The highest value of WFPS, 70%, occurred immediately after several consecutive days of rain, corresponding to gravimetric water content of 0.36 $g \cdot g^{-1}$ which is higher than the water content at field capacity. Towards the end of the wheat growing season, the WFPS was about 28%, close to the water content at permanent wilting point.

The mean concentration of soil WEOC was higher in the unfertilized treatment than in the fertilized treatment (**Table 1**), but the differences were only significant ($p < 0.05$) in November. Trends toward lower WEOC in the fertilized treatment than in the unfertilized treatment suggest that WEOC in fertilized soils might have been metabolized at a greater rate. The WEOC concentrations remained relatively unchanged and low during the plant growth, ranging between 12.52 and 32.13 mg C kg^{-1} (**Table 1**). Similar results were found by Elmi *et al.* [14] who reported that WEOC concentration was relatively uniform, ranging from 10 to 30 $mg \cdot kg^{-1}$ (15 cm depth) in a corn field. The WEOC is a dynamic pool C, controlled by several physical and biological mechanisms. Sorption and desorption are two key processes for WEOC stabilization and production in soils, soluble root exudates and microbial processes can also contribute with dissolved C, and plants and microorganisms can consume WEOC [35]. Then, the variety of factors determining the concentration of WEOC indicates that is difficult to explain its variation over the time in the soil.

3.3. Daily Denitrification Rates

Denitrification rates varied greatly throughout the wheat growing season, being more variable in the fertilized treatment (**Figure 4**). Daily rates of denitrification ranged from 0.776 to 472.83 ng N_2O-N $cm^{-2} \cdot day^{-1}$ in the unfertilized treatment and from 0.789 to 4472.00 ng N_2O-N $cm^{-2} \cdot day^{-1}$ in the fertilized treatment. Most of the time, denitrification rates were relatively low but on occasions increased to high values as 4472.00 ng N $cm^{-2} \cdot day^{-1}$ on 13 October. This irregular pattern of denitrification rates has been shown by other experiments [36] [37] and was often the result of complex interactions among weather conditions, soil properties and management practices, each having specific effects on denitrification. In fact, the coefficients of variation (CV) of daily measurements ranged from 9 to 157% indicating a high spatial and temporal variability. Several studies showed a large variability of denitrification rates with coefficients of variation ranging from 70% to 379% when they were measured using intact cores and under field conditions [38]-[40].

Throughout August the fertilized treatment emitted more N_2O-N than the unfertilized treatment but the daily denitrification rate was only significantly different on 24 August ($p < 0.05$). During this month, denitrification rate was relatively low (<500 ng N_2O-N $cm^{-2} \cdot day^{-1}$) even though the NO_3-N content was high, particularly in the fertilized treatment, and the WFPS ranged from 50% to 70%. This means that other factors such as temperature and/or availability of C, independent to inorganic N and water content, controlled N_2O-N losses by denitrification. Previous studies indicate that C availability is the most important limiting factor controlling denitrification, even in soils with low NO_3-N contents [41] [42]. According to Burton and Beauchamp [43], soluble organic

Table 1. Water-Extractable Organic Carbon (WEOC) concentration in soil from the control and fertilized treatments in the wheat growing season.

Date	WEOC (mg C kg^{-1})	
	Control	Fertilized
August	23.80 a	22.45 a
September	32.13 a	21.73 a
October	13.20 a	12.52 a
November	17.89 a	14.13 b
December	22.20 a	20.25 a

Different letters within each row indicate significant differences (p < 0.05) between treatments.

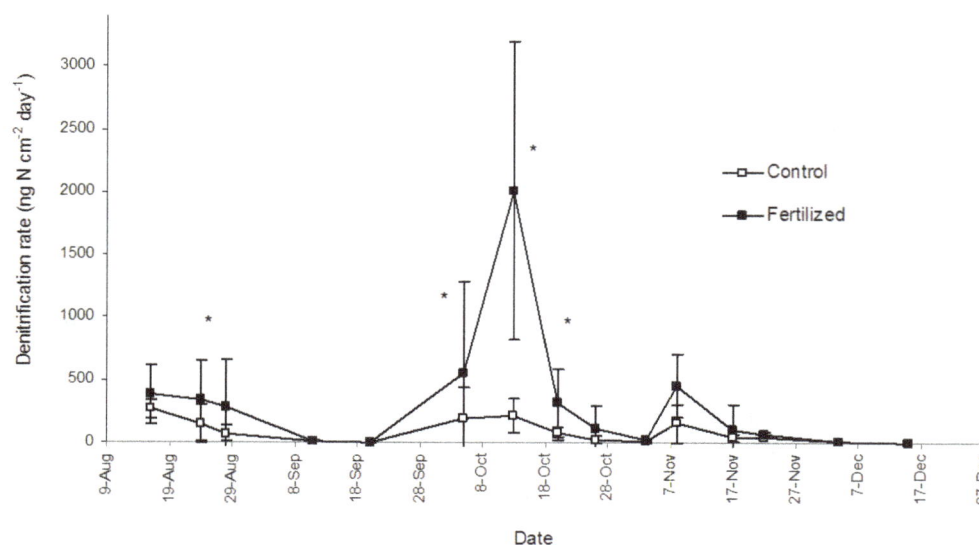

Figure 4. Daily denitrification rates (ng N_2O-N cm^{-2}·day^{-1}) measured during the growth period of spring wheat. Vertical bars represent standard errors. Asterisks indicate significant differences between fertilized and nonfertilized treatments (p < 0.05).

C contents greater than 60 - 80 mg C kg^{-1} are required for denitrification. Since the WEOC levels measured during the growing season were less than 32.13 mg C kg^{-1}, it is possible that C was limiting this process during the crop growth. However, because the WEOC was measured each month and no more frequently, its concentration may not reflect accurately the effect on denitrification rate.

After this initial period, the daily denitrification rate decreased and was very low (<20 ng N_2O-N cm^{-2}·day^{-1}) during September. Soil water content was low, approaching 33% WFPS during the dry phase from 28 August through 20 September, when the total rainfall was 9.6 mm. However, a second peak of N_2O occurred on 13 October following an increase in WFPS. The highest peak was recorded in the fertilized soil, after several rainfall events reflected in a WFPS of 70%, and coinciding with an average NO_3-N content of 25.35 mg N kg^{-1}. The denitrification rate of the fertilized soil was more than nine times the rate of the control soil, suggesting that N_2O-N production was driven mainly by NO_3-N availability since all the other measured soil properties (available C and water content) were similar in both treatments. Previous studies have shown that NO_3-N would not be the limiting factor unless all other parameters were optimized [44]. Then, NO_3-N could be limiting when concentrations are lower than 5 - 10 mg N kg^{-1} in a clay loam soil [45], 20 mg N kg^{-1} under grassland soil [46] or 40 mg N kg^{-1} in maize under no-tillage [10]. A small N_2O-N peak was observed on 8 November in the fertilized treatment, but the emissions were not statistically different (p > 0.05) between both treatments. Towards the end of the wheat growing season, WFPS and NO_3-N concentration dropped to about 30% and 6 mg N kg^{-1}; respectively, while the corresponding denitrification rate decreased to less than 12.7 ng N_2O-N cm^{-2}·day^{-1}. This result was probably consequence to depletion of soil NO_3-N pool by plant uptake in combination with low water content leading to cessation of denitrification.

The active denitrification during spring appears to have been associated mainly with high soil moisture contents. It has been recognized that O_2 concentrations can affect both synthesis and activity of denitrifying enzymes [47]. Denitrification increases with increases in WFPS and the maximum emission of N_2O-N by denitrification occurs at WFPS values >60% [34] [48]. A high correlation between the logarithm of daily denitrification rate and WFPS (r = 0.67, n = 90) was found, although nearly all WFPS values are below 60% during the measurement period. On the other hand, although NO_3-N concentration has been recognized to play a significant role in the regulation of denitrification process [49], it was not a good indicator of the logarithm of the denitrification rate because the relationship between both variables was weak (r = 0.21, n = 90). Denitrification is maybe more influenced by the diffusion of the NO_3-N to the active denitrification sites even at a high NO_3-N concentration [45] than by the concentration of available NO_3-N.

3.4. Cumulative N_2O Losses

Cumulative losses of N_2O-N for the 4-month period were lower for the unfertilized treatment compared with the fertilized treatment, but this difference was not statistically significant (p > 0.05). The high variability in N_2O-N rates probably overrides the statistical significances of the means. During the wheat season, the fertilized soil had a cumulative N_2O-N loss of 3.5 kg N_2O-N ha^{-1} whereas it was 0.9 kg N_2O-N ha^{-1} in the unfertlized soil (**Figure 5**). The results agree with those of Skiba *et al.* [11], and Granli and Bockman [50] who found higher total N_2O losses from the fertilized treatment compared with the control. The emissions of N_2O-N by denitrification expressed as a percentage of applied N fertilizer were low, 2%, after subtraction of N_2O-N emissions attributable to the control. For the non-fertilized plots, total N_2O-N losses were also low but this confirms that there are emissions even without N fertilization due to the potential for N mineralization of this soil.

Most cumulative N_2O losses in the wheat crop, about 82%, occurred early in the spring probably because of high precipitation, NO_3-N availability and higher air temperature. This indicates the importance of this period for the evaluation of total N_2O-N losses from wheat crop, in temperate climate zones.

3.5. Denitrification and Its Relationship to Other Loss Processes

In this study, denitrification measurements were performed on a site where other aspects of the N cycle were measured. A previous study reported that ammonia volatilization losses from this soil were low, on average 1.69 kg N ha^{-1}, which represented 1.5% of the N fertilizer when 120 kg N ha^{-1} of urea was applied at sowing of wheat

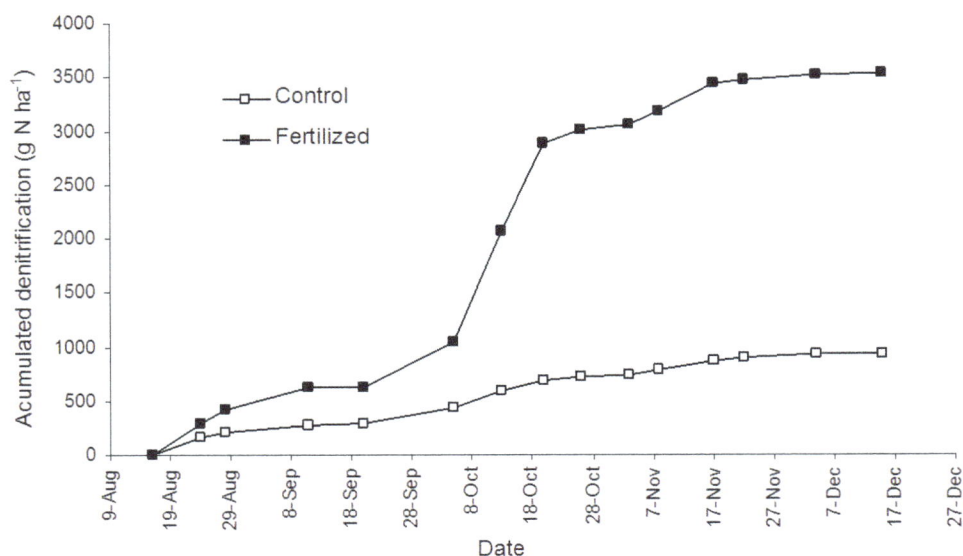

Figure 5. Cumulative N_2O emissions (g N_2O-N ha^{-1}) by denitrification throughout the growing season of wheat (August-mid December) from the control and fertilized treatments.

crop [28]. Using Ceres-Wheat model, it was predicted that NO_3-N losses by leaching ranged from 12 to 62 kg N ha^{-1} for rates of 0 to 175 kg N ha^{-1}; respectively, while denitrification losses fluctuated between 1.2 and 3.9 kg·N·ha^{-1} depending on the N rate at sowing of wheat [17]. According to this information, the denitrification process was expected to be a minor pathway of N loss. In fact, the results of this research support this conclusion since N_2O-N rates measured by acetylene were comparable to those predicted by the model, 0.9 kg N ha^{-1} and 3.5 kg N ha^{-1} in the control and fertilized plot; respectively. However, the lack of measurements during the first days immediately after N fertilizer application may have resulted in an underestimation of the cumulative N_2O-N loss as emissions could have occurred during that period. In addition, weather conditions during the study were drier (350 mm) than the median over the 1971-2003 period (489 mm). This means that a median year could lead to higher N_2O emissions than those measured in this study due to a higher rainfall.

4. Conclusions

The peaks of denitrification rate in the wheat crop were observed when N fertilizer was applied and rain was more frequent and intensive. A high correlation was found between daily denitrification rate and WFPS (r = 0.67, n = 90), however, the NO_3-N soil concentration was not a good indicator of denitrification rate (r = 0.21, n = 90). The accumulative N_2O loss under spring wheat was not significantly different between the fertilized and control treatments. This is due to the high measured variability of denitrification rates. Denitrification process does not appear to be a major pathway for loss of N from this crop.

Denitrification rates were small considering that they represent 2% of the N added to the soil. Most cumulative N_2O-N losses occurred early in the spring (82%). Therefore sampling schedules need to focus during this period for wheat crop, in temperate climate regions.

References

[1] IPCC (2006) Guidelines for National Greenhouse Gas Inventories Third Authors/Experts Meeting: Industrial Processes and Product Use, Washington D.C., 27-29 July 2004.

[2] Firestone, M.K. and Davidson, E.A. (1989) Microbiological Basis of NO and N_2O Production and Consumption in Soils. In: Andreae, M.O. and Schimel, D.S., Eds., *Exchange of Trace Gases between Terrestrial Ecosystems and the Atmosphere*, John Willey and Sons, New York, 7-21.

[3] Bateman, E.J. and Baggs, E.M. (2005) Contributions of Nitrification and Denitrification to N_2O Emissions from Soils at Different Water-Filled Pore Space. *Biology and Fertility of Soils*, **41**, 379-388. http://dx.doi.org/10.1007/s00374-005-0858-3

[4] Drury, C.F., Zhang, T.Q. and Kay, B.D. (2003) The Non-Limiting and Least Limiting Water Ranges for Soil Nitrogen Mineralization. *Soil Science Society of American Journal*, **67**, 1388-1404. http://dx.doi.org/10.2136/sssaj2003.1388

[5] Parton, W.J., Mosier, A.R. and Schimel, D.S. (1988) Rates and Pathways of Nitrous Oxide Production in a Shortgrass Steppe. *Biogeochemistry*, **6**, 45-48. http://dx.doi.org/10.1007/BF00002932

[6] Skiba, U., Smith, K.A. and Fowler, D. (1993) Nitrification and Denitrification as Sources of Nitric Oxide and Nitrous Oxide in a Sandy Loam Soil. *Soil Biology and Biochemistry*, **25**, 1527-1536. http://dx.doi.org/10.1016/0038-0717(93)90007-X

[7] Coyne, M.S. (2008) Biological Denitrification. In: Schepers, J.S. and Raun, W.R., Eds., *Nitrogen in Agricultural Systems*, American Society of Agronomy, Crop Science Society of America. Soil Science Society of America, Madison, 202-254. http://dx.doi.org/10.2134/agronmonogr49.c7

[8] Firestone, M.K. (1982) Biological Denitrification. In: Stevenson, F.J., Ed., Nitrogen Agricultural Soils. Agron. Monogr. 22. American Society Agronomy, Crop Science Society of America and Soil Science Society of American, Madison, 289-326.

[9] Grant, R.F., Pattey, E., Goddard, T.W., Kryzanowski, L.M. and Puurveen, H. (2006) Modeling the Effects of Fertilizer Application Rate on Nitrous Oxide Emissions. *Soil Science Society of American Journal*, **70**, 235-248. http://dx.doi.org/10.2136/sssaj2005.0104

[10] Sainz Rozas, H., Echeverría, H.E. and Picone, L.I. (2001) Denitrification in Maize No-Tillage: Effect of Nitrogen Rate and Application Time. *Soil Science Society of American Journal*, **65**, 1314-1323. http://dx.doi.org/10.2136/sssaj2001.6541314x

[11] Skiba, U., Fowler, D. and Smith, K. (1994) Emissions of NO and N_2O from Soils. *Environmental Monitoring and Assessment*, **31**, 153-158. http://dx.doi.org/10.1007/BF00547191

[12] Stehfest, E. and Bouwman, L. (2006) N_2O and NO Emission from Agricultural Fields and Soils under Natural Vegeta-

tion: Summarizing Available Measurement Data and Modeling of Global Annual Emissions. *Nutrient Cycling in Agroecosystems*, **74**, 207-228. http://dx.doi.org/10.1007/s10705-006-9000-7

[13] Zhang, Y.M., Chen, D.L., Zhang, J.B., Edis, R., Hu, C.S. and Zhu, A.N. (2004) Ammonia Volatilization and Denitrification Losses from an Irrigated Maize-Wheat Rotation Field in the North China Plain. *Pedosphere*, **14**, 533-540.

[14] Elmi, A.A., Astatkie, T., Madramootoo, C., Gordon, R. and Burton, D. (2005) Assessment of Denitrification Gaseous End Products in the Soil Profile under Two Water Table Management Practices Using Repeated Measures Analysis. *Journal of Environmental Quality*, **34**, 446-454. http://dx.doi.org/10.2134/jeq2005.0446

[15] Satorre, E.H. and Slafer, G.A. (1999) Wheat Production Systems of the Pampas. In: *Wheat: Ecology and Physiology of Yield Determination*, Food Products Press, New York, 333-348.

[16] Reussi Calvo, N., Sainz Rozas, H., Echeverría, H. and Berardo, A. (2013) Contribution of Anaerobically Incubated Nitrogen to the Diagnosis of Nitrogen Status in Spring Wheat. *Agronomy Journal*, **105**, 321-328. http://dx.doi.org/10.2134/agronj2012.0287

[17] Barbieri, P.A., Rozas, H.S. and Echeverría, H.E. (2008) Time of Nitrogen Application Affects Nitrogen Use Efficiency of Wheat in the Humid Pampas of Argentina. *Canadian Journal of Plant Science*, **88**, 849-857. http://dx.doi.org/10.4141/CJPS07026

[18] Calviño, P.A. and Sadras, V.O. (2002) On-Farm Assessment of Constraints to Wheat Yield in the South-Eastern Pampas. *Field Crops Research*, **74**, 1-11. http://dx.doi.org/10.1016/S0378-4290(01)00193-9

[19] Gobierno Argentino (2007) Segunda Comunicación Nacional del Gobierno Argentino a la Convención Marco de las Naciones Unidas sobre Cambio Climático. Proyecto BIRF No. TFO51287. http://www.fundacionbariloche.org.ar

[20] Taboada, M.A. and Cosentino, V.R.N. (2012) Emisión de óxido nitroso (N_2O) desde suelos agrícolas. *XIX Congreso Latino Americano de la ciencia del suelo, XXIII Congreso argentino de la Ciencia del suelo*, 16-20 abril 2012, Resumen en actas.

[21] Yoshinari, T., Hynes, R. and Knowles, R. (1977) Acetylene Inhibition of Nitrous Oxide Reduction and Measurement of Denitrification and Nitrogen Fixation in Soil. *Soil Biology and Biochemistry*, **9**, 177-183. http://dx.doi.org/10.1016/0038-0717(77)90072-4

[22] Tiedje, J.M. (1982) Denitrification. In: Page, A.L., Ed., *Methods of Soil Analysis: Part 2. Microbiological and Biochemical Properties*, Agron. Monogr. 9, American Society of Agronomy and Soil Science Society American, Madison, 1011-1026.

[23] Keeney, D.R. and Nelson, D.W. (1982) Nitrogen Inorganic Forms. In: Page, A.L., Ed., *Methods of Soil Analysis Part 2 Chemical and Microbiological Properties*, American Science of Agronomy and Soil Science Society of America, Madison, 643-693.

[24] Mebius, L.J. (1960) A Rapid Method for the Determination of Organic Carbon in Soil. *Analytica Chimica Acta*, **22**, 120-124. http://dx.doi.org/10.1016/S0003-2670(00)88254-9

[25] Blake, G.R. and Hartge, K.H. (1986) Bulk Density. In: Klute, A., Ed., *Methods of Soil Analysis, Part 1*, 2nd Edition, Agron. Monogr. 9, American Society of Agronomy and Soil Science Society of America, Madison, 363-375.

[26] Shapiro, S.S. and Wilk, M.B. (1965) An Analysis of Variance Test for Normality (Complete Samples). *Biometrika*, **52**, 591-611. http://dx.doi.org/10.1093/biomet/52.3-4.591

[27] SAS Institute Inc. (1988) User's Guide: Statistics. Version 6.03 ed. SAS/STAT. SAS Inst., Cary.

[28] Videla, C.C., Ferrari, J.L., Echeverria, H.E. and Travasso, M.I. (1996) Transformaciones del nitrógeno en el cultivo de trigo. *Ciencia del suelo*, **14**, 1-6.

[29] García, F. and Fabrizzi, K. (2001) Dinámica del nitrógeno en ecosistemas agrícolas: Efecto de la siembra directa. Siembra directa en el Cono Sur. Programa cooperativo para el desarrollo tecnológico agroalimentario y agroindustrial del Cono Sur. Instituto Interamericano de Cooperación para la Agricultura, Procisur, Montevideo, 299-323.

[30] Ferrari, M.H. (1995) Hidrólisis de la urea en suelos de la region pampeana. Tesis Ing. Agr. Facultad de Ciencias Agrarias, UNMdP, Balcarce.

[31] Navarro, C.A., Echeverría, H.E., Gonzalez, N.S. and Iglesias, M.A. (1980) Cinética de las reacciones de amonificación y nitrificación en algunos suelos de Argentina. *IX Reunión Argentina de la Ciencia del Suelo*, 15 al 20 septiembre 1980, 431-437.

[32] Abbate, P.E. and Andrade, F.H. (2006) Los nutrientes del suelo y la determinación del rendimiento de los cultivos de grano. In: Echeverría, H.E. and García, F.O., Eds., *Fertilidad de suelos y fertilización de cultivos*, Editorial INTA, 43-65.

[33] Baethgen, W.E. and Alley, M.M. (1989) Optimizing Soil and Fertilizer Nitrogen Use for Intensively Managed Winter Wheat. I. Crop Nitrogen Uptake. *Agronomy Journal*, **81**, 116-120.

[34] Linn, D.M. and Doran, J.W. (1984) Effect of Water Filled Pore Space on Carbon Dioxide and Nitrous Oxide Produc-

tion in Tilled and Non-Tilled Soils. *Soil Science Society of America Journal*, **48**, 1267-1272. http://dx.doi.org/10.2136/sssaj1984.03615995004800060013x

[35] Neff, J.C. and Asner, G.P. (2001) Dissolved Organic Carbon in Terrestrial Ecosystems: Synthesis and a Model. *Ecosystems*, **4**, 29-48. http://dx.doi.org/10.1007/s100210000058

[36] Hénault, C., Grossel, A., Mary, B., Roussel, M. and Léonard, J. (2012) Nitrous Oxide Emission by Agricultural Soils: A Review of Spatial and Temporal Variability for Mitigation. *Pedosphere*, **22**, 426-433. http://dx.doi.org/10.1016/S1002-0160(12)60029-0

[37] Mathieu, O., Lévêque, J., Hénault, C., Milloux, M.J., Bizouard, F. and Andreux, F. (2006) Emissions and Spatial Variability of N_2O, N_2 and Nitrous Oxide Mole Fraction at the Field Scale, Revealed with ^{15}N Isotopic Techniques. *Soil Biology and Biochemistry*, **38**, 941-951. http://dx.doi.org/10.1016/j.soilbio.2005.08.010

[38] Folorunso, O.A. and Rolston, D.E. (1984) Spatial Variability of Field Measured Denitrification Gas Fluxes. *Soil Science Society of America Journal*, **48**, 1214-1219. http://dx.doi.org/10.2136/sssaj1984.03615995004800060002x

[39] Parkin, T.B., Kaspar, H.F., Sextone, A.J. and Tiedje, J.M. (1984) A Gas-Flow Soil Method to Measure Field Denitrification Rates. *Soil Biology and Biochemistry*, **16**, 323-330. http://dx.doi.org/10.1016/0038-0717(84)90026-9

[40] Parsons, L., Scott Smith, M. and Murray, R.E. (1991) Soil Denitrification Dynamics: Spatial and Temporal Variations of Enzyme Activity, Populations and Nitrogen Gas Loss. *Soil Science Society of America Journal*, **55**, 90-95. http://dx.doi.org/10.2136/sssaj1991.03615995005500010016x

[41] Drury, C.F., McKenney, D.J. and Findlay, W.I. (1991) Relationships between Denitrification, Microbial Biomass and Indigenous Soil Properties. *Soil Biology and Biochemistry*, **23**, 751-755. http://dx.doi.org/10.1016/0038-0717(91)90145-A

[42] Weier, K.L., Doran, J.W., Power, J.F. and Walters, D.T. (1993) Denitrification and the Dinitrogen/Nitrous Oxide Ratio as Affected by Soil Water, Available Carbon and Nitrate. *Soil Science Society of America Journal*, **57**, 66-72. http://dx.doi.org/10.2136/sssaj1993.03615995005700010013x

[43] Burton, D.L. and Beauchamp, E.G. (1985) Denitrification Rate Relationship with Soil Parameters in the Field. *Communications in Soil Science and Plant Analysis*, **16**, 539-549. http://dx.doi.org/10.1080/00103628509367626

[44] Vinther, F.P. (1984) Total Denitrification and the Ratio between N_2O and N_2 during the Growth of Spring Barley. *Plant and Soil*, **76**, 227-232. http://dx.doi.org/10.1007/BF02205582

[45] Ryden, J.C. (1983) Denitrification Loss from a Grassland Soil in the Field Receiving Different Rates of Nitrogen as Ammonium Nitrate. *Journal of Soil Science*, **34**, 355-365. http://dx.doi.org/10.1111/j.1365-2389.1983.tb01041.x

[46] Myrold, D.D. and Tiedje, J.M. (1985) Establishment of Denitrification Capacity in Soil: Effects of Carbon, Nitrate and Moisture. *Soil Biology and Biochemistry*, **17**, 819-822. http://dx.doi.org/10.1016/0038-0717(85)90140-3

[47] Dendooven, L. and Anderson, J.M. (1994) Dynamics of Reduction Enzymes Involved in the Denitrification Process in Pasture Soil. *Soil Biology and Biochemistry*, **26**, 1501-1506. http://dx.doi.org/10.1016/0038-0717(94)90091-4

[48] Davidson, E.A. (1991) Fluxes of Nitrous Oxide and Nitric Oxide from Terrestrial Ecosystems. In: Rogers, J.E. and Whitman, W.B., Eds., *Microbial Production and Consumption of Greenhouse Gases: Methane, Nitrogen Oxide, and Halomethanes*, ASM Press, Washington DC, 219-235.

[49] Tiedje, J.M. (1988) Ecology of Denitrification and Dissimilatory Nitrate Reduction to Ammonium. In: Zehnder, A.J.B., Ed., *Biology of Anaerobic Microorganisms*, John Wiley and Sons, New York, 179-244.

[50] Granli, T. and Bøckman, O.C. (1994) Nitrous Oxide from Agriculture. *Norwegian Journal of Agricultural Sciences*, **12**, 1-128.

7

Testing of Decision Making Tools for Village Land Use Planning and Natural Resources Management in Kilimanjaro Region

Anthony Z. Sangeda[1*], Frederick C. Kahimba[1], Reuben A. L. Kashaga[1], Ernest Semu[1], Christopher P. Mahonge[1], Francis X. Mkanda[2]

[1]Sokoine University of Agriculture, Morogoro, Tanzania
[2]Sustainable Land Management Project, Regional Commissioner's Office, Moshi, Tanzania
Email: [*]sangedaaz@gmail.com

Abstract

This paper focuses on participatory testing of decision making tools (DMTs) at village level to assist in development of land use plans (LUPs) for sustainable land management (SLM) in Kilimanjaro Region, Tanzania. Data were collected using conditional surveys through key informant interviews with the project's district stakeholders in each district, focused group discussions with selected villagers and participatory mapping of natural resources. Soil health, land degradation, carbon stock, and hydrological conditions were assessed in the seven pilot villages in all seven districts using DMTs as part of testing and validation. Results indicated soils of poor to medium health, and land degradation as portrayed by gullies and wind erosion in lowlands and better in uplands. Carbon and forest disturbance status could not be assessed using one-year data but hydrological analysis revealed that water resources were relatively good in uplands and poor in the lowlands. Challenges with regard to land use include increased gully erosion, decreased stream flow, reduced vegetation cover due to shifting from coffee with tree sheds to annual crops farming, cultivation near water sources, and overgrazing. Empowering the community with decision making tools at village level is essential to ensure that village land uses are planned in a participatory manner for sustainable land and natural resources management in Kilimanjaro and other regions in Tanzania.

Keywords

Decision Making Tool, Land Use Planning, Sustainable Land Management, Natural Resources Management, Kilimanjaro

---

1. Introduction

Tanzania experiences deficiency in capacity, particularly on knowledge, information and decision making for land use planning at all levels [1]. The links between research and extension service are weak and research findings remain in shelves without being put into practice. These problems are compounded by the absence of a comprehensive monitoring system and decision making tools (DMTs) to guide village land use planning for natural resources management (NRM) at local level. Although some districts in Tanzania have been conducting isolated monitoring of environmental variables, there are no specific tools at village level to assess land degradation dynamics, carbon stock, soil health, or hydrological information, thereby limiting the application of adaptive management based on early detection of negative impacts. The lack of tools has limited the usefulness of planning and decision making to correct or mitigate the impact of current practices and to define land use policies at village level, where land degradation is experienced the most [2].

The government of Tanzania, with support from the global environment facility (GEF), through United Nations Development Programme (UNDP), is implementing a 4-year project aimed at reducing land degradation on the highlands of the Kilimanjaro Region. The project is in response to the fact that despite its local and global significance, the Kilimanjaro ecosystem is experiencing extensive degradation and deforestation driven by a set of complex interrelated factors, such as rapid population increase, land-use change, poor land-management practices, unsustainable harvesting of natural resources, declining commodity prices, and climate change [3]. Examples of degradation include completely deforested patches, removed vegetation cover and intense soil erosion in many areas that has resulted in big gullies. Such areas require some form of intervention to promote a favourable environment for the establishment of plants and to increase soil protection, thereby arresting further soil/land degradation. The project's goal is to ensure that "sustainable land management (SLM) provides the basis for economic development, food security, and sustainable livelihoods while restoring the ecological integrity of the Kilimanjaro Region's ecosystems". Its purpose is to provide local land-users and managers with the enabling environment (policy, financial, institutional, capacity) necessary for the widespread adoption of sustainable land-management practices.

As part of capacity building, the project is facilitating integrated land-use planning at landscape and village levels because it is a necessary vehicle for institutionalization of SLM. Although land-use planning is mandated by the constitution, it is only being done to a very limited extent due to resource problems [2]. The project has, therefore, designed decision making tools that have been successfully tested with community groups in project pilot villages. This paper describes the testing of the decision making tools and preliminary findings of the tests that were conducted after the tools were designed. This paper presents findings of decision making tools for use at village and landscape levels that we designed and tested as part of the sustainable land management project in the Kilimanjaro Region.

2. Materials and Methods

2.1. Description of Study Area

The study was conducted in seven pilot project villages, each village drawn from one district within the Kilimanjaro Region. Three villages of Shighatini (Mwanga District), Ushiri (Rombo District), and Manio (Siha District) were from the uplands (1200 - 1900 m.a.s.l). Two villages of Roo (Hai District) and Sango (Moshi Rural District) were from midlands with moderate altitude (900 - 1200 m.a.s.l). The remaining two villages of Mnazi (Moshi Urban) and Mabilioni (Same District) were from the lowland areas (500 - 900 m.a.s.l). The location of the test sites allowed researchers to test the tools in all the heterogeneity within the Kilimanjaro Region. Due to differences in topo-sequence, these three categories of villages had different characteristics in terms of agro-ecology and socio-economic activities of the communities. For example, there were more conservation agriculture and less free grazing in the uplands compared to the lowlands.

The Kilimanjaro Region is located in the north-eastern part of Tanzania (**Figure 1**). It lies between latitudes 2°25'S and 4°15'S, and between longitudes 36°25'30"E and 38°10'45"E. The region is bordered by the country of Kenya on the north, Arusha Region on the north-west, Manyara Region on the west, and Tanga Region to the south-east [4]. Main relief features in the region include Mt. Kilimanjaro with a snow covered peak at (5895 m.a.s.l) and the Pare Mountain ranges running from north-west of Mwanga District to south-east of Same District.

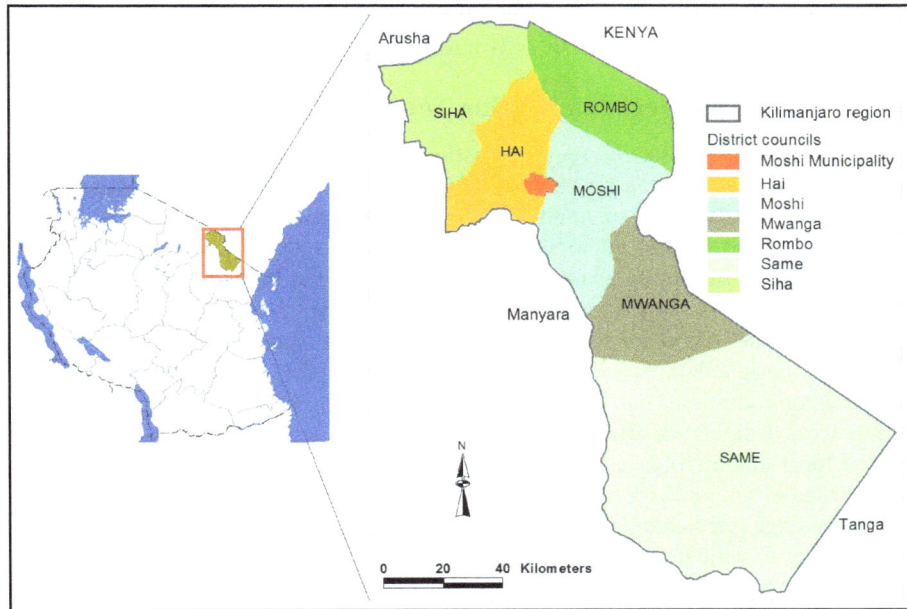

Figure 1. Location map showing sustainable land management (SLM) project area in Kilimanjaro Region, Tanzania.

The region has a population of 1,640,087 people, with an average annual growth rate of 1.8% [5]. About 48.7% of the total land area of 13,209 km^2 or 1,320,900 ha is arable, 21.3% is under game reserves, 15.3% under grasslands and rangelands, 12.4% under forest reserves, and 2.3% are water bodies (lakes, dams, and rivers). This amount of land gives an arable density of about 0.8 ha per capita, which is higher than the national figure of 0.21 per capita. Arable density is the number of people per unit area of arable land and is a measure of self-sustainability in terms of food production [6]. This result implies that self-sustainability in terms of food production in Kilimanjaro is lower than in the rest of Tanzania. The implication is that cultivation intensity in the region is higher than for the whole nation, which could be a factor contributing to land degradation.

The climate of the region is modified by the presence of the mountain ranges, with mean annual precipitation ranging from 400 mm in the lowlands to more than 1200 mm in the highlands, and more than 2000 mm in mountainous areas. The temperature of Kilimanjaro Region is variable depending on topo-sequence. Highlands are much cooler than lowlands with temperatures ranging between 15°C and 30°C. Lowland maximum temperatures go as high as 40°C during the dry season. January to March are the warmest months. The region faces significant risks from climate change, with a rise in mean temperature, decrease in annual precipitation, and increase in variability with the rainfall patterns becoming increasingly more unimodal. These occurrences have increased the vulnerability as farmers may have less water available for crop production.

Kilimanjaro Region is also characterized with higher wind speeds. On the eastern part of the region in Same District, mid-day winds range between 15.2 - 24.5 km/hr from September to March, and calms down to 13.3 - 16.3 km/hr from April to August. On the western part in lowlands of Hai District wind speeds are higher in the months of September to March (15.2 - 18.0 km/hr), and become lower from April to August (11.4 - 12.2 km/hr). The central part in Moshi town normally has mid-day wind speeds in ranges of 14.6 - 17.1 km/hr from September to March, and 8.3 - 11.0 km/hr from April to August [7]. This is an indication that the lowlands of Kilimanjaro Region are very vulnerable to wind erosion during the dry season, especially if the land is loosened e.g. by overgrazing and left bare without vegetation/crop cover as a result of harvesting all crop residues for animal feeds.

With regard to soils in Kilimanjaro Region they vary depending on the landscape. Generally the soils are volcanic in the uplands around Mount Kilimanjaro, and ferralistic soils in the Pare Mountains areas with alluvial plains in all the lowlands, mainly affected by salinity. Due to steep slopes in the region even the middle and lower altitude areas have suffered extensive soil erosion. Previous studies showed that soil erosion by water is a major problem in the region [8]. Although farmers in the area have traditionally practiced soil conservation as evidenced by the remnants of terraces, these practices are no longer institutionalized within the community groups, and hence they are being practiced by only a few farmers.

2.2. Methods Used to Test Decision Making Tools

To design tools that are suitable for the region, existing decision making tools (DMTs) for assessment of land degradation [9], estimation of carbon stocks [10] [11], soil health [12] [13], and hydrology [14] [15] were reviewed. These four tool kits were deemed adequate to enable framers make decisions about their land. The vigorous literature review was done by authors to familiarize various decision making tools used in other places of the world. However, many of the reviewed tools were those used at higher levels for decision makers. In contrast, the designed tools in this study aimed to be used at local level; hence a number of modifications and simplifications had to be made. Although some of the reviewed tools from the literature were in terms of graphical models, all the new designed tools were standardized into tabular frameworks, which are user friendly to semi-illiterate local communities.

The draft tools were designed in office after data collection, which was done prior to testing in the field using beneficiaries (the local community) in a participatory approach. During validation, the communities were involved to assess whether an indicator of the DMT was appropriate and relevant or not. All irrelevant indicators pointed in the first village were changed before testing the tool in the next village. In this case, researchers were meeting every evening to validate the tool based on the inputs and comments raised by villagers after which an updated version of the tool had to be printed for the next village and so on. All the tools used for testing and validation were translated into Kiswahili language which is the *lingua franca* in Tanzania.

A total of seventy (70) land users, ten (10) in each village were involved in testing the applicability of the tools. The villagers involved were selected based on their membership in a particular village committee. Representatives from land use, environment/natural resources, water and women committees, village extension officer, village executive officer, and village chairperson were involved in testing the tools.

The testing was conducted in a form of training where the researchers explained the importance of participatory land use planning and management of natural resources in the village. Communities were also told of the importance of using specific tools in data collection and monitoring with time. Additionally, they were introduced to various simplified equipment (like tape measure, infiltrometer, penentrometer etc.) for use, how they can be used to collect information, and how that information could help in land use and natural resources management within the village.

The work started in a classroom session for one hour where communities were introduced to all the four decision making tools and copies of each tool were distributed to them. The session was followed by a transect walk into the field. At each site, a nearby degraded land, forest, farm and water source was identified to be used as a medium for practical testing of the respective toolkits. While at each site (e.g. degraded area) the team was asked to open the land degradation tool kit and assess the extent of degradation using the indicators in the tool kit. In most cases researcher introduced the assessment methods in the first place and then communities in groups replicated the assessment thereafter. On each site and for each toolkit, the first test was done by the researcher as demonstration and the subsequent tests were then carried by communities in groups and later the results were compared.

2.3. Topo-Sequence Effects on Land Use for Kilimanjaro Region

The districts of Kilimanjaro are located in series along the Dar es Salaam-Arusha highway with the exception of Rombo District. Considering the longitudinal (east-west) cross section, the region was divided into two main parts; the eastern part covering the districts of Same, Mwanga and part of Moshi rural district, and the western part covering districts of Siha, Hai, Moshi urban, and western part of Moshi rural districts (**Figure 1**).

The climatic conditions of the eastern part districts were mainly modified by the presence of the Pare mountain ranges, characterized by low rainfalls especially in the lowlands (<400 mm) along the Pangani River areas. The central and western part climate was modified by the ecology of the Kilimanjaro Mountain, making it much cooler, covered with clouds most of the time, and experiencing higher rainfalls. Considering the topo-sequence, all the districts were divided into highlands, midlands, and lowland areas (**Figure 2(a)** and **Figure 2(b)**). The upland areas were characterized by permanent crop plantations, forest reserves, and steep slopes, some with soil conservation measures such as bench terraces (**Figure 3**), and higher population density in settlement areas.

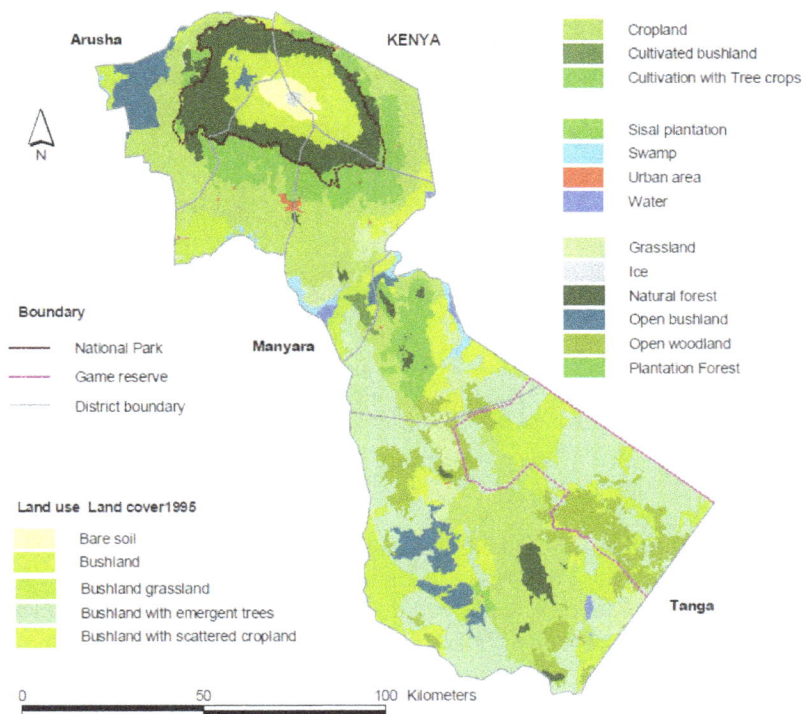

(a)

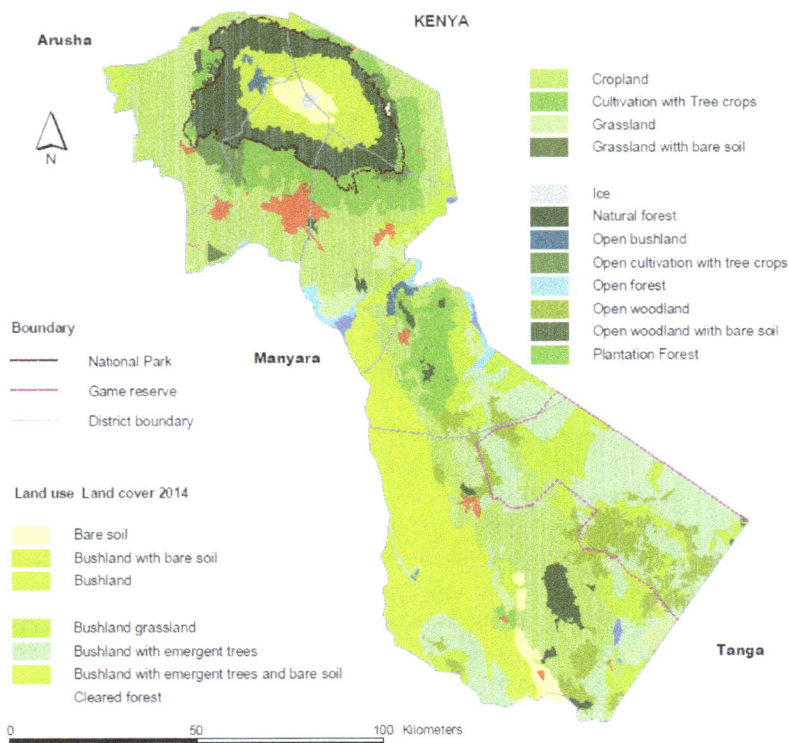

(b)

Figure 2. (a) Kilimanjaro Region land use map for 1995; (b) Kilimanjaro Region land use map for 2014.

Figure 3. Terraces with elephant grass in the uplands of Kilimanjaro Region.

Gully erosion was not profound in this zone due to lesser accumulation of stream flows and existence of good forest cover as opposed to midlands and lowlands (**Figure 4**). However, the paradigm shift from perennial coffee plantations to maize farming has resulted in removal of tree sheds in the fields to create favorable environment for annual crops such as maize. This shift has caused most of the land in the uplands and midlands to remain bare during dry season as crop residues are harvested for animal feeds as it can be depicted in the temporal land use change maps (**Figure 2(a)** and **Figure 2(b)**).

The midland and lowland areas were characterized mainly by both permanent (mainly midlands) and seasonal crops. Lowland areas were also occupied by pastoral communities and seasonal cropping systems. Most of the fields in the lowlands were used as grazing fields making the soil loose and prone to both water and wind erosion.

2.4. The Selection of Decision Making Tools

The land degradation tool was designed to assess the extent and severity of land degradation in a village. It set indicators of land degradation and mitigation measures to prevent further degradation. Indicators for land degradation included soil erosion (sheet, rill and gully), loss of vegetation cover, time series of poor performance of crops in terms of growth and yield, and emergence of invasive plant species. Others included cultivation along steep slopes without conservation measures and development of settlements in the slopes of water catchment. Change of greenish colour of natural vegetation cover in water catchment areas was also identified as an indicator of land degradation. The community used the tool to monitor natural resources potential and existing land degradation conditions, and appreciated appropriate measures to take for different land degradation scenarios to ensure sustainable land management. The measures were categorized into short-term, medium-term and long-term based on the stage of land degradation (initial, intermediate and later stage).

The soil health assessment tool aimed at guiding the local community in mapping village soils based on soil health suitability. It assisted in allocating various village land uses based on soil fertility criteria and suitability. For example, fertile soils to be designated for agriculture while poor soils could be left for pasture and therefore demarcated for livestock and other village investments. The aim of soil health assessment was to inform the land users (communities) on characteristics of the soil, hence its ability to render relevant ecosystem functions commensurate with its environment and expectations. An example of assessment tool is given in **Table 2**.

The Carbon stock assessment tool was a simplified tool to assess changes in carbon stock. It used simple and readily available tools to measure amount of tree biomass/increment and assessing levels of forest disturbances. Measurements were easily done by local communities, but data analysis needed to be carried out by carbon experts using established allometric models. The assessment led to knowing the condition of the forest, which reflected the amount of carbon gained or lost. The information was regarded as important for planning *i.e.* if there were more loss in carbon then communities would be required to enforce rules and bylaws in managing forest resources, do enrichment planting or involve more village members in the management.

Figure 4. Land degradation in the lowlands of Kilimanjaro Region.

The hydrological/water resources assessment tool involved a participatory water sources and water points' mapping as a decision management tool for village land use planning. During the hydrological mapping various water sources such as rivers, springs, traditional canals, micro-dams (*ndiva/nduwa*) were mapped and their hydrological characteristics documented. Mapping water points involved locating water points for domestic uses, livestock use and water sources and points used specifically for agriculture/irrigation purposes. Examples of water points for domestic water uses included boreholes with water lifting devices, public water taps (DPs), and private water taps in a house with domestic water service. Water points and sources for livestock included cattle troughs, and chaco-dams (*malambo*) exclusive for animal drinking. Mapping was also done on points with multipurpose water uses such as rivers where livestock were sent to drink water at a specific location. Specific water sources for agriculture were such as micro-dams (*ndiva/nduwa*). Water reservoir (*Nduwa*) in Sango village in Moshi District council was an example of project pilot village with multipurpose water uses.

3. Results and Discussion

3.1. Outcome of Tool Testing

3.1.1. Land Degradation Assessment

The land degradation tool has provided a guideline on how to identify whether or not land degradation has occurred and at what extent. It focused on three indicators namely soil erosion, vegetative cover and change in greenish colour of crops/plants. Regarding soil erosion indicator, this decision making tool has covered three types of soil erosion namely sheet erosion, rill erosion and gully erosion. For all the three indicators, four levels for measuring the extent or degree of land degradation were used namely "no degradation", "low degradation", "medium degradation" and "high degradation". The no degradation level was represented by the numeral value "0", low degradation level was operationalized by the scale of 1-3, medium degradation, 4-6, and high degradation, 7-9. All the three indicators were operationalized into the four levels and thereafter for every row of an indicator space was provided for farmer's scoring at the extreme right side. Results showed that, there was more degradation in lowlands (Mabilioni and Mnazi Villages) in terms of both soil and wind erosion as compared to mid and uplands.

Having identified the extent of degradation, the tool gave an opportunity for the community to recommend strategies to be undertaken under different statuses of land degradation based on the visual evidences observed by farmers from selected indicators of soil erosion, vegetative cover and crop/plant color (**Table 1**). Hence the tool enabled farmers and other resource users and local managers to understand land degradation status by observing visual evidences, and their severity scales and therefore to facilitate the farmer to make appropriate decision. The farmers were also provided with opportunity of judging from multiple indicators (soil erosion, vegetative cover, and crop colour) thereby triangulating evidence of land degradation, and therefore enhancing the validity of observation (**Table 1**).

Table 1. Suggested land degradation remedial actions.

Type of land degradation	Remedial actions
Sheet erosion	1) Construct soil erosion control structures such as ridges, *fanyajuu* and terraces on steep slopes. 2) Apply manure/use leguminous plants in the field to replenish the lost nutrients. 3) Plant trees/leguminous trees/agroforestry trees to serve as windbreak in the field especially for the affected area.
Rill erosion	1) Plant grass or encourage restoration of natural vegetation by abandoning human activities on the area. 2) Use contour ridges to harvest and prevent violent flow of water during rains.
Gully erosion	1) Construct gabions. 2) Planting tress e.g. castrol oil trees (*Minyonyo*) and sisal to help hold the soil. 3) Stop anthropogenic use of land surrounding the gullies to encourage natural vegetation restoration.
Reduction of vegetation cover	1) Plant trees/leguminous trees/agroforestry trees in the affected area. 2) Use of bench terraces. 3) Use of water diversion structures.
Loss of greenish colour	1) Construct water harvesting structures such as ridges, *fanyajuu* and terraces on steep slopes. 2) Plant trees and use leguminous plants/crops around contours on steep slopes to replenish the lost nutrients. 3) Use of fertilizers such as manure and inorganic fertilizers to replenish the lost nutrients.

3.1.2. Soil Health Assessment

The soil health assessment was performed with the help of soil health score sheet in farmlands. Key soil health aspects that farmers learned and practiced include ease of penetration using penetrometer, ease of infiltration, diversity of macro-life, presence of earthworms, soil structure, root development, aggregate stability of the soil samples at the 10 - 20 cm depth, and plant size and leaf color (**Table 2** and **Table 3**). After the testing, farmers were able to describe types of nutrient deficiencies in their farmlands and appropriate agronomic and soil fertility measures needed to restore the soil health. The villagers were able to replicate on their own the procedures for evaluating soil health using the soil health score sheet.

A few challenges were encountered while testing with the soil health score sheet. The test needed to be conducted in cropped fields to give a real picture of the farmland soil health status. However, in fields with seasonal crops when crop residues are removed the score sheet may indicate poor vegetation cover, implying poor soil health, contrary to the real farm conditions. Amount of moisture content also need to be assessed during penetration tests because wet fields gave results that were different from very dry fields within the same area. Hence on the same field, testing during dry season would give results that are different from those taken during the rainy season. As a corrective measure, there is a need for taking into consideration season of the year, soil wetness and crop cover. In a dry soil, as happens during the dry season, this is taken care of by first pouring water into the soil and letting it infiltrate through for a day or two before test is carried out to stabilize the moisture at about field capacity.

The results of the testing in seven villages are given in **Table 3**.

Figure 5 shows differences in the soil health status of the different villages. Mabilioni village was lowest in ranking and Roo was highest. The health parameters that contributed most to the lower scores were, macrolife diversity, earthworms, and ground cover (**Table 3**). Other parameters gave somewhat average scores for many villages, with Roo village showing overall higher scores. Soil physical properties (ease of penetration, ease of infiltration, soil structure, root development and aggregate stability) were generally medium to high for most villages. Plant colour/size, as proxy for chemical soil fertility was high for Roo, Mnazi and Shigatini villages. The abnormal colours that were widely mentioned as occurring on crops, thus contributing to low soil health scores, were yellowish/pale green and purple colours, implying low levels of nitrogen and phosphorus, in the soils respectively. Generally, Mabilioni village soils were of poor health while Roo village had good health soils. All other villages had soils of medium health.

3.1.3. Carbon Stock Assessment

Carbon assessment is normally a technical exercise that needs to be done at required standards and accuracy. However, communities in the seven pilot villages were able to use the simplified tools prepared for assessing carbon stock (**Figure 6**). The tape measure that was used for measuring circumference (to be converted to diameter) was a normal standard fiber tape measure, which could easily be available in villages. Tape measures

Table 2. The soil health assessment tool: the soil health card results score sheet.

Date: _____ Village/Ward/District: _____ (draw a sketch map of tested area overleaf)
Agricultural activity: _____ Soil Type/Description: _____ Productivity: _____
Days since 20 mm rain: ____ Soil moisture condition at testing time (tick): dry ...., moist ...., water logged ....

S/N	Test/score	Poor (1-3)	Fair (4-6)	Good (7-9)	Scores of individual tests runs from same soil type (1-9)
1	Ease of penetration (using a penetrometer)	Wire probe will not penetrate	Wire probe penetrates with difficulty to less than 20 cm	Wire probe easily penetrates to 20 cm	
2	Ease of infiltration	More than 7 minutes	3 to 7 minutes	Less than 3 minutes	
3	Diversity of macro-life	Fewer than two types of soil animals	Two to five types of soil animals	More than five types of soil animals	
4	Earthworms	0 - 3	4 - 6	More than 6	
5	Soil structure	Mostly in clods or with a surface crust, few crumbs	Some clods but also many 10 mm crumbs	Friable, readily breaks into 10 mm crumbs	
6	Root development	Few fine roots only found near the surface	Some fine roots mostly near the surface	Many fine roots throughout	
7	Aggregate stability, 10 - 20 cm depth	Aggregate broke apart in less than one minute	Aggregate remained intact after one minute	Aggregate remained intact after swirling	
8	Soil pH, 5 - 20 cm depth	pH 5 or lower	pH 5.5	pH 6 to pH 7	
9	Plant size and leaf colour	Stunted plants, leaf discolouration	Some variation in growth and colour	Appropriate leaf colour and uniform plant growth	

Table 3. Possible causes of low test scores on soil health.

S/N	Test result	Situation indicated	Possible causes
1	Low ground cover	Ground plants absent or plant growth is poor	Unsuitable plant type(s), soil compaction, soil erosion, shading, overstocking (animals eating most of the plant residues)
2	Low variety of soil fauna	Lack of food for fauna, poor soil structure, presence of harmful (agro)chemicals	Sparse litter, low soil organic matter, lack of soil spaces and channels, frequency or intensity of tillage has been excessive, mortality from recent use of insecticides or regular use of cumulative chemical(s) such as copper fungicides
3	Low earthworm count	pH unfavourable, poor food supply, lack of soil spaces, predators or parasites present, presence of harmful chemical	Soil pH naturally low, sparse litter and/or ground cover (and roots), low organic matter content, loss of topsoil, soil compaction, poor structure, predators (such as flatworms) and parasites (e.g. parasitic flies, mortality from recent use of insecticides or regular use of cumulative chemical(s) such as copper fungicides
4	Low probe penetrability	Soil is generally hard at the surface only, hard layer at greater depth	Low organic matter content, compacted by traffic or livestock due to overstocking) especially if soil is wet at the time, compacted by heavy vehicles or "hard pan" formed by soil inverting cultivators
5	Slow water infiltration	High proportion of clay particles and lack of spaces, channels or burrows in soil	Naturally high clay content of soil type, possible loss of topsoil through erosion, soil compaction, poor soil structure, lack of earthworms, surface crusting
6	Poor root development	Hard soil lacking spaces, poor plant nutrition, root disease or attack	Loss of topsoil, poor soil structure, soil compaction, soil pH not suitable for crop, lack of major or minor nutrients, presence of soil-borne plant pathogens, root-feeding nematodes or root-feeding insects
7	Poor soil structure	Powdery soil, few crumbs, excessive clods	Lack of soil-binding substances and processes, low soil organic matter content (sparse ground cover), few worms, topsoil loss, soil compaction, "puddling" of wet soil by livestock, excessive cultivation
8	Low aggregate stability (leading to high slaking of soil aggregates)	Soil particles disperse when wet	Topsoil loss (0 - 10 cm), compaction, low organic matter, excess tillage down to 20 cm, poor mixing of soil by soil animals, acidic soil conditions
9	Poor leaf color	Unthrifty (poorly-growing) plant	Deficiency or unavailability of one or more essential nutrients (e.g. P, N, K, Ca, etc.), (can be confirmed by soil or leaf analysis), presence of some plant diseases, water logging

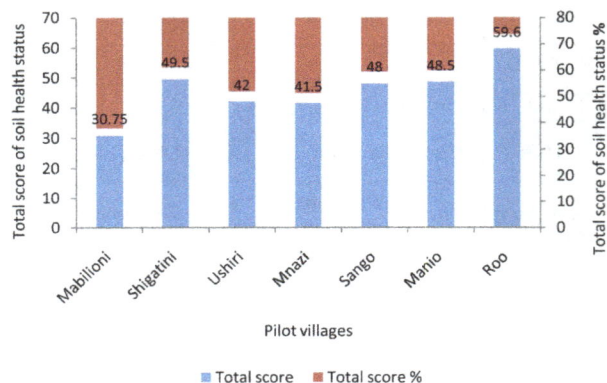

Figure 5. Soil health status in the pilot villages as captured by soil health assessment tool. Key: low = 0% - 33 %; medium = above 33% - 66%; high = above 66% - 100%.

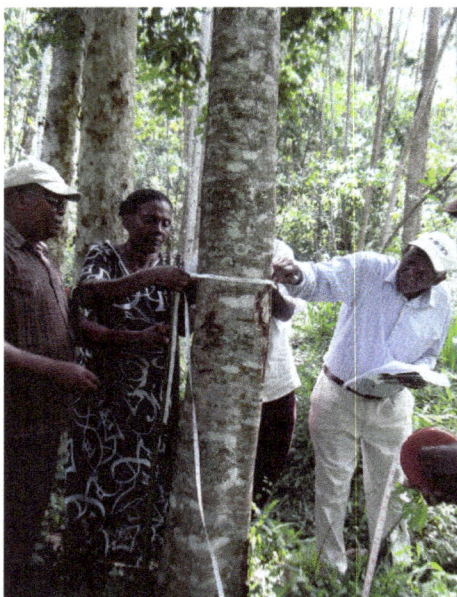

Figure 6. Participatory carbon assessment by communities in Shighatini Village, Mwanga District in Kilimanjaro Region, Tanzania.

were common to villagers doing tailoring, carpentry and masonry. Conversion of circumferences to diameters was easily done by communities using their mobile phone calculators. Complex equipment such as GPS for locating position, compass for direction and hypsometer for tree height estimation were not common to almost all villagers in all the seven villages. In this case, their use was minimized or rather used by professional foresters. That means, for carbon assessment, communities will need further detailed training or will need to be accompanied by respective district or ward forest officers.

The assessment involved hands-on exercises done by communities themselves. After explanation and demonstration on how to demarcate plots and how to take measurements, tools were given to a group of communities and they were asked to undertake the exercise by themselves. The trainers were around to assist communities where they went wrong. Some of group members were taking measurements and some recording the data. Important instrument that was fabricated locally for use was the 1.3 m stick, which was used to determine the height where the diameter at breast height (DBH) measurement was to be taken. The rest of the tools were the ones that could easily be bought in village shops.

The carbon assessment toolkit was simplified in a tabular form for easy of following and recording of data. Communities had sufficient knowledge of reading, writing, and calculating basic statistics. Most of them knew basic information about trees and the role of forests for environmental benefits and livelihoods. The most interesting activity in the testing exercise for the community was the hands-on practical testing of measuring the forest condition. They actively participated to measure the diameter of trees; count and measure tree stamps and assess forest disturbance. Results revealed very few and small trees in lowlands as compared to uplands where there were good protected forests. Unmanaged forests in lowlands were diminishing and converted to bush lands probably due to double effects of climate change and overgrazing. These changes also reflected low carbon storage potential.

The main challenge in pilot villages was the fact that most villages did not have natural forests that could qualify in carbon trading. Most forest patches were small mainly of plantations, which do not qualify in the REDD+ mechanism by now.

3.1.4. Hydrological Assessment

The hydrological assessment involved determination and quantification of the available water resources within the village, mapping of their location, and allocating various uses to available water resources. Upland and midland villages were good in hydrological values as compared to lowlands. Using this tool, the villagers were able to quantify the amount of water (discharge in l/sec) flowing from various sources such as rivers, canals/streams, water pipe and boreholes. The villagers also were able to evaluate the physical water quality, perform reconnaissance survey of potential areas for drilling boreholes, and played a part in participatory mapping and allocation of the water resources to come up with sustainable village water resources use plans. The community appreciated the simplicity in determining the required information using the hydrological assessment tool.

Key challenges observed were availability of a flowing water source in the proximity of the training area. In some villages such as Mabilioni in Same District the trainees (and trainers) had to walk quite a distance (2 - 3 km) to reach the Pangani River water source. Locating key features that signify groundwater potential was a challenge in some places, especially in villages that are located in mountain areas such as Manio Village in Siha District. Based on the hydrological analysis it was also observed that the community had a good understanding on how to preserve the water sources through planting trees around the water sources and banning the cultivation and cutting of trees along the water courses.

Key water uses emphasized were for domestic, livestock, agriculture, environment, construction and other uses. The documentation of the location and hydrological conditions of the water resources within the village were concluded by mapping of the available water resources (**Figure 7**) and locating areas for potential future development using the existing water resources. Existing land uses and water sources and points mapping prepared by the community is presented in **Figure 7**.

4. Conclusion

The developed decision making tools have proven to be useful tools in assisting the community and other stakeholders to plan for their present and future land uses on sustainable manner. The tested tools were on land degradation, soil health, carbon stock and hydrology. Soil health was poor in lowlands and good in midlands and uplands. Land degradation was more pronounced in lowland areas as a result of water and wind erosion of the soils. Carbon stock was assumed to be lower in lowlands because of more shrubs and less trees as compared to mid and uplands. Upland and midland villages were good in hydrological values as compared to lowlands. Generally, the tools have shown that upland village communities are better in land management than lowland communities. That means: more resources are required to be channeled to manage the more degraded lands in lowland areas of Kilimanjaro as compared to uplands. Therefore, empowering the communities with decision making tools at village level and landscape level is essential. While the region is making efforts to ensure that the land management and use is sustainable, there is a need to enforce existing laws (e.g. banning harvest of trees) and make environment a permanent agenda in local government authority meetings.

Acknowledgements

The authors appreciate the financial support from the United Nations Development Programme (UNDP), and the Regional Administrative Secretary (RAS) Kilimanjaro Region through a 4-year project on reducing land

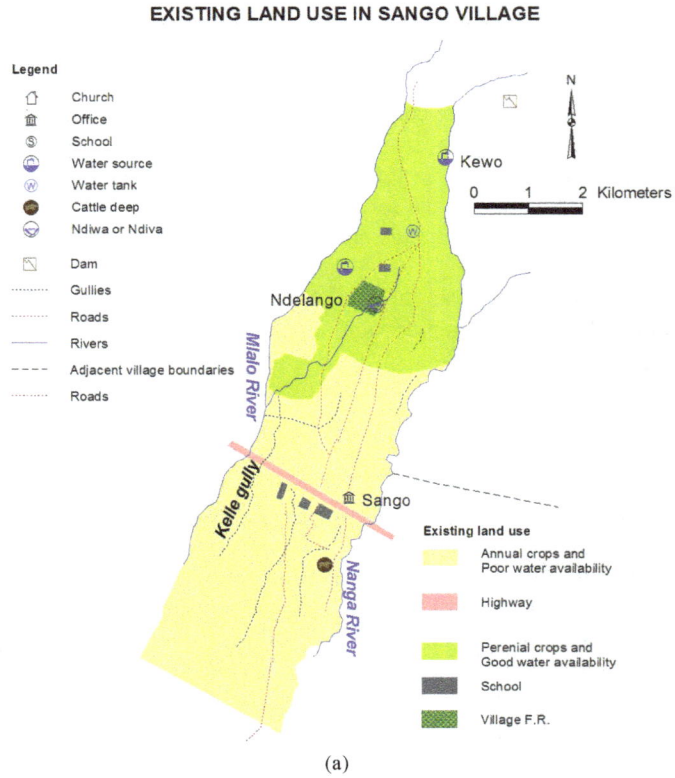

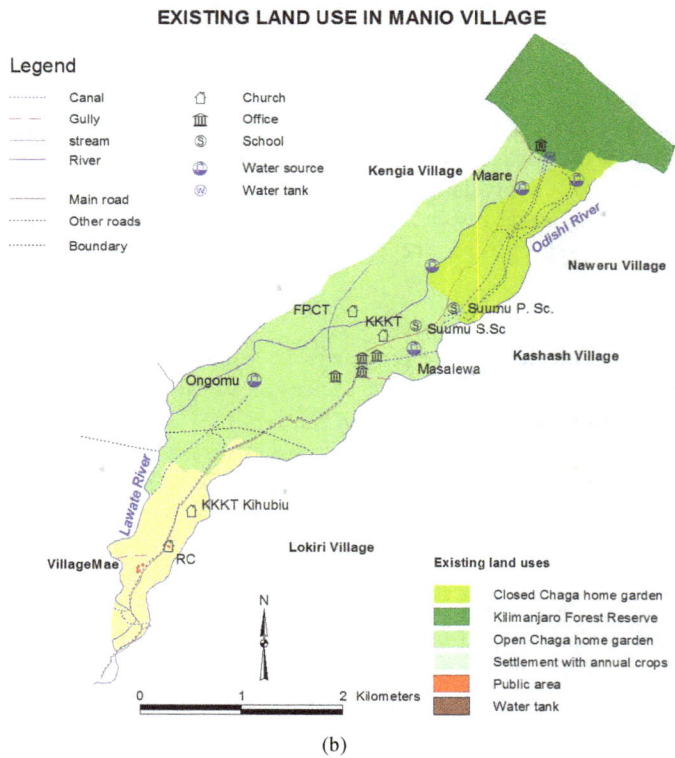

Figure 7. Participatory mapping of (a) land uses in Sango Village, Moshi rural district and (b) water resources/water points in Manio Village, Siha District, Kilimanjaro, Tanzania.

degradation on the highlands of the Kilimanjaro Region. We are also thankful to district executive directors, district facilitation team members (DFTs), district focal persons, extension staff, pilot village leaders and all our respondents in all seven districts and villages for their cooperation during data collection exercises and validation of DMTs. Lastly, we are grateful to SLM technical team and all people who were involved in proof reading and commenting on this paper.

References

[1]   FAO (2000) Land Resources: Sustainable-Land-Management/Field-Farm-Level.

[2]   NLUPC (1998) Guidelines for Participatory Village Land Use Management in Tanzania. National Land Use Planning Commission, Ministry of Lands and Human Settlements Development. Dar es Salaam, Tanzania.

[3]   UNDP (2010) Reducing Land Degradation on the Highlands of Kilimanjaro Region. United Nations Development Program, Tanzania LRM Project Document. http://www.tz.undp.org/content/dam/tanzania/Sustainable%20Land%20Management%20in%20Kilimanjaro.pdf

[4]   URT (1998) Kilimanjaro Region Socio-Economic Profile. The Planning Commission Dar es Salaam and Regional Commissioner's office, Kilimanjaro Region.

[5]   URT (2013) Population and Housing Census. Population Distribution by Administrative Areas. National Bureau of Statistics, Ministry of Finance, Dar es Salaam, and Office of Chief Government Statistician, President's Office, Finance, Economy and Development Planning, Zanzibar, 244 p.

[6]   Plane, D.A. and Rogerson, P.A. (1994) The Geographical Analysis of Populations: With Applications to Planning and Business. John Wiley & Sons, New York.

[7]   Tanzania Meteorological Agency (TMA) (2005) Annual Report. Government Printer, Dar es Salaam, Tanzania.

[8]   Maro, G.P., Mrema, J.P., Msanya, B.M., Janssen, B.H. and Teri, J.M. (2014) Developing a Coffee Yield Prediction and Integrated Soil Fertility Management Recommendation Model for Northern Tanzania. *International Journal of Plant & Soil Science*, **3**, 380-396. http://dx.doi.org/10.9734/IJPSS/2014/6883

[9]   Stocking and Murnaghan (2001) Handbook for Field Assessment of Land Degradation. Earthscan, London.

[10]   UNFCC (2013) Estimation of Carbon Stocks and Change in Carbon Stocks of Trees and Shrubs in A/R CDM Project Activities. http://cdm.unfccc.int/methodologies/ARmethodologies/tools/ar-am-tool-14-v4.1.pdf

[11]   NAFORMA (2011) Forest Inventory Systems and the FAO/NAFORMA Program.

[12]   Jenkins, A. (2006) Northern Rivers Soil Health Card: A Monitoring Tool for Farmers Developed by Farmers. NSW Department of Primary Industry, Wollongbar Agricultural Institute, Bruxner Hwy Wollongbar. www.dpi.nsw.gov.au

[13]   Jenkins, A. (2010) Striking a Match: How to Ignite a Passion for Soils. World Congress of Soil Science, Soil Solutions for a Changing World. Brisbane, 26-29.

[14]   Mango, G. and Kalenzi, D. (2011) The Study to Develop a Strategy for Establishing Cost Effective Land Use Plans in Iringa and Njombe Regions. URT, National Land Use Planning Commission Report, Dar es Salaam, 47 p.

[15]   IIED (2010) Participatory Land Use Planning as a Tool for Community Empowerment in Northern Tanzania. Ujamaa Community Resource Team, Gatekeeper 147: December 2010. International Institute for Environment and Development. http://pubs.iied.org/pdfs/14608IIED.pdf

8

Effect of Petroleum Products on Soil Catalase and Dehydrogenase Activities

Fidelis Ifeakachuku Achuba*, Patrick Nwanze Okoh

Department of Biochemistry, Delta State University, Abraka, Nigeria
Email: *achubabch@yahoo.com

Abstract

The effect of refined petroleum products on the activities of selected enzymes (catalase and dehydrogenase) was studied. There was a significant decrease ($p < 0.01$) in catalase activity. Catalase activity was higher in diesel and engine-oil treated soil after twelve days relative to petrol and kerosene. These observations indicate that the enzyme activity is the order of petrol > kerosene > diesel > engine oil. However, a significant increase ($p < 0.01$) was observed in dehydrogenase activity after twelve days relative to control values. Although, the refined petroleum products caused a similar pattern in the alteration of soil dehydrogenase activity, as they affected catalase activities, the general results indicate that the toxic effect is in the order of kerosene > diesel > petrol > engine oil. On the whole the results reveal that refined petroleum products alter soil biochemistry.

Keywords

Catalase Activity, Dehydrogenase Activity, Kerosene, Diesel, Engine Oil, Petrol

1. Introduction

Petroleum compounds have multifarious industrial and domestic uses. They are extensively employed in the making of solvents, dry cleaning fluids quick-products, automobile fuel, household solvents, and lubricants, cosmetics, water proofing agents, cleaning agents, and specialty chemicals [1]. These activities have led to the widespread contamination of the environment. Oil spill is the major cause for the high influx of petroleum to the biosphere [2] [3]. However, other points of soil pollution with refinery products are petrol stations, car and tractor servicing parks and, seaport areas [4]. Other areas of concern are mining and distribution of petroleum-based products [5] [6]. Besides, heavier use of machinery in agriculture together with unsatisfactory care while disposing of old or used petroleum products leads to considerable pollution of the natural environment [7].

---

*Corresponding author.

The soil is a key component of natural ecosystem because environmental sustainability depends largely on sustainable soil ecosystem [8]. When soil is polluted, the physiochemical properties are affected which may decrease its productive potentials [9]-[12]. In Nigeria, most of the terrestrial ecosystem and shoreline in oil-producing areas are important agricultural land under cultivation [13]. Any contact with petroleum and/or refined petroleum products causes damage to soil conditions of these agricultural lands, which culminates in loss of soil fertility [11] [12] [14]-[16]. Soil enzymes are important biotic components which are responsible for soil biochemical reactions [17]. The effects of petroleum hydrocarbon on soil enzyme activities are well documented [18]-[21].

Catalase activity, alongside with the dehydrogenase activities, gives information on the microbial activities in soil. Both catalase and dehydrogenase activity are very sensitive to heavy metal pollution [22]-[24]. Their values can therefore be used as a simple toxicity testing tool [25]. The aim of this study was to determine the effect of petroleum products (kerosene, diesel, engine oil and petrol in soil on soil catalase and dehydogenase activities.

2. Materials and Methods

Refined petroleum products of known physical properties were obtained from Warri Refining and Petrochemical Company, Warri, Nigeria. The soil (sand 84%, silt 5.0%, clay 0.4% and organic matter 0.6%, pH 6.1) was obtained from a fallow land in Delta State University, Abraka. Some of the chemical properties of the soil are listed in **Table 1**.

Soil (1600 g) was taken in each of small size planting bags (1178.3 cm^3, 15 cm deep) and divided into six groups of five replicates. Groups 1 to 5 contained 0.1%, 0.25%, 0.5%, 1.0% and 2.0% (v/w) respectively of each of the petroleum products while group six served as control (0.0%). To the first bag, 1.0 ml of kerosene, corresponding to 0.1%, was added. The petroleum products treated soil samples were mixed vigorously with the hand to obtain homogeneity of the mixture. The procedure was repeated for 0.25%, 0.5%, 1.0%, 1.5% and 2.0%. Each treatment, including control, was replicated five times.

2.1. Preparation of Extract and Assay of Catalase Activity

One hundred ml of phosphate buffer, pH 7.4, was added to 10 g of soil and stirred vigorously. The soil suspension was filtered using cheesecloth. The filtrate was centrifuged at maximum speed of 7000 g for 10 min to obtain supernatant (S1). Catalase activity was determined as described by Rani *et al.* [26]. Catalase breaks down hydrogen peroxide to give oxygen that oxidises potassium dichromate. The oxidation of chromate gives a chromophore that absorbs maximally at 610 nm. The enzyme extract (0.5 ml) was added to the reaction mixture containing 1 ml of 0.05 M phosphate buffer (pH 7.5), 0.5 ml of 0.2 MH$_2$O$_2$, 0.4 ml H$_2$O and incubated for different time period t$_1$, t$_2$ and t$_3$ for 1 minute, 2 minutes and 3 minutes respectively. The reaction was terminated after each time interval by the addition of 2 ml of acid reagent (dichromate/acetic acid mixture) which was prepared by mixing 5% potassium dichromate with glacial acetic acid (1:3 by volume). To the control, the enzyme was added after the addition of acid reagent. All the tubes were heated for 10minutes in boiling water and the absorbance was read at 610 nm. Catalase activity was expressed in terms of moles of H$_2$O$_2$ consumed/min.

Table 1. Physicochemical properties of test soil.

Parameters	Value
pH	6.09
Total organic carbon, %	2.90
Phosphorus, mg/kg	<0.01
Nitrogen, mg/kg	8.47
Nitrate, mg/kg	9.86
Cation exchange capacity, meq/100g	0.74
Sodium, mg/kg	9.06
Potassium, mg/kg	6.72
Calcium, mg/kg	2.98
Magnesium, mg/kg	0.31

2.2. Assay of Soil Dehydrogenase Activity

Dehydrogenase activity was determined using the method described by Tabatabai [27]. Dehydrogenases convert 2, 3, 5-triphenyl tetrazolium chloride to formazan. The asbsorbance of formazan was read spectrophotometrically at 485 nm. Sieved soil (1 gm) was placed in test tubes (15 × 100 mm), mixed with 1 ml of 3% aqueous (w/v) 2, 3, 5-triphenyl tetrazolium chloride and stirred with a glass rod. After 96 h of incubation (27°C) 10 ml of ethanol was added to each test tube and the suspension was vortexed for 30 s. The tubes were then incubated for 1 h to allow suspended soil to settle. The resulting supernatant (5 ml) was carefully transferred to clean test tubes using Pasteur pipettes. Absorbance was read spectrophotometrically at 485 nm. Extinction coefficient of 15433 Mol/cm [28] was used for evaluating the concentration of formazan formed.

2.3. Statistical Analysis

The results were expressed as mean + SEM. All results were compared with respect to the control. Comparisons between the test and control were made by using analysis of variance (ANOVA). Differences at $p < 0.01$ were considered as significant.

3. Results

The results of the effect of refined petroleum products on soil catalase activites after four , eight and twelve days soil treatment are presented in **Figure 1** The activities of catalase in soil samples were decreased significantly ($p < 0.01$) after four days of the treatment of soil samples with kerosene, diesel, engine oil and petrol. At the highest concentrations petrol decreased catalase activity significantly ($p < 0.01$) more than did engine oil, and engine oil decreased catalase significantly ($p < 0.01$) activity more than did either kerosene or diesel. After eight days of treatment of soil with peroleum produts, kerosene treatment of soil resulted in significant ($p < 0.01$) decrease of catalase activity at 2% concentration compared with control. Diesel treatment of soil gave rise to a significant ($p < 0.01$) decrease of catalase activity at 1% and 2% concentrations. However, engine oil treatment of soil resulted in significant ($p < 0.01$) decrease in catalase activity in all the concentrations tested (0.25% - 2%). Finally, petrol treatment of soil brought about a significant increase of soil catalase activity at 0.25%, 0.5%, 1.5% and 2% compared with control. Comparing the activities of soil catalase in kerosene, diesel, engine oil and petrol treated soil; it was found to be significantly lower in all the tested concentrations in engine oil treatment of soil except at 0.25% concentration than the other petroleum products. But petrol treatment resulted in significantly higher catalase activity at 0.25% and 0.5% concentrations, than in other petroleum products treated soils

The activities of catalase after twelve days following the treatment of soil with kerosene significantly ($p < 0.01$) decreased catalase activity at a concentration of 0.25% but not at higher concentrations. Diesel treatment of soil resulted in decrease in catalase activity that was significant ($p < 0.01$) only at 0.25% level of contamination. Engine oil treatment of soil resulted in significant ($p < 0.01$) increase of catalase activity from 0.5% up to 1.5%, thereafter, an additional increase in concentration brought about a significant ($p < 0.01$) decrease of catalase activity relative to the control. Petrol treatment of soil resulted in a significant ($p < 0.01$) decrease of catalase activity only at 0.25%, 1% and 1.5% concentrations compared with control. When the activities of soil catalase in petrol, kerosene, and diesel and engine oil treatment of soil are compared; it was found to be significantly higher in engine oil treatment of soil at 1% and 1.5% concentrations than the other petroleum products. However, catalase activity was significantly lower in petrol treatment of soil at 1% concentration than the other petroleum products.

The activities of soil dehydrogenase after four days following the treatment of soil samples with kerosene, diesel, engine oil and petrol are presented in **Figure 2**. Kerosene treatment of soil resulted in significant increase of soil dehydrogenase activity from 0.5% up to 2%. Similarly, diesel treatment of soil led to a significant increase of soil dehydrogenase activity at 0.5% up to 2%. Engine oil and petrol treated soils had an all round significant ($p < 0.01$) increase of soil dehydrogenase activity in all the concentrations tested compared with control. When the activities of soil dehydrogenase in engine oil treatment of soil were compared with other petroleum products treated soil samples, they were found to be significantly ($p < 0.01$) higher at 0.25%, 0.5%, 1.5% and 2% concentrations over other petroleum products. Petrol treatment of soil recorded a significantly higher activity of soil dehydrogenase only at 1% than other petroleum products. The activities of soil dehydrogenase after eight days following treatment of soil samples with kerosene, diesel, engine oil and petrol are shown in **Figure 2**. In all the concentrations tested, all petroleum products, treated soil samples resulted in significant increase (p < 0.01) of soil dehydrogenase activity compared to control. When the activities of soil dehydrogenase in soils treated with kerosene, diesel, engine oil and petrol are compared, it was found to be significantly higher in engine oil at all levels of concentrations than the other petroleum products. The activities of soil dehydrogenase after twelve days of following treatment of soil with kerosene, diesel, engine oil and petrol are shown in **Figure 6**. After twelve days, kerosene treated soil resulted in significant increase (p < 0.01) of dehydrogenase activity at 1%, 1.5% and 2% levels of soil contamination. Diesel treatment of soil in all the concentrations tested, gave rise to significant increase (p < 0.01) of dehydrogenase activity from 1% to 2%, while, engine oil and petrol treatment of soil samples resulted in significant (p < 0.01) increase of dehydrogenase activity at 0.5% to 2% concentrations compared to the control. when the dehydrogenase activities of soils treated with kerosene, diesel, engine oil and petrol were compared, the activities were found to be to almost the same at each level of soil contamination except that of kerosene, at 2%, which was significantly lower than the other petroleum products. On the whole, the activities of soil dehydrogenase, in all the concentrations were significantly (p < 0.01) higher compared to control.

4. Discussion

The refined petroleum products altered soil catalase activity. The activity of the enzyme decreased after four days of treatment of soil with petroleum products (**Figure 1**), increased by the eight day (**Figure 2**) at lower

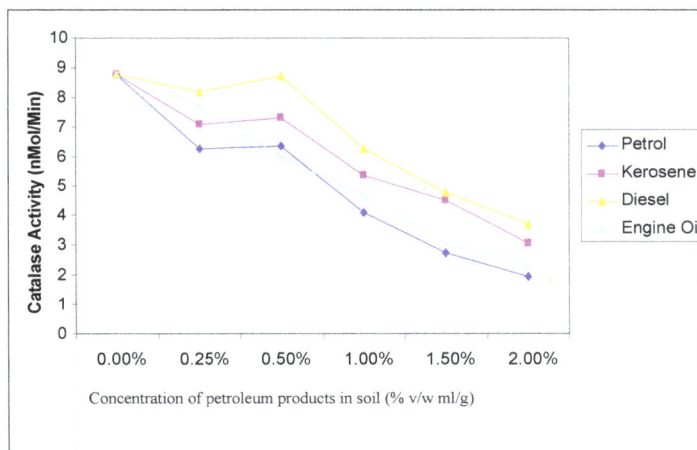

Figure 1. Dependence of catalase activity on concentration of four petroleum products in soil after four days.

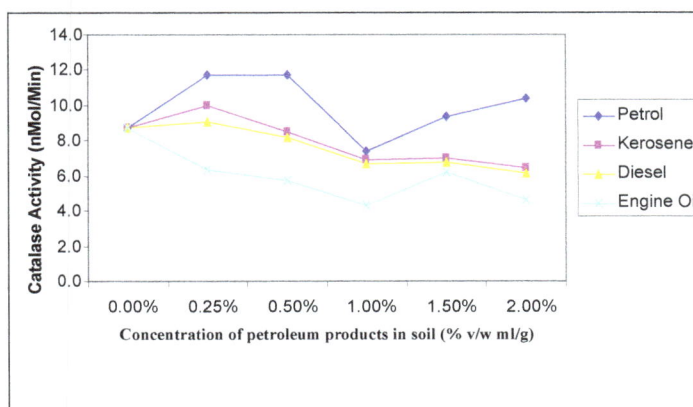

Figure 2. Dependence of catalase activity on concentration of four petroleum products in soil after eight days.

concentrations of kerosene, diesel and petrol in soil and then started to decrease after twelve days (**Figure 3**) of treatment of soil. Petroleum products, when present in soil, creates an unsatisfactory condition for soil organisms, mainly due to poor aeration, immobilization of soil nutrients and lowering of soil pH which culminates in petroleum mediated reduction in the number of hydrocarbon degrading microorganisms [11] [15] [17] [19] [29]. This may be the basis for decrease in catalase activity after four days of treatment. Decrease in catalase activity when soil is exposed to petroleum was previously reported [19]. However, the increase in the activity of the enzyme after eight days (**Figure 2**) could be predicated on increased microbial activity towards biodegradation of available petroleum hydrocarbon. This might explain why the activity of the enzyme started to decrease after twelve days of incubation with petrol and kerosene (**Figure 3**). Earlier reports indicated that there is a decrease in catalase activity after biodegradation has decreased [30]-[32]. However, the enzyme activity was higher in diesel and engine oil treated soil samples after twelve days relative to petrol and kerosene. These observations indicate that biodegradation was in the order of petrol > kerosene > diesel > engine oil.

Earlier reports indicated that soil polluted with petroleum hydrocarbon experienced increased dehydrogenase activity [18] [19]. The result of the current investigation agrees with this fact. The activity of the enzyme increased after four (**Figure 4**) and eight days of post treatment (**Figure 5**) with the four refined petroleum products (petrol, kerosene, and diesel and engine oil). However, the enzyme activity decreased after twelve days (**Figure 6**) in all the soil treated with the four refined petroleum products. The increase in enzyme activity has been attributed to the involvement of certain microorganisms in the metabolism of hydrocarbons whereas; the

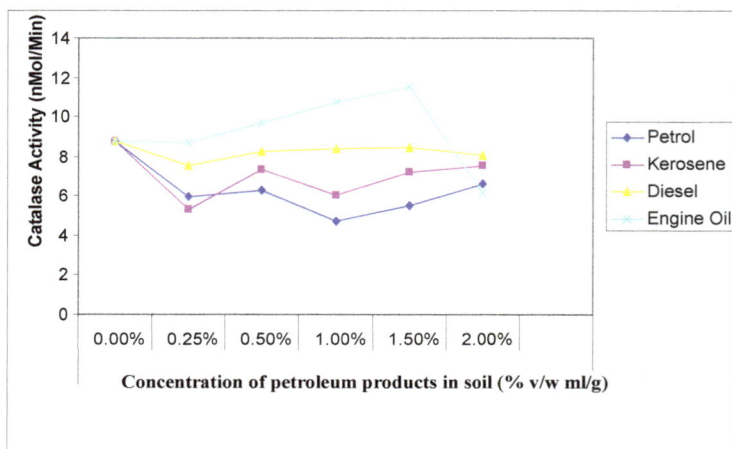

Figure 3. Dependence of catalase activity on concentration of four petroleum products in soil after twelve days.

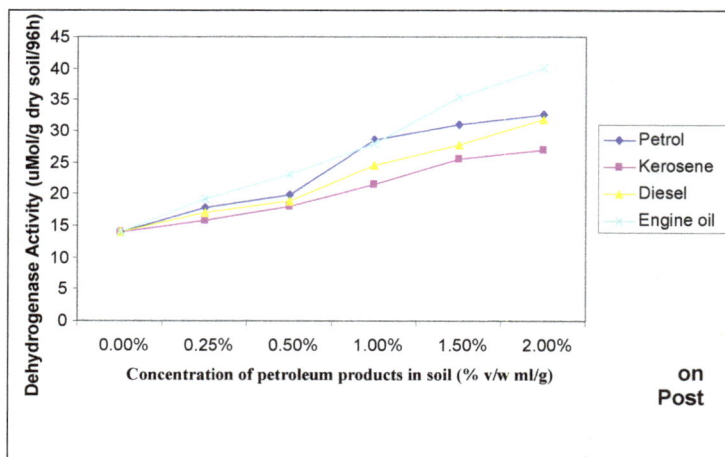

Figure 4. Dependence of dehydrogenase activity on concentration of four petroleum products in soil after four days.

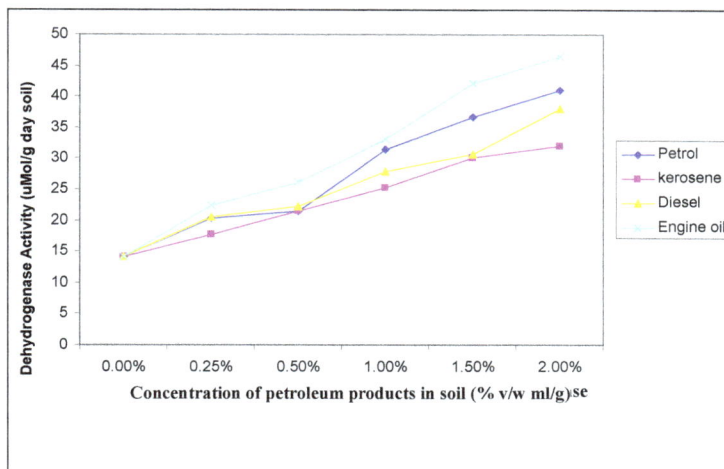

Figure 5. Dependence of dehydrogenase activity on concentration of four petroleum products in soil after eight days.

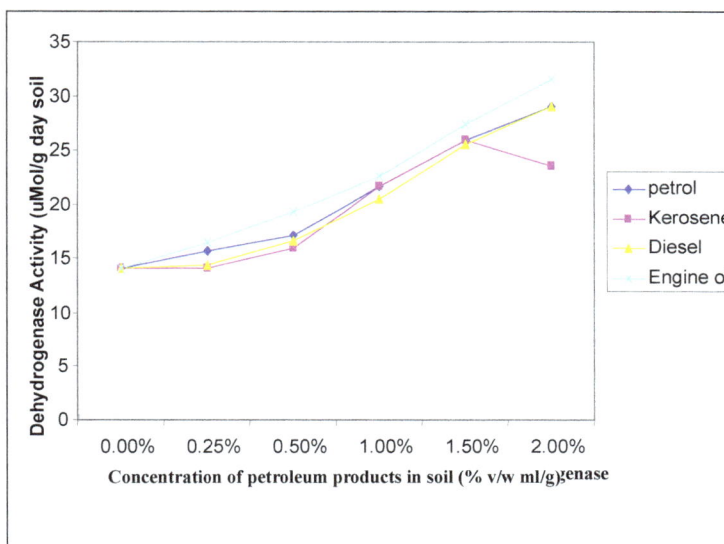

Figure 6. Dependence of dehydrogenase activity on concentration of four petroleum products in soil after twelve days.

decrease in the enzyme activity after twelve days of incubation could be due to decease in petroleum content of the soil. Decrease in soil dehydrogenase activity after biodegradation of petroleum has been reported [33] [34]. Although, the refined petroleum products caused a similar pattern, as they affect catalase activity, in the alteration of soil dehydrogenase activity, the general results seem to reveal that the toxicity effect is in the order of kerosene > diesel > petrol > engine oil. The toxicity of kerosene and diesel was reported earlier [20] [35].

5. Conclusion

In conclusion, it is pertinent to say that kerosene is more toxic to soil microorganisms, by decreasing the soil enzyme activities, in short term. However, engine oil stayed longer in the soil compared to the other three refined petroleum products, as reflected by sustained dehydrogenase after four days of post treatment.

References

[1]   Kenny, J., Kutcherov, V., Bendeliani, N. and Alekseev, V. (2002) The Evolution of Multi-Component System of High

Pressures: VI. The Thermodynamic Stability of the Hydrogen Carbon System: The Genesis of Hydrocarbon and the Origin of Petroleum. *Proceeding of the National Academic of Sciences USA*, **99**, 10976-10981.

[2]   Achi, C. (2003) Hydrocarbon Exploration, Environmental Degradation and Poverty. The Nigeria Delta Experience. *Diffuse Pollution Conference*, Dublin.

[3]   Tolulope, A.O. (2004) Oil Exploration and Environmental Degradation. The Nigerian Experience. *Environmental Informatics Archives*, **2**, 387-393.

[4]   Michalcewicz, W. (1995) Wply is oleju napadewo-godo silnikow Dielsa m liczbnosc batkeri gryzbow promiienowciow oraz biomase mikoorg-anizmow glebowych. *Rocz. Panstw. Zakl. Hig.*, **46**, 91-97.

[5]   Song, H. and Bartha, R. (1990) Effects of Jet Fuel Spills in the Microbisl Community of Soil. *Applied and Environmental Microbiology*, **56**, 646-651.

[6]   Jorgensen, K.S., Purstinen, J. and Sourti, A.M. (2000) Bioremediation of Petroleum Hydrocarbon-Contaminated by Compositing in Biophiles. *Environmental Pollution*, **107**, 245-254. http://dx.doi.org/10.1016/S0269-7491(99)00144-X

[7]   Odjegba, V.J. and Sadiq, A.O. (2002) Effect of Spent Engine Oil on the Growth Parameters, Chlorophyll and Protein Levels of *Amaranthus hybridus* L. *The Environmentalist*, **22**, 23-28. http://dx.doi.org/10.1023/A:1014515924037

[8]   Adriano, O.C., Chopecka, A. and Kaplan, K.I. (1998) Role of Soil Chemistry in Soil Remediation and Ecosystem Conservation. Soil Science Society of America, Special Publication, Madison, 361-386.

[9]   Osuji, L.C., Adesiyan, S.O. and Obute, G.C. (2004) Post Impact Assessment of Oil Pollution in the Agtada West Plain of Niger Delta Nigeria: Field Reconnaissance and Total Extractable Hydrocarbon Contact. *Chemistry & Biodiversity*, **1**, 1569-1577. http://dx.doi.org/10.1002/cbdv.200490117

[10] Osuji, L.C., Inioborg, O.I. and Ojinnata, C.M. (2006) Preliminary Investigation of Mgbede-20 Oil Polluted Site in Niger Delta Nigeria. *Chemistry & Biodiversity*, **3**, 568-577. http://dx.doi.org/10.1002/cbdv.200690060

[11] Osuji, L.C. and Nwoye, L. (2007) An Appraisal of the Impact of Petroleum Hydrocarbons on Soil Fertility; the Owaza Expectation. *African Journal of Agricultural Research*, **2**, 318-324.

[12] Wyszkowska, J., Kuncharki, J., Jastrazabska, E. and Hlasko, A. (2001) The Biological Properties of the Soil as Influenced by Chromium Contamination. *Polish Journal of Environmental Studies*, **10**, 37-42.

[13] Egborge, A.B.M. (1994) Water Pollution in Nigeria: Bio-Diversity and Chemistry of Warri River. Ben Miller Publication, Warri.

[14] Sztompka, E. (1999) Biodegradation of Engine Oil in Soil. *Acta Microbiologica Polonica*, **489**, 185-196.

[15] Atuanya, E.I. (1987) Effects of Waste Engine Oil Pollution on Physical and Chemical Properties of the Soil. *Nigerian Journal of Applied Science*, **55**, 155-176.

[16] Amadi, A., Abbey, S.D. and Nma, A. (1996) Chronic Effect of Oil Spill on Soil Properties and Michroflora of Rainforest Ecosystem in Nigeria. *Water, Air, and Soil Pollution*, **86**, 1-11. http://dx.doi.org/10.1007/BF00279142

[17] Zahir, A.Z., Malik, M.A.R. and Arshad, M. (2001) Soil Enzymes Research: A Review. *Journal of Biological Sciences*, **1**, 299-301. http://dx.doi.org/10.3923/jbs.2001.299.307

[18] Li, H., Zhang, Y., Zhang, C.G. and Chen, G.X. (2005) Effect of Petroleum-Containing Wastewater Irrigation on Bacterial Diversities and Enzymatic Activities in a Paddy Soil Irrigation Area. *Journal of Environmental Quality*, **34**, 1073-1080. http://dx.doi.org/10.2134/jeq2004.0438

[19] Achuba, F.I. and Peretiemo Clarke, B.O. (2008) Effect of Spent Engine Oil on Soil Catalase and Dehydrogenase Activities. *International Agrophysics*, **22**, 1-4.

[20] Wyszkowska, J., Kuncharski, J. and Waldowska, E. (2002) The Influence of Diesel Oil Contamination on Soil Microorganism and Oat Growth. *Rostlinna Vyroba*, **48**, 58-62.

[21] Wyszkowska, J. and Kuchaarski, J. (2000) Biochemical Properties of Soil Contaminated by Petrol. *Polish Journal of Environmental Studies*, **9**, 479-485.

[22] Naplekova, N.N. and Bulavko, G.I. (1983) Enzyme Activity of Soils Polluted by Lead Compounds. *Soviet Soil Science*, **15**, 33-38.

[23] Perez, M.M. and Gonzalez, C.S. (1987) Effect of Cadmium and Lead on Soil Enzyme Activity. *Review of Ecology and Biology Solutions*, **1**, 11-18.

[24] Wilkes, B.M. (1991) Effects of Single and Successive Addition of Cd., Ni and Zn on Carbon Dioxide Evolution and Dehydrogenase Activity in Sandy Soil. *Biology and Fertility of Soils*, **11**, 34-37. http://dx.doi.org/10.1007/BF00335831

[25] Rogers, J.C. and Li, S. (1985) Effect of Metals and Other Inorganic Ions on Soil Microbial Activity. Soil Dehydrogenase Assay as a Simple Toxicity Test. *Bulletin of Environmental Contamination and Toxicology*, **34**, 858-865. http://dx.doi.org/10.1007/BF01609817

[26] Rani, P., Meena Unni, K. and Karthikeyan, J. (2004) Evaluation of Antioxidant Properties of Berries. *Indian Journal of*

Clinical Biochemistry, **19**, 103-110. http://dx.doi.org/10.1007/BF02894266

[27] Tabatabai, M.A. (1982) Soil Enzymes, Dehydrogenases. In: Miller, R.H. and Keeney, D.R., Eds., *Methods of Soil Analysis. Part* 2. *Chemical and Microbiolgical Properties*, Agronomy Monograph, No. 9, ASA and SSSA, Madison.

[28] Dushoff, I.M., Payne, J., Hershey, F.B. and Donaldson, R.C. (1965) Oxygen Uptake Tetrazolium Reduction during Skin Cycle of Mouse. *American Journal of Physiology*, **209**, 231-235.

[29] Maila, M.P. and Cloete, T.E. (2005) The Use of Biological Activities to Monitor the Removal of Fuel Contaminants —Perspectives to Monitoring Hydrocarbon Contamination: A Review. *International Biodeterioration & Biodegradation*, **55**, 1-8. http://dx.doi.org/10.1016/j.ibiod.2004.10.003

[30] Frankenberger, W.T. and Johansson, J.B. (1982) Influence of Crude Oil and Refined Petroleum Products on Soil Dehydrogenase Activity. *Journal of Environmental Quality*, **11**, 602-235. http://dx.doi.org/10.2134/jeq1982.00472425001100040010x

[31] Van der Waarde, J.J., Dijkhuis, E.J., Henssen, M.J.C. and Keuing, S. (1995) Enzyme Assays as Indicators for Bioremediation. In: Hinchee, R.E., Douglas, G.S. and Ong, S.K., Eds., *Monitoring and Verification of Bioremediation*, Batelle Press, Columbus, 59-63.

[32] Schinner, F., Ohlinger, R. and Margesin, R. (1996) Methods in Soil Biology. Springer Press, Berlin. http://dx.doi.org/10.1007/978-3-642-60966-4

[33] Janke, S., Schamber, H. and Kunze, C. (1992) Beeinflussung der Biodenbiologischen Aktivat durch Heizol. *Angewandte Botanik*, **66**, 42-45.

[34] Margesin, R. and Schinner, F. (1997) Bioremediation of Diesel-Oil Contaminated Alpine Soil at Low Temperatures. *Applied Microbiology and Biotechnology*, **47**, 462-468. http://dx.doi.org/10.1007/s002530050957

[35] Wemedo, S.A., Obire, O. and Ijogubo, O.A. (2002) Myco-Flora of Kerosene Polluted Soil in Nigeria. *Journal of Applied Sciences and Environmental Management*, **6**, 14-17.

9

Analysis of Staffing and Training Needs for Effective Delivery of Extension Service in Sustainable Land Management in Kilimanjaro Region, Tanzania

John F. Kessy

Department of Forest Economics, Sokoine University of Agriculture, Morogoro, Tanzania
Email: jfkessy2012@gmail.com

Abstract

An assessment of staffing and training needs for effective delivery of extension services in mainstreaming sustainable land management (SLM) practices in Kilimanjaro Region was conducted in June/July 2013. Data collection methods included discussions with key informants at the regional and district levels, consultations with village level stakeholders and potential collaborators, review of human resources data both at regional, district and ward levels and collection of individual staff bio-data including capacity deficiencies. The staffing situation at the regional and district levels was considered to be adequate for effective mainstreaming of SLM interventions in the region. Staffing at ward and village levels was very poor and largely inadequate for sustainable execution of extension services. It is optimistically estimated that on average the staffing at ward level needs to be increased by at least 50%. In some districts the deficiency of extension staff at ward level was as high as 80%. Training needs exist at all levels from the region down to community level. At the regional and district levels both long and short term training programs were required. At the community level required training is more practical and purely focused in mainstreaming SLM interventions at individual households and community lands. Potential collaborators with local government were identified in four main categories namely, NGOs/CBOs, private sector, government departments and faith-based organizations. The study recommends a capacity building program on specific knowledge gaps identified at regional, district, ward and village levels. The study further recommends that immediate measures need to be taken by the district authorities to address the staffing problem at ward level including recruitment of volunteers and developing collaboration framework with identified potential partners.

Keywords

Staffing, Training, Extension Services, Sustainable Land Management

1. Introduction

The Global Environment Facility (GEF) through UNDP has committed US $2.63 million over a period of 4 years to contribute to reducing land degradation on the highlands of Kilimanjaro through implementation of actions targeting removal of barriers for Sustainable Land Management (SLM) in a multi-level approach. The project goal is to ensure "Sustainable Land Management provides the basis for economic development, food security and sustainable livelihoods while restoring the ecological integrity of the Kilimanjaro Region's ecosystems".

The Kilimanjaro ecosystem is experiencing an extensive process of degradation and deforestation, with serious consequences on its ability to continue providing ecosystem services and declining land productivity [1]-[3]. Degradation is driven by a set of complex and interrelated factors, such as rapid increase of a population largely dependent on natural resources, land use change, poor land management practices, unsustainable harvesting of natural resources, declining commodity prices and climate change. As in other parts of the country, local communities in Kilimanjaro are faced with a number of shocks for which they must develop coping strategies. Among the documented shocks [4] for example include climate change causing crop failure, economic hardships, health related shocks causing diseases and deaths, land use conflicts as well as policy related challenges.

The project intends to address four key barriers to sustainable land management in the region which are: limited livelihood opportunities outside the natural resources, weak incentives for adoption of SLM, weaknesses in the policy, planning and institutional environment that influence SLM, and inadequate skills at all levels required for promoting and/or adopting SLM. The project has a set of outputs linked to activities and interventions aimed to address the identified challenges. However, for the project outputs to be realized the capacities of the implementing agents on the ground particularly the regional and district technical teams need to be enhanced through a systematic capacity building program. The reported study aimed to assess the staffing and training needs for effective delivery of SLM related extension services in the region. Specifically, the study aimed to:

- Assess existing capacity for extension service delivery at regional, district and community level;
- Assess the required capacity and gaps for effective extension delivery;
- Identify best partnership practices between local government staff, NGOs and the private sector in delivering of extension services;
- Recommend a capacity bridging program involving partnerships and synergies with existing collaborators in the districts.

2. Methodology Deployed

A range of data-collection methodologies was deployed during the assessment to facilitate effective participation of key stakeholders and triangulate the findings objectively [5] [6]. The methodologies deployed are briefly described in the following sub-sections.

2.1. Discussions with Key Informants in the Region

This approach was used to collect information from the project management team, regional and district officials, and selected professional and technical staff using a specified data-collection checklist. At the regional level the key informants, among others included the Regional Technical Team, responsible for implementing the project, and other relevant administrators. At the district level discussions involved the District Facilitating Teams, technical officers, and other extension agents in the district. The data collection process was structured to capture comparable information from both the region and districts for triangulation purposes.

2.2. Village/Community Level Discussions

From the design stage of the assessment it was strongly felt that community needs and perceptions in relation to capacity building should be given attention. In each district one representative village was selected. In the selected village a consultative meeting was organized consisting of representatives from the community. Representatives included 5 members from the village government, 5 from village environmental committees, 3 members representing influential farmers in the village and 2 representatives who were female household heads.

2.3. Collection of Bio-Data of Employees and Deficiencies

Current data on employees was collected from the regional, district and ward offices. A special form was designed to capture the entire list of existing extension staff in the region and districts, their capacities and deficiencies.

2.4. Reviews of Relevant Documents

Various documents provided by the project, local government authorities and other stakeholders were reviewed. From these documents secondary information on the capacity status in the target districts was obtained. Particularly, human resources management data was reviewed from the personnel department files and planning documents at regional and district levels.

2.5. Data Analysis

Data was analyzed with the purpose of establishing existing capacities and gaps [7] [8] in relation to the provision of extension services for mainstreaming SLM activities at regional, district and ward/village levels. For qualitative information collected through checklists, content analysis was deployed to extract valuable pieces of information that add value to the assessment. Quantitative data especially on staffing levels was analyzed using standard statistical procedures.

3. Results and Discussions

3.1. Analysis of Staffing Needs

In the context of this assessment, staffing was contextualized to entail the process of acquiring, deploying, and retaining human resources of sufficient quantity and quality to create positive impacts on effective implementation of project interventions in Kilimanjaro Region [9] [10]. For purposes of providing extension services at the district, ward and village level as well as coordination at regional level, staffing is perceived as a management function of the regional and district authorities within the Local Government system in Tanzania. In this context it involves manning the regional, district, ward and village level administrative structures with the needed technical and extension officers for effective service delivery to the target population. This is supposed to be achieved through proper and effective selection, appraisal and development of the personnel in their respective positions to fulfill the roles assigned to them by employers [11]. It is in this context that the assessment was done to establish the staffing deficiencies at all levels in order to propose mechanisms through which the SLM project in collaboration with the local government structure in Kilimanjaro could improve the situation for a more effective delivery of extension services.

3.1.1. Staffing at Regional Level

Two categories of staff are responsible for mainstreaming SLM activities at the regional level. They include staff of the Economic and Productive Section within the regional administration structure and project staff employed by UNDP. Assessment of staffing needs considered both categories of staff, who actually from the Regional Technical Team. According to the establishment of the Economic and Productive Section (**Table 1**) the section is supposed to have an administrative secretary, 4 agricultural officers, 1 irrigation engineer, 2 agro-engineers, 2 cooperative officers, 2 trade officers, 2 livestock officers, 1 veterinary officer, 1 forest officer, 1 game officer and 1 beekeeping officer.

Assessment of actual strength in the section shows that the section has deficiency of nine technical staff. Examining the staffing deficiency in relation to the task of mainstreaming SLM in the region it was apparent that lack of an irrigation engineer, two agro-engineers, one veterinary officer, and a beekeeping officer can adversely affect the mainstreaming of SLM activities in the region. This is because SLM interventions in the region require knowledge and professional skills from all disciplines including engineering, veterinary science and beekeeping. Discussions with the Regional Technical Team confirmed the fact that the shortage, especially of engineers has a bearing on delivery of extension services related to rehabilitation of degraded lands, and improvement of irrigation systems.

It was observed that most of the vacant positions at the regional level were already budgeted for (**Table 1**). What remained to be done was to fill the positions through normal recruitment procedures. Discussion with the regional team during the assessment revealed further that team spirit and holistic approaches have been promoted in delivering extension services to the communities. The inter-disciplinary approach that has been

Table 1. Analysis of staffing and capacity deficiencies at regional level.

S/No	Position	Establishment	Actual	Deficiencies	Comment
		Staffing at regional level			
1	Assistant administrative secretary	1	1	0	Staffing as per requirement
2	Agriculture officer	4	4	0	Staffing as per requirement
3	Irrigation engineer	1	0	1	Approved in budget 2012/2013
4	Agro-engineer	2	0	2	Approved in budget 2012/2013
5	Cooperative officer	2	1	1	Approved in budget 2012/2013
6	Trade officer	2	1	1	Budgeted for the year 2013/2014
7	Livestock officer	2	2	0	Staffing as per requirement
8	Veterinary officer	1	0	1	Approved in budget 2012/2013
9	Fisheries officer	1	0	1	Approved in budget 2012/2013
10	Forestry officer	1	1	0	Staffing as per requirement
11	Game officer	1	0	1	Budgeted for the year 2013/2014
12	Beekeeping officer	1	0	1	Budgeted for the year 2013/2014
	Total	19	10	9	
		Regional staff employed by UNDP			
S/No	Position	Establishment	Actual	Deficiency	Comment
1	Technical advisor	1	1	0	
2	National project coordinator	1	1	0	
3	Assistant project coordinator	1	1	0	
4	Socio-economic specialist	1	1	0	

adopted has minimized the impact of staff deficiency. Regarding the project staff that were employed by UNDP the assessment revealed that all established project positions at the regional level have actually been staffed. There is no deficiency in terms of project staff because the recruitment process by UNDP closely matched the requirements of the project and qualifications of the employed staff.

3.1.2. Staffing at District Level

Extension services for mainstreaming SLM activities at the district level were facilitated by the District Facilitation Team (DFT) under the leadership of a district focal person. The composition of the DFT is normally determined by the District Executive Director (DED) who appoints members of a DFT. During the appointment, the DED ensures that all the needed technical competences for handling sustainable land management issues are included in the team. Most common technical officers who constitute the DFT are drawn from relevant disciplines including agriculture, forestry, beekeeping, livestock development, community development, planning, land management and engineering. Composition of a DFT depends on existing professionals in a respective district. In total, the DFT is made up of eight members. The role of DFT is to facilitate the process of mainstreaming SLM interventions through delivery of extension services to the communities.

The assessment of staffing levels in the districts examined whether the DFTs had adequate staff for effective delivery of extension services. Since the DFT membership was drawn from relevant departments for SLM, it was implied that if the DFT was in place, adequately staffed and operational, extension service delivery in the districts would be smooth. It was revealed that in most of the districts staffing for implementation of SLM activities was adequate. It was only in two districts namely Mwanga and Rombo where deficiencies of 1 and 2 technical staff were respectively encountered. In terms of qualifications of the technical staff within the districts there was much variation. Most of the encountered staff either had a degree, an advanced diploma or an ordinary diploma. For implementing and sustaining SLM interventions in the districts, these qualifications were adequate. However, some capacity building in some specific areas of intervention shall be necessary as presented in the Section 3.2 of this paper.

3.1.3. Staffing at Ward and Village Levels

Eventually implementation of planned SLM intervention takes place in the target villages with facilitation from extension officers based either at the ward of village levels. During the assessment, it was therefore considered an issue of prime importance to assess existing staffing situation in the wards in order to establish staffing deficiencies which can adversely affect mainstreaming and sustaining SLM interventions in the region.

The ward level was purposely selected as the focus of analysis as opposed to village level for a number of reasons. In all the districts assessed, extension services at the community level were provided by extension offices based in the wards but serving a number of villages. In some instances, an extension officer served more that one ward. In very few incidences, some villages have village level extension officers. However, such villages were too few to direct the focus of the assessment at village level. The implication here is that it is more realistic to consider staffing at ward rather than village level if one needs to make an impact in terms staffing. As such, it was assumed that provision of extension services at the community level would largely depend on existing extension offers at the ward level.

The assessment took an inventory of existing extension officers in all the wards where the project is operational. The inventory took into consideration staffing requirement for each ward in terms of various disciplines that are relevant to SLM interventions. The assessment further established the number of existing staff in various disciplines and established staffing gaps or deficiencies based on the optional requirement of the wards. Findings indicate that in all the districts serious staffing deficiencies exist at ward level. As shown in **Table 2** below, in terms of percentages the highest recorded staff deficiency was 80.5% from Mwanga District followed by Siha (77%), Same (71.4%) Moshi Municipality (70%), Hai (67%), Rombo (64%) and Moshi Rural (55%).

It should be noted that the relatively lower percentage of deficit in Moshi Rural was actually a reflection of the fact that in several cases one extension officer was serving more than one ward. It should also be noted that the degree of deficiency depended much on the number of wards within the district which are involved in SLM interventions. For example Mwanga has a total of 11 wards and mostly has extension offices in agriculture and livestock production with very few community development officers and without foresters, beekeeping officers or water services attendants at ward level. This makes the staffing deficiency relatively high. When all the target wards were considered together staffing deficiency at ward level was found to be about 54%. It can, therefore, be implied that the wards where SLM interventions are implemented, have only about half of the required extension staff. Yet, this is an optimistic estimate. Depending on available resources, the local government authorities in collaboration with partners can target to reduce this deficiency by a percentage say (e.g., 50%) and apportion the recruitments based on demand in various wards. The staffing, can either be in terms of employed project extension staff, volunteers supported by the project or combination of both categories.

3.2. Analysis of Training Needs

In the context of this assessment training was considered in contemporary terms as an integral part of a comprehensive and systematic program addressing capacity issues, on how to use available technology best suited to the project's goals and embedded in a personnel development plan [9] [10]. It was further perceived that training should constitute build-in incentives on the part of project implementation staff to apply the new skills and should empower trainees to train others in using the technology. This is expected to influence staff performance in delivering extension services. As such, training was not considered to be stand alone and one-off interventions but rather a continuous process of capacity building in the region [7] [8]. New skills and competences are

Table 2. Staffing and capacity deficiencies at ward level in all the districts.

District	Number of ward	Ward staff deficiencies %			
		Required	Present	Deficiency (number)	Deficiency (%)
Mwanga	11	77	15	62	80.5
Same	3	21	6	15	71.4
Siha	2	13	3	10	77
Hai	6	42	14	28	67
Mosh Rural	13	91	41.7	50.3	55
Moshi Urban	10	70	21	49	70
Rombo	2	14	5	9	64
Overall		328	106	176	54

developed responding to specific individual and community needs. This conceptualization guided the assessment process. The assessment of training needs was done for all the levels of SLM implementation including regions, districts, and ward/village levels. In each level the assessment was guided by the roles and responsibilities of staff and other stakeholders at that level in provision of extension services for mainstreaming SLM interventions in the region.

3.2.1. Regional Level

The regional team is responsible for overall coordination of the SLM interventions for the whole region. It is more strategic in nature with leadership and management roles including overall project cycle management. The assessment revealed that two categories of training needs exist at the regional level. They include both long and short term training programs. Long term training is required for some staff who would like to increase their professional competences. They include regional staffs who are interested to increase their qualifications and attain either graduate or post-graduate trainings. Short term training needs are very much related to acquisition of knowledge and skills in order to increase the capacity of regional staff to comfortably handle the coordination role. These are needed for almost all staff at the regional level. It was expressed during the assessment process that short training covering one to two weeks should be organized to provide knowledge and skills on some thematic areas of interest to SLM. The most relevant thematic areas include:

- Leadership and management skills;
- Strategic planning skills;
- Result based monitoring and evaluation skills;
- Project cycle management skills;
- Team building and communication skills;
- Conflict management skills.

It is proposed that the trainings should be formal in nature facilitated by resource person from within the country mostly to be drawn from existing institutions of higher learning and competent NGOs/private sector.

3.2.2. District Level

At the district level the assessment revealed that this is where intensive training and capacity building is needed. This is because the district team is responsible not only for the coordination and supervision of implementation of SLM activities in the district but also because it is envisaged that trainings at ward and village level should be facilitated by the district teams. Additionally the responsibility of integrating SLM interventions in the district planning process and ensuring acceptability by political structures in the districts rests with the District Facilitating Teams (DFTs). It is therefore important to ensure that the district teams are competent in most of the SLM technologies and practices. The assessment revealed that similar to the regional team, the district teams need both long and short duration training programs. A number of needed thematic training areas of training were identified to be relevant to the district teams. These include the following:

- Agroforesty technology;
- Alternative energy technologies;
- Alternative livelihood options and income generation activities at local level;
- Beekeeping skills;
- Tree nursery techniques and technologies;
- Soil erosion control mechanisms;
- Socio-economic survey methods;
- Survey and mapping skills (including the use of GPS);
- Water-harvesting techniques;
- Non-timber forest products harvesting and value addition;
- Fish farming techniques;
- Computer applications and database management;
- Project cycle management including M & E;
- Geographical information systems;
- Contemporary livestock production technologies;
- Team building and communication skills;
- Lobbying and advocacy skills;

- Business administration and financing mechanism (options at local level);
- PES options including REDD+;
- Efficient irrigation technologies.

It is proposed that training at the district level be provided to all relevant technical staff using the Training of Trainers (ToT) approach, the implication being that the district teams have to train those at lower levels. The training should combine theoretical orientation, practical skills and resources allowing some study tours to concretize the practical skills. The proposed duration of the training at the district level is two weeks whereby the first week is spent on theoretical orientation while the second week is used for practical skills including a study tours. The training should be organized within the region but rotating among the districts. Selection of the district where a particular training can be hosted should be guided by the possibility of carrying out practical training in that particular district on the thematic training area. For example, soil-erosion control training should be conducted in a district where soil erosion is a serious problem. As for the study tours, they can be organized within the region or outside depending particular skills required.

3.2.3. Ward and Village Levels

As pointed out earlier it expected that training at ward and village levels should be facilitated by the district teams. In situations where specific practical skills are needed and an NGO or institution in the private sector or FBO partner has the needed skills, collaboration between the district team and the partners should be sought to optimize learning. This may apply for example in areas of alternative energy technologies, nursery techniques, beekeeping, and water harvesting and efficient irrigation systems where a number of partners working in the districts have knowledge and experience as summarized in Section 8 of this report.

The assessment reveled that relevant training at ward and village levels should be on practical skills and adoption of technology as opposed to theoretical concepts. The most relevant thematic areas were identified and they include:
- Alternative energy technologies;
- Livelihood options and micro-financing mechanisms;
- Tree nursery and agroforestry technologies;
- Soil erosion control mechanisms;
- Beekeeping skills;
- Non-timber forest product collection and value addition;
- Water harvesting and efficient irrigation technologies.

It is proposed that the training approach should focus on demonstrations, practical orientation and study tours for practical learning. The trainings should best be organized at ward level and should rotate from one ward to another within the district taking advantage of peculiarities of different wards in relation to required SLM interventions.

3.3. Analysis of Community Perceptions on Existing Capacities

During the assessment process, deliberate efforts were made to get community-level opinion in relation to existing capacities to implement and sustain SLM interventions. In each district, a representative village was selected and a consultative meeting arranged. The consultations aimed first to establish whether villagers are aware of the SLM introduced interventions and how community members participate. It was revealed that all the involved villagers were aware of a project. Villagers reported to participate in project activities including tree planting, soil-erosion control, forest protection, protection of water resources, beekeeping, improved agriculture practices, reviving of traditional irrigation systems, alternative energy sources, and engaging in income generation activities. Villagers were aware of what the project expected from the community which they summarized to be active participation in project activities including practicing improved land-management practices in their own farms.

When asked about their ability and capacity to actively and effectively participate in project activities, the trend in all the villages was to request for capacity building arguing that they did not have enough knowledge and resource for effective participation. Among the areas which they requested for training included:
- Improved agriculture and livestock keeping practices;
- Beekeeping;
- Soil erosion control and conservation;
- Establishment of tree nurseries and agroforestry;

- Alternative energy sources;
- Income generation activities and micro-financing.

They also reported to need support in terms of improved crop variety seeds and livestock herds. When asked about their perceptions on the ability of extension officers to deliver good services, all the consulted villagers reported that extension officers are not enough. They requested for increased number of extension officers particularly in agriculture, livestock and forestry professionals. It was also pointed out by villagers that some extension officers need to be trained on recent development in agriculture.

Shortage of extension officers was reported to result into delays in terms of getting services as one extension officer was serving many villagers in the ward. Inquiries were made to find out whether villagers were willing, ready and capable of sustaining SLM interventions as introduced by extension officers in their localities. Findings indicated that in all the villages surveyed willingness existed. Most of the interviewed villagers were ready to adopt introduced SLM practices in their farms but expressed their need for capacity building and support in terms of:

- Marketing of farm produce;
- Trainings and employment of extension officers;
- Access to credit facilities;
- Access to seeds, seedlings and planting materials;
- Alternative incomes and sources of energy.

3.4. Assessment of Potential Collaborators

The possibility of collaborating with other actors in pursuing SLM interventions was examined. The other actors in this context included like-minded organizations such as NGOs, FBOs, the private sector and other government departments whose activities complement SLM interventions in the region. The assessment was done in every district and the findings are summarized in the following matrix (**Table 3**).

It should be noted that different modalities of engaging the potential collaborators exist. In some instances they can be sub contracted in others they can take part in project activities using their resources but also they can fundraise in collaboration with districts teams.

4. Major Conclusions and Recommendations

Through the assessment it has become possible to understand the existing situation in terms of staffing and training needs in the districts and wards where the SLM project is operational as well as at the regional secretariat where project coordination is affected. The major conclusions from the assessment include the following:

- The existing staffing situation at the regional and district levels is considered to be adequate for effective delivery of extension services. Where gaps exist for example at the regional level, mechanisms are already in place to fill the gaps including inclusion of the recruitment needs in annual budgets.
- Staffing at ward and village levels is very poor and largely inadequate for sustainable execution of project interventions. It is optimistically estimated that average the staffing at ward level needs to be improved by at least 50%. In fact in some districts the shortage of extension staff at ward level is as high as 80%. This is an area where the project can significantly make an impact in terms of staffing.
- Training needs exist at all levels from the region down to community level. At the regional and district levels both long term and short term trainings are required. While the former is for improving the professional skills of the staff the later is largely focused in ensuring that the needed knowledge and practical skills to handle planned project interventions are in place. At the community level required training is more practical and purely focused in mainstreaming SLM interventions at individual households and community lands.
- A number of potential collaborators were identified. They fall in four main categories namely, NGOs/CBOs, private sector, government departments and faith based organizations. Possible collaborating mechanisms included sub-contracting, joint ventures in fundraising, provision of training services and using partner's activities/interventions to complement SLM interventions in areas where the partner is working. However, deliberate efforts to initiate the envisaged collaboration should be taken by the project management team.
- A training program for the project covering all the levels has been proposed.

It is recommended that the SLM local government authorities in collaboration with the project management team should:

Table 3. Identified potential partners, their activities and roles in SLM.

District	Potential partner	Activities of the partner	Areas of collaboration
Mwanga	MIFIPRO	Irrigation, tree planting and beekeeping	Irrigation & tree planting
	TIP	Traditional irrigation systems	Irrigation systems
	SMECAO	Energy saving stoves, Tree nurses and conservation of water sources	Stoves & tree planting
Same	SAIPRO	Irrigation, tree planting, income generation activities and dry land agriculture	Irrigation, tree planting and income generation
	FIDE	Biogas and alternation energy sources	Alternation energy
	SMECAO	See Mwanga above	
	FBOS	Tree plants, irrigation, income generation activities and alternative energy	Tree planting, stoves, irrigation
Siha	TATEDO	Alternative energy, efficient stoves, Tree planting	Stoves and tree planting
	OHINYI	Micro financing	Micro-financing
	IVAENYI	Credit facilities and biogas	Micro financing, biogas
Hai	TATEDO	See above	
	TACRI	Improved coffee varieties, shade trees	Improved agriculture
	FBOs	See above	
	Shoot & Roots	Tree planting	Tree planting
Moshi Rural	TATEDO	See above	
	TACRI	See above	
	FLORESTA	Conservation of water sources & tree planting as well as promotion of indigenous crops?	
	Pangani Water Basin	Management of water resources	Water resources
	MUWASA	Management of water resources	Water resources
Moshi Municipal	Pangani Water Basin	See above	
	FITI	Tree planting, training and beekeeping	Trainings, tree planting
	Tanzania Forest Services	Tree planting, forest management	Tree planting & forest management
	Green Garden	Tree nursing and efficient stoves	Tree planting
	KIWAMA	Tree nursing	Tree planting
	CARMATEC	Energy serving technologies, biogas	Alternative energy
Rombo	TATEDO	See above	
	ENVIRONCARE	Tree planting and energy saving stoves	Stoves, tree planting
	TACRI	See above	
	KEDA	Tree planting	Tree planting
	Seliani Agric Research	Agroforesty research	Agroforesty promotion

- Take immediate measures to implement a capacity building program for mainstreaming SLM activities in the region based on the identified knowledge gaps.
- Decide on immediate measures to be taken to address the identified staffing problem at ward level in all the districts.
- Continue their efforts to develop memorandum of understanding with identified potential collaborators in the region for collaborative delivery of extension services.

Acknowledgements

The author wishes to acknowledge the financial support provided by the UNDP Tanzania Office through the Sustainable Land Management project in Kilimanjaro for conducting this study. The support provided by the regional and district levels teams of experts to the author during the assessment period is highly appreciated.

References

[1] Kessy, J.F., Oktingati, A. and Solberg, B. (1993) The Economics of Rehabilitating Denuded Areas in Tanzania: The

Case of Legho Mulo Moshi. Faculty of Forestry Record No. 60, Sokoine University of Agriculture.

[2] Kessy, J.F. and Oktingati, A. (1994) An Analysis of Some Socio-Economic Factors Affecting Farmer's Involvement in Agroforestry Extension Projects in Tanzania. *Annals of Forestry*, **2**, 26-32.

[3] Oktingàti, A. and Kessy, J.F. (1991) The Farming Systems on Mount Kilimanjaro. In: Newmark, W.D., Ed., *Conservation of Mount Kilimanjaro*, Chapter 8, IUCN, Gland.

[4] Flora, K., Mashindano, O., Rweyemam, D. and Charles, P. (2011) Poverty Escape Routes in Central Tanzania: Copping Strategies in Singida and Dodoma Regions. Volume III, The Economic and Social Research Foundation (ESRF), Dar es Salaam.

[5] Kessy, J.F. (1998) Conservation and Utilization of Natural Resources in the East Usambara Forest Reserves: Conventional Views and Local Perspectives. Tropical Forest Management Papers No. 18, Wageningen.

[6] UNDP (2007) Capacity Assessment Methodology. Users Guide, Capacity Development Group, Bureau for Development Policy, UNDP.

[7] Wignaraja, K. (2009) Capacity Development: A UNDP Primer. UNDP, USA.

[8] Miller, J.A. and Osinski, D.M. (2002) Training Needs Assessment.
http://www.ispi.org/pdf/suggestedReading/Miller_Osinski.pdf

[9] Chris, B., Sparrow, P., Vernon, G. and Houldsworth, E. (2011) International Human Resources Management. 3rd Edition, Chartered Institute of Personnel and Development, London.

[10] Itika, J. (2011) Fundamentals of Human Resources Management: Emerging Experiences from Africa. African Public Administration and Management Series, Vol. 2, African Studies Centre, Leiden.

[11] Robert, L. and Hendon, J. (2012) Human Resources Management: Functions, Applications, Skill Development. Sage Publications, Thousand Oaks.

10

Genetic Differentiation Caused by Chromium Treatment in *Leersia hexandra* Swartz Revealed by RAPD Analysis

X. W. Cai*, Y. Shao, Z. M. Lin

College of Earth Science, Guilin University of Technology, Guilin City, China
Email: *monkeycxw@tom.com

Abstract

Randomly amplified polymorphic DNA (RAPD) technique was applied to assess the genetic variations and phylogenetic relationships in genetic differentiation within 4 Chromium-treatment *Leersia hexandra*. The fresh leaves of *Leersia hexandra* cultivated on the condition of chrome pollution and exogenous organic acids were used as experimental material. The genomic DNA of *Leersia hexandra* was extracted by using CTAB method. The results showed that different samples of *Leersia hexandra* exhibited DNA polymorphism when using the random primer S43, S51 and S55 as the primers in the RAPD reaction. One specific DNA band about 1000 bp was found in the sample which treated with 10 mmol/L concentration EDTA when used the S43 primer to RAPD. The obvious differences between different EDTA-treatment levels suggest that EDTA has certain effects on enrichment to heavy metals of *Leersia hexandra*, it will be more favored to *Leersia hexandra* accumulation of chromium when EDTA concentration increased.

Keywords

Chromium Treatment, Genetic Differentiation, Randomly Amplified Polymorphic DNA (RAPD), *Leersia hexandra* Swartz

1. Introduction

Leersia hexandra Swartz, a perennial marshy plant, has been reported to be a Cr-accumulating plant with high tolerance to Cr. Under nutrient solution culture, it did not show any obvious symptoms of Cr toxicity when Cr concentrations in the leaves reached 5608 mg·kg^{-1} dry weight [1]

---

*Corresponding author.

Previous and continuous researches show that environmental factors have a significant impact on the plants [2]. Excessive Cr in soil has negative impact on plant growth. At the same time, it can also accumulate in plants by roots, and enter human and animal bodies and harm them through the food chain [3].

Random Amplified Polymorphic DNA (RAPD) is a kind of molecular marker based on PCR. It uses 10 bp random primers amplifying the different DNA fragments in genome to show the polymorphism [4] [5]. One RAPD amplification is actually a simple PCR reaction, and it suits for a large number of samples for rapid analysis. The required DNA template is very small amounts, generally an amplification of only 10 ng to 50 ng DNA [6]. RAPD was used to determine genetic variability of ten populations of alfalfa [7] and the relationships between RAPD markers and 22 quantitative traits of caraway (*Carum carvi* L.) were analyzed [8]. The RAPD-PCR method was used to describe the pattern of DNA band variation in the samples influenced by the environmental pollution, to describe the level of pollution in an area contaminated with smoke and waste from an iron-steel factory, and to reveal the level of potential [9].

At present, there are not molecular biology reports on heavy metal chromium enrichment of *L. hexandra*, and little molecular ecology study on plants to heavy metal pollutants under the long -term effect. Reports on using RAPD to analyze heavy metal pollution on plant population genetic diversity are increasing recently, Li *et al.* (2007) used RAPD to analyze genetic diversity and genetic differentiation of *Dicranopteris dichotoma* populations and lead-zinc mine tailings in population which grow in nature potential [10]. Wen *et al.* (2001) planted 4 *Datura* seeds in different regions and do analysis of this 4 *Datura* habitats potential [11]. Gu *et al.* (2008) used RAPD to analyze genetic diversity of *clethroides* populations which grow in lead zinc tailings in the storing time of 10 years and 20 years in natural and contrast soil potential [12]. *L. hexandra* is a kind of important heavy metal chromium enrichment plant, so it is important to study its molecular enrichment mechanism. Therefore, this study used *L. hexandra* treated with heavy metal pollution in the laboratory as experimental materials, analyzed the genetic diversity of *L. hexandra* under different EDTA treatment of heavy metal chromium pollution through RAPD, and tried to explore the differentiation and evolution of *L. hexandra* population under long-term persistence of toxic heavy metals pollution from molecular level.

2. Material and Methods

2.1. Materials

Plant material handling
L. hexandra collected from Guilin city, China, and artificial cultivated in sunlight greenhouse. 400 mg chrome metal Cr^{3+} ($CrCl_3$) to per kg were added to soil when *L. hexandra* planted. EDTA solution of 0 mmol/L, 2.5 mmol/L, 5 mmol/L and 10 mmol/L concentration were prepared to handle *L. hexandra* plant, 3 repeats for every kind of concentration gradient, joined the appropriate EDTA solutions to *L. hexandra* plant every other week, and added a total of 3 times.

Reagents and solutions
2 × CTAB extracting buffer (2% CTAB, 1.4 mol/L NaCl, and 20 mmol/L EDTA, and 100 mmol/L Tris-Cl, pH 8); phenol/chloroform/isoamyl alcohol (25:24:1), chloroform/isoamyl alcohol (24:1), β-mercaptoethanol; 3 mol/L NaAc, (pH 5.2) isopropyl alcohol; TE solution (10 mmol/L Tris-Cl, 1 mmol/LEDTA, pH 8); 75% alcohol; alcohol; 1 × TAE buffer (40 mmol/L Tris, and 20 mmol/L HAc, 1 mmol/L EDTA, pH 8.0), Taq DNA polymerase; agarose; Lambda DNA/EcoR I + Hind III Marker.

Instruments and equipments
General refrigerators; HVE-50 high pressure sterilization pot (Japan Hirayama), LEGEND MICRO17 high speed centrifuge (United States saimofei Fisher); trace moving liquid (Japan Nichipet EX); T1Thermocyler amplification apparatus (Germany BIOMETRA); DYY-12 Sanheng Multiple use Electrophoresis apparatus (Beijing Liuyi); DYCP-34A Electrophoresis tanks (Beijing Liuyi); MRS-1200048U scanner (Shanghai, Zhongjing technology); Furi FR-980 gel imager (Shanghai, Furi technology).

The primer sequences
The primer sequences used in this study are shown in **Table 1**.

2.2. Experimental Methods

Extraction and purification for L. hexandra genomic DNA

Table 1. RAPD random primer of *Leersia hexandra* Swartz

NO	(5′→3′)	NO	(5′→3′)	NO	(5′→3′)	NO	(5′→3′)
S41	ACCGCGAAGG	S51	AGCGCCATTG	S82	GGCACTGAGG	S94	GGATGAGACC
S42	GGACCCAACC	S52	CACCGTATCC	S83	GAGCCCTCCA	S95	ACTGGGACTC
S43	GTCGCCGTCA	S53	GGGGTGACGA	S84	AGCGTGTCTG	S96	AGCGTCCTCC
S44	TCTGGTGAGG	S54	CTTCCCCAAG	S85	CTGAGACGGA	S97	ACGACCGACA
S45	TGAGCGGACA	S55	CATCCGTGCT	S86	GTGCCTAACC	S98	GGCTCATGTG
S46	ACCTGAACGG	S56	AGGGCGTAAG	S87	GAACCTGCGG	S99	GTCAGGGCAA
S47	TTGGCACGGG	S57	TTTCCCACGG	S88	TCACGTCCAC	S60	ACCCGGTCAC
S48	GTGTGCCCCA	S58	GAGAGCCAAC	S89	CTGACGTCAC	S81	CTACGGAGGA
S49	CTCTGGAGAC	S59	CTGGGGACTT	S92	CAGCTCACGA	S93	CTCTCCGCCA

Applied an improved ctab method potential [13] [14], to extract total DNA from leaves of *L. hexandra* dried by silica gel, and used phenol/chloroform/isoamyl alcohol (25:24:1) to extract total DNA, anhydrous ethanol to purified total DNA, 0.8% agarose gel electrophoresis to detect DNA extraction and effects.

Primers screening

Used *L. hexandra* genomic DNA extracted as DNA templates for RAPD random primers screening. Primers are used in this research process all synthesized by Shanghai bio-engineering technology services company limited. Random primers with clear RAPD amplification and stability response were screened for final primers of *L. hexandra* RAPD molecular markers.

RAPD amplification reaction of L. hexandra

We used *L. hexandra* genomic DNA extracted as DNA templates for RAPD amplification reaction. Optimized reaction system: 25 μl total volume, 18.5 μl double distilled water, 2.5 μl 10 uffer liquid, 2 μl Mg^{2+} concentration for 20 mmol/L, 0.5 μl dNTP concentration for 200 μmol/L, 0.5 μl primer concentration for 0.4 μmol/L, 1 μl template DNA originated from DNA extraction diluted 4 times with TE buffer liquid, 2U/25 μl Taq enzyme. Circulation system after optimized: for 5 min 94°C predegeneration, then for 40 cycles, followed by 1 min 94°C predegeneration, 1 min 36°C annealing, 1 min 72°C stretch. Last, 7 min 72°C complete the extension potential [15]. All RAPD reactions were in T1 Thermocyler amplification a paratus (Germany BIOMETRA). Results of PCR amplified products detected by 0.8% agarose gel electrophoresis, and used Lambda DNA/Hind III + EcoR I Markers as molecular standard, and electrophoresis in 120 V constant-voltage for 40 min then do EB dyeing for 10 min, electrophoresis buffer of 1 × TAE, observed by Furi FR-980 gel imaging device, and then taken pictures by the gel image analysis system.

3. Results and Discussion

3.1. *L. hexandra* DNA Extraction Results

Agarose gel electrophoresis results of *L. hexandra* leaves genomic DNA is shown in **Figure 1**. Test combined with common CTAB method for genomic DNA extraction, and extracted genomic DNA respectively for 12 samples, high quality DNA have great influence on stability and reliability experiment results. Seen from **Figure 1**, the primary belt of *L. hexandra* genomic DNA is at about 21,226 bp. After agarose gel electrophoresis for DNA extracted from *L. hexandra* leaves, its band isn't very bright and clear but with much tail, which showed that process of DNA extraction degradated in severe, proteins and RNA degradation was not very thorough. To illustrate, extraction of DNA purity in this experiment was not high, but in little amount of DNA in RAPD experiment, generally, expansion just need 10 ng to 50 ng DNA, purity requirements were not very high. The RAPD results showed that the extraction genomic DNA from *L. hexandra* can be applied to RAPD reaction.

3.2. *L. hexandra* RAPD Primers Screening Results

Used *L. hexandra* genomic DNA as a template, for screening of the 46 UBC random primers, results showed that 29 UBC hadn't amplified products or obscured and couldn't tell, other 17 UBC random primers amplified a strip of clear and stable, that account for 36.9% of total primers. They varied approximately 200 bp or 3000 bp size, each primer amplification of fragments between changes in 1 - 7. Part *L. hexandra* leaves DNA RAPD primers screened results are as shown in **Figure 2**.

3.3. *L. hexandra* RAPD Amplification Results

Screened 46 random primers (see **Table 1**) preliminary, and used primers that can amplify stability and clear DNA bans as DNA amplification primers of *L. hexandra* polluted by chromium, and screened random primers S43, S51, S55 from tests results for RAPD reaction in this study. RAPD analysis was used for different samples of *L. hexandra* which under heavy metals chromium of 400 mg/kg concentration pollution and then added different concentration exogenous EDTA, sample RAPD results detected by agarose gel electrophoresis as shown in **Figure 3**, **Figure 4** and **Figure 5**.

3.4. RAPD Amplification Results of Primer S43

Figure 3 shows the amplification results of S43 RAPD random primers of *L. hexandra* samples. EDTA concentration of the first sample was 0 mmol/L (contrast). EDTA concentration of the second and third sample was 2.5

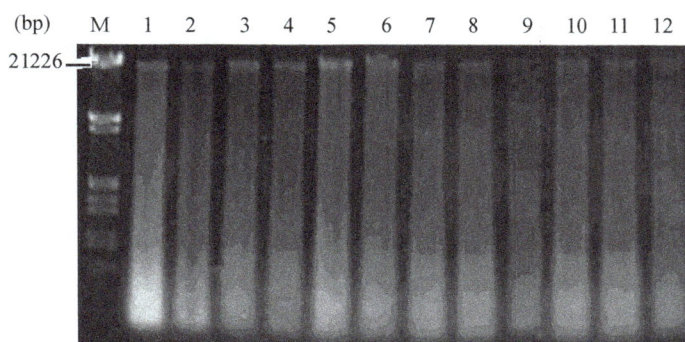

Figure 1. Leaves general DNA electrophoresis spectrums of *L. hexandra*, 1 - 3: EDTA 0 mmol/L; 4 - 6: EDTA 2.5 mmol/L; 7 - 9: EDTA 5.0 mmol/L; 10 - 12: EDTA 10 mmol/L. M is Lambda DNA/Hind III + EcoR I Markers.

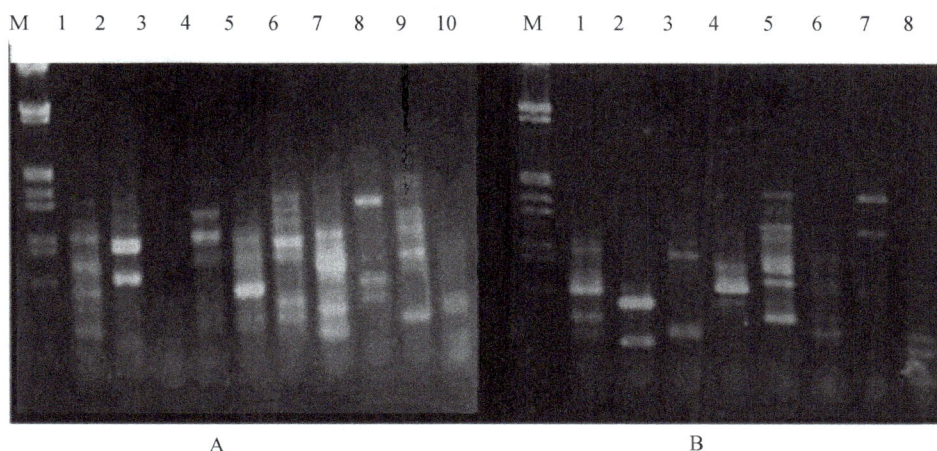

Figure 2. Filtered results of RAPD part primers in *L. hexandra* leaves DNA, A: Lane 1 - 10 Respectively Express Primer S51, S52, S57, S59, S60, S93, S94, S95, S96 and S97 amplification results; B: Lane 1 - 8 Respectively Express Primer S4, S9, S98, S99, S43, S44, S55 and S56 amplification results; M is Lambda DNA/Hind III + EcoR I Markers.

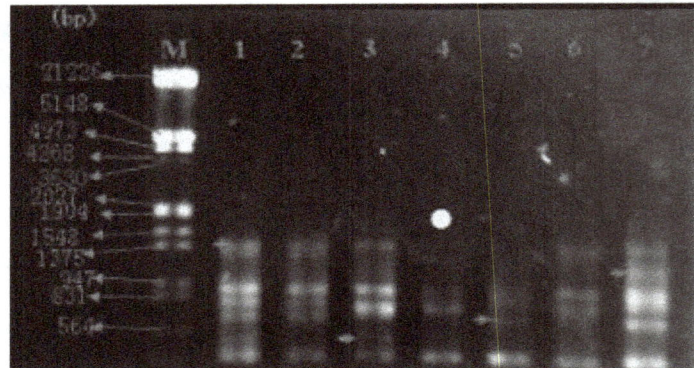

Figure 3. Amplification results in primer 43(GTCGCCGTCA) RAPD, 1: EDTA 0 mmol/L (Contrast); 2,3: EDTA 2.5 mmol/L; 4,5: EDTA 5.0 mmol/L; 6,7: EDTA 10 mmol/L; M is Lambda DNA/Hind III + EcoR I Markers.

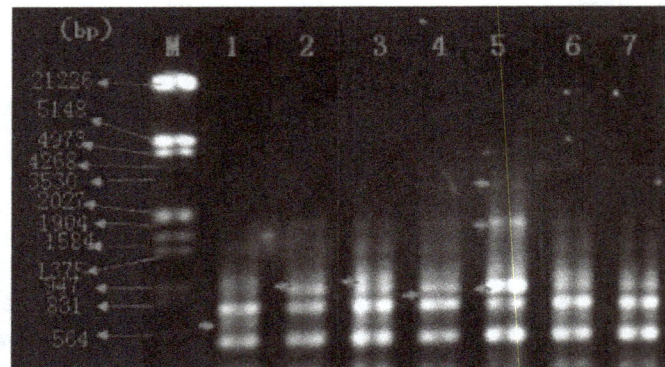

Figure 4. Amplification results in primer 51(GTCGCCGTCA) RAPD. The treatment of samples 1 - 7 is the same as **Figure 3**.

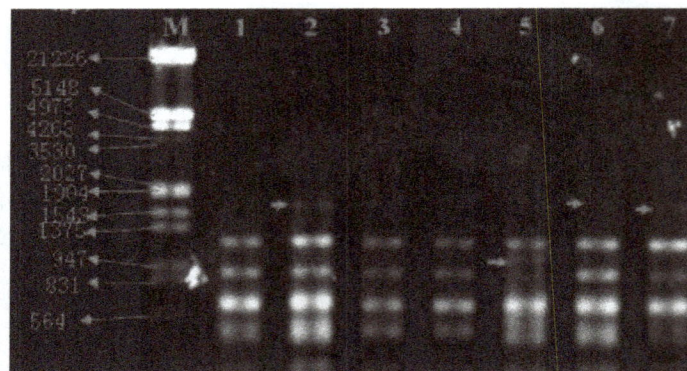

Figure 5. Amplification results in primer 55(GTCGCCGTCA)RAPD. The treatment of samples 1 - 7 is the same as **Figure 3**.

mmol/L and the third sample appeared DNA polymorphism at about 564 bp (as shown by the arrow in the 3rd sample). EDTA concentration of the fourth and fifth sample was 5 mmol/L, and the fifth sample appeared DNA polymorphism at about 700 bp (as shown by the arrow in the 5th sample). EDTA concentration of the sixth and seventh sample was 10 mmol/L, and the 7th appeared DNA polymorphism at about 1000 bp (as shown by the arrow in the 7th sample).

EDTA concentration of the first was 0 mmol/L (contrast). Comparison between the second and third and the first samples, the third appeared DNA polymorphism at about 564 bp (as shown by the arrow in the 3rd sample). Comparison between the fourth and fifth sample and the first sample, both appeared strip absence at about 1375 bp (absence strip as shown by the arrow in the first sample), in addition, the fourth sample appeared strip absence at about 700 bp (absence strip as shown by the arrow in the first sample). Comparison between the sixth and seventh and the first sample, the sixth sample appeared strip absence at about 700 bp (as shown by the arrow in the first sample), and the seventh sample appeared DNA polymorphism at about 1000 bp (as shown by the arrow in the 7th sample).

RAPD amplification results of primer S51

Figure 4 shows the amplification results of S51 RAPD random primers of *L. hexandra* samples. EDTA concentration of the first was 0 mmol/L (contrast). EDTA concentration of the second and third sample was 2.5 mmol/L and compared with each other, the third sample appeared DNA polymorphism at about 1000 bp (as shown by the arrow in the second sample). EDTA concentration of the fourth and fifth sample was 5 mmol/L, and compared with each other, the fifth sample appeared DNA polymorphism at about 3530bp. EDTA concentration of the sixth and seventh sample was 10 mmol/L, and compare result shown that both were almost no difference.

EDTA concentration of the first was 0 mmol/L (contrast). Comparison between the second and third and the first sample, both appeared DNA polymorphism at about 947 bp (as shown by the arrow in the 2nd and 3rd sample). Comparison between the fourth and fifth sample and the first sample, both appeared DNA polymorphism at about 947 bp (as shown by the arrow in the 4th and 5th sample), in addition, the 5th sample appeared DNA polymorphism at about 1904 bp and 3530 bp (as shown by the arrow in the 5th sample), and both the samples appeared strip absence at about 700 bp (strip absence as shown by the arrow in the first sample). Compared to the sixth and seventh sample, the contrast showed no difference.

RAPD amplification results of primer S55

Figure 5 shows the amplification results of S55 RAPD random primers of *L. hexandra* samples. EDTA concentration of the first was 0 mmol/L (contrast). EDTA concentration of the second and third sample was 2.5 mmol/L and compared with each other, the 2nd sample appeared DNA polymorphism at about 1548 bp (as shown by the arrow in the 2nd sample). EDTA concentration of the fourth and fifth sample was 5 mmol/L, and the 5th sample appeared DNA polymorphism at about 947 bp (as shown by the arrow in the 5th sample). EDTA concentration of the sixth and seventh sample was 10 mmol/L, and compared with each other, result shown that both were almost no difference.

EDTA concentration of the first was 0 mmol/L (contrast). Comparison between the second, the third and the first sample, the second sample appeared DNA polymorphism at about 1548 bp (as shown by the arrow in the 2nd sample). Comparison between the fourth, the fifth sample and the first sample, the fifth sample appeared DNA polymorphism at about 947 bp (as shown by the arrow in the 5th sample). Comparison between the sixth and seventh and the first sample, both samples appeared DNA polymorphism at about 1548 bp (as shown by the arrow in the 6th and 7th sample).

Study on the evolution of pollution can help to understand how plants adapt to the environment and evolve, to understand biological adaptation mechanism under pollution. Knowledge on the genetic diversity can be used in future breeding programs potential [16] [17]. Students concluded that DNA polymorphism detected by RAPD analysis in conjunction with other biochemical parameters could be a powerful eco-toxicological tool in biomonitoring arsenic pollution potential [18].

Heavy metal pollution is an important form of soil pollution, the past researches focused on heavy metal toxic effect and mechanism of plants, and study on migration, accumulation, distribution and other aspects on heavy metals in the plant's tissues, organs and ecological systems potential [19]. The accumulating ability of *Leersia hexandra* Swartz may be due to they live in electroplating waste water pollution in streams for a long time, in the process of adapting to this environment produced a variation, some genetic characteristics have changed, been the emergence of new forms of Chromium patience and Enrichment capacity also increased. And there are few studies on molecular ecology about plant under long-term effects of heavy metals. Through RAPD analysis of genetic diversity of *L. hexandra* under the heavy metals pollution, we can understand fundamental changes that is DNA level changes in the cells under tolerance in heavy metals pollution, so as to reform *L. hexandra* gene by using modern molecular techniques, to make greatly ability on enrichment and accumulation to heavy metals, so as to develop ability of restoring soil .

This study first screen primers of *L. hexandra* RAPD polymorphicing, and screen random primers S43

(GTCGCCGTCA), S51 (AGCGCCATTG), S55 (CATCCGTGCT) from the 46 random primers to be random primers of *L. hexandra* RAPD molecular markers preliminary. And then used S43, S51, S55 as primers on the polymorphism of different samples for research. The combined use of both RAPD and ISSR markers for the genetic analysis of the cultivar varieties have been performed earlier potential [20] [21] (Gupta, *et al.*, 2008, Lu, *et al.*, 2009). Our study utilizes a total of 55 primers to disclose the genetic similarity between the selected varieties.

Compared EDTA concentration of 5 mM/kg, 10 mM/kg, 2.5 mM/kg with the control, the former expansion is larger than that of the latter. Different degrees of difference in comparison between different EDTA concentrations and control of treated RAPD results of *L. hexandra*, this shows that EDTA has certain effects during enrichment of *L. hexandra* to heavy metals. But perhaps, in a certain scope, it will be more favourable to *L. hexandra* accumulation of chromium metal when EDTA concentration increased. *L. hexandra* in the accumulation of chromium metal is also the process of chromium metal tolerance, and cause changes in gene eventually, to make it more resistant chromium metal against.

4. Conclusion

Different *L. hexandra* samples exhibited DNA polymorphism when using the random primer S43, S51 and S55 as the primers in the RAPD reaction. One specific DNA band about 1000 bp was found in the sample which was treated with 10 mmol/L concentration EDTA when used the S43 primer to RAPD. The obvious differences between different EDTA-treatment levels suggest that EDTA has not much effects on enrichment to heavy metals of *L. hexandra*, so we should do more work to research the *L. hexandra* accumulation of chromium mechanism when EDTA concentration increased.

Acknowledgements

This research was funded by Guangxi Nature Science of Foundation (2011GXNSFA018012), Key Discipline Physical Geography Established in Guilin University of Technology and Guangxi Key Laboratory of Hidden Metallic Ore Deposits Exploration Guilin University of Technology, Guilin, China.

References

[1]   Zhang, X.H., Liu, J., Huang, H.T., Chen, J., Zhu, Y.N. and Wang, D.Q. (2007) Chromium Accumulation by the Hyperaccumulator Plant *Leersia hexandra* Swartz. *Chemosphere*, **67**, 1138-1143. http://dx.doi.org/10.1016/j.chemosphere.2006.11.014

[2]   Wang, T., Zhang, Q.B. and Ma, K.P. (2006) Treeline Dynamics in Relation to Climatic Variability in the Central Tianshan Mountains, Northwestern China. *Global Ecology and Biogeography*, **15**, 406-415. http://dx.doi.org/10.1111/j.1466-822X.2006.00233.x

[3]   Zhang, X.W. and Liu, B. (2010) Effect of Chromium on Crops Growth. *Environmental Science*, **23**, 48-51.

[4]   Hu, Y.Q. and Zhao, S.J. (2010) RAPD Technology and Its Application in Plant Research. *Biotechnology Bulletin*, **5**, 74-77.

[5]   Williams, J.G.K., Kubelik, A.R., Licak, J.A., *et al.* (1990) DNA Polymorphisms Amplified by Arbitrary Primers Are Useful as Genetic Marker. *Nucleic Acids Research*, **18**, 6531-6535. http://dx.doi.org/10.1093/nar/18.22.6531

[6]   Wen, C.H., Wang, Z.H. and Duan, C.Q. (1998) Pollution Resistant Differentiation and Evolution in Plants and the Application of Molecular Biology Technique. *Ecological Science*, **17**, 19-24.

[7]   Živković, B., Radović, J., Sokolović, D., Šiler, B., Banjanac, T. and Štrbanović, R. (2012) Assessment of Genetic Diversity among Alfalfa (*Medicago sativa* L.) Genotypes by Morphometry, Seed Storage Proteins and RAPD Analysis. *Industrial Crops and Products*, **40**, 285-291. http://dx.doi.org/10.1016/j.indcrop.2012.03.027

[8]   Bocianowski, J. and Seidler-Łożykowska, K. (2012) The Relationship between RAPD Markers and Quantitative Traits of Caraway (*Carum carvi* L.). *Industrial Crops and Products*, **36**, 135-139. http://dx.doi.org/10.1016/j.indcrop.2011.08.019

[9]   Cansaran-Duman, D., Atakol, O. and Aras, S. (2011) Assessment of Air Pollution Genotoxicity by RAPD in *Evernia prunastri* L. Ach. from around Iron-Steel Factory in Karabük, Turkey. *Journal of Environmental Sciences*, **23**, 1171-1178. http://dx.doi.org/10.1016/S1001-0742(10)60505-0

[10]   Li, J.M., Jin, Z.X. and Zhu, H.C. (2007) Heavy Metal Pollution Fern Population Genetic Differentiation Analysis of RAPD. *Journal of Ecology*, **26**, 171-176.

[11] Wen, C.H., Duan, C.Q., Xiu, C.X., *et al.* (2001) Heavy Metal Pollution of Differentiation of RAPD Analysis of Datura Stramonium L. *Journal of Ecology*, **21**, 1238-1245.

[12] Gu, Q.P., Jin, Z.X. and Li, J.M. (2008) Under Heavy Metal Stress Clethroides Populations Genetic Diversity Analysis of RAPD. *Jiangsu Agricultural Sciences*, **2008**, 54-58.

[13] Yu, Z.X., Ou, G.Z., Chen, Q.X., *et al.* (2010) Pitaya Total DNA Extraction Method Comparison Research. *Agricultural Science Bulletin*, **26**, 300-303.

[14] Cheng, S.P., Shi, J., Shi, G.A., *et al.* (2010) *Pistacia chinensis* Leaf Genomic DNA Extraction Methods. *Journal of Henan University of Science and Technology*, **31**, 71-73.

[15] Wang, J.J., Wu, S.L., Zhang, F.C., *et al.* (2010) Grape Genome DNA Extraction and RAPD Reaction System Optimization. *Agricultural Sciences of Xinjiang*, **47**, 1066-1070.

[16] Lin, G.M., Sui, F.F., Lin, N., *et al.* (2010) Wilfordii Leaf DNA Extraction and RAPD Reaction System. *Chinese Agricultural Science Bulletin*, **26**, 40-43.

[17] Singh, S., Panda, M.K. and Nayak, S. (2012) Evaluation of Genetic Diversity in Turmeric (*Curcuma longa* L.) Using RAPD and ISSR Markers. *Industrial Crops and Products*, **37**, 284-291.

[18] Ahmad, M.A., Gaur, R. and Gupta, M. (2012) Comparative Biochemical and RAPD Analysis in Two Varieties of Rice (*Oryza sativa*) under Arsenic Stress by Using Various Biomarkers. *Journal of Hazardous Materials*, **217-218**, 141-148. http://dx.doi.org/10.1016/j.jhazmat.2012.03.005

[19] Tang, Y., Wang, P.P. and Zhang, N. (2006) Research in Heavy Metal Toxicity Mechanism in Plant. *Journal of Shengyang Agricutural University*, **37**, 551-555.

[20] Tao, J.X., Zhang, X.H., Luo, H., *et al.* (2010) *Leersia hexandra* Swartz for Electroplating Sludge Contaminated Soil Chromium Copper Nickel Uptake and Accumulation. *Journal of Guilin University of Technology*, **30**, 144-147.

[21] Gupta, S., Srivastava, M., Mishra, G.P., Naik, P.K., Chauhan, R.S., Tiwari, S.K., Kumar, M. and Singh, R. (2008) Analogy of ISSR and RAPD Markers for Comparative Analysis of Genetic Diversity among Different *Jatropha curcas* Genotypes. *African Journal of Biotechnology*, **7**, 4230-4243.

11

The Use of Plants and Wildflowers as Bioremediation for Contaminated Soils in the Hong Kong S.A.R.

Angelo Indelicato

Dragages Hong Kong Ltd., Hong Kong, China
Email: angelo.indelicato@dragageshk.com

Abstract

Heavy metal contamination of the biosphere has increased sharply over the last century. Anthropogenic activities such as industrialisation and demographic growth can be considered as the main causes of it. Soil contamination affects every organism and poses major environmental and human health problems worldwide. The issue has been addressed in the past and a few methodologies have been developed in order to effectively clean up the contaminated areas. However, many of these remedies are very aggressive and can damage the soil. This paper focuses on the use of gentler techniques, which take advantage of the properties of several plants and wildflowers that absorb heavy metals and polycyclic aromatic hydrocarbons, and their potential application in megacities such as Hong Kong.

Keywords

Hong Kong, Bioremediation, Wildflowers, Contaminated Land

1. Introduction

Hong Kong (**Figure 1**) is located at the south-eastern tip of the Chinese mainland. With a population of over 7 million inhabitants and a total area of approximately 1100 km^2, it is one of the most densely populated cities in the world. The rapid economic and demographic growth in the region during the last few decades has had a significant environmental impact. Hong Kong's mountainous landscape and limited flat land compresses much of the residential and commercial activities on the hillsides and in the coastal areas surrounding Victoria Harbour. These estates are located close to roads and highways, making them susceptible to various sources of pollution.

Urban soil is remarkably contaminated by heavy metals and other pollutants. Elevated concentrations of

Figure 1. Hong Kong (©Google Earth 2014).

heavy metals are known to have an adverse effect on human health, especially on that of children, as they have a high absorption rate due to their active digestion and sensitivity to haemoglobin [1]. According to a survey conducted in 1981 by Lau and Wong, the highest amount of Cadmium (54 mg/kg) was found in a recreational park, and the highest amount of Copper (205 mg/kg) was identified in an industrial estate in Aberdeen.

Rural expanses are also subject to soil contamination. To ease the densely populated centre, the so-called "satellite" cities in the New Territories have been developed and expanded so now housing estates are interspersed with agricultural and industrial areas. The highest concentration of Lead and Zinc, 229 mg/kg and 259 mg/kg respectively, were found in the agricultural area of Man Uk Pin [1].

Pollution in Hong Kong not only accumulates in Kowloon and around Victoria Harbour but it affects every part of the city and its surroundings. As the concentration of these pollutants increases, it is vital to find new methods to clean the soil and the water. This study will focus on effective bioremediation treatments to soil pollution.

2. Soil Composition

Soil is a complex mixture of organic and inorganic material. The organic components derive from the decayed remains of plants and animals [2]. The inorganic parts are made of rock fragments that were formed by bedrock weathering over thousands of years.

Typical soils are made of several main layers, or horizons. The uppermost (topsoil) is humus-rich and comprises of decaying organisms, microorganisms, insects and worms. Below this layer lies the subsoil, created predominated by inorganic particles, minerals and organic substances leached from the horizon above. The deepest horizon consists of fragmented bedrock mixed within a matrix of silt, sand and clay, also known as saprolitic soil.

Organic matter and anthropogenic compounds such as pesticides and their degradation products can react with humus. Water-soluble humus, like fulvic acid, can form complexes with organic pesticides at the soil surface. Percolating water can transport these compounds deep into the soil.

Soil is an innate transmitter of pollutants to surface water, groundwater, the atmosphere and food, thus posing a substantial threat to human life. When air and water are polluted, switching off the source of pollution, followed by dilution and self-purification, can reverse the problem. However, these techniques cannot be used to eliminate heavy metal contamination in the soil, making remediation expensive and time-consuming [3].

In Hong Kong, above the saprolite lie two deposits from the Pleistocene. The Middle Pleistocene Colluvial Deposit (Po Chu Tam Formation) is stiff to very stiff, slightly clayey sandy silt with 30% to 70% highly to completely decomposed sub-angular to sub-rounded cobbles and boulders [4]. A similar composition is also found in the Late Pleistocene Colluvial Deposits (Shum Wan Formation).

Two deposits from the Holocene have been identified: the colluvial and alluvial deposits. The colluvial deposit is soft to firm clayey sandy silt to silty fine sand with angular to sub-angular gravel and cobbles. The latter (Fan Ling Formation), occurring mainly along the course of narrow streams, is made of clayey silty sand with a thin layer of organic mud [4].

The urban soil has undertaken drastic modification due to burial, mixing, truncation or wholesale removal. Contamination by building debris is common, with cement, mortar, concrete and occasional brick fragments and other inert foreign materials [5]. There is no organic-matter rich topsoil and the release of alkaline leachate from calcareous construction wastes can raise the pH level beyond the normal tolerance [5].

3. Types of Contaminants Present in the Soil

Contaminants present in the soil can be quite different. Heavy metal contamination refers to the excessive deposition of toxic heavy metals in the soil caused by human activity. Some examples of metals with significant biological toxicity are Mercury (Hg), Cadmium (Cd), Lead (Pb), Chromium (Cr) and Arsenic (As). Other heavy metals like Zinc (Zn), Copper (Cu), Nickel (Ni) and so on [3], are also present. With the exception of Cu and Cd, most of heavy metal contamination is both colourless and odourless and it does not explicitly damage the environment in the short term. However, when it exceeds environmental tolerance, heavy metals in the soil can be activated and cause serious ecological damage [3].

Atmospheric deposition, sewage irrigation, improper stacking of industrial solid waste, use of pesticides and fertilizers, and so on, are among the many sources of soil contamination. Heavy metals in the atmosphere, on the other hand, are mainly from gas and dust produced by energy and transport, with burning leaded gasoline and the dust produced by automobile tyre wear as the leading causes. In a city like Hong Kong plagued with extreme traffic congestion, high-rise buildings form a man-made canyon landscape where the dust collects at the bottom and the wind has zero to slim chances of clearing it up, thus aggravating this pollution. Areas such as Mongkok and Prince Edward have more serious concentrations of Pb than others and, considering the amount of people living there, many are exposed to toxic gases on a daily basis, with debilitating consequences to their health. Heavy metal pollutants tend to persist in the soil, although impermeable concrete paving can stop and even seal water infiltration and prevent the soil from otherwise leaching and spreading the pollutants [5]. Another main source comes from energy production, such as thermal power stations that produce coal ashes (containing Cd, Cu, Pb and Zn), which are dispersed on land and at sea. Animal waste, spent litter and compost which may contain high levels of trace-metals such as Cu and Zn, are traditionally used in local farms and can also contribute to soil contamination [6].

Inorganic forms of As were once used extensively as agricultural pesticides and high concentrations of Lead are due to the heavy use of Pb-containing agrochemicals. Agricultural wastes are among the most serious causes of pollution in Hong Kong. Pig effluents and poultry droppings highly contribute to the total quantity of readily putrescible matter entering the streams but also in the soil as it is used as fertilizer. It contains substantial amount of Pb, Cu, Zn and Mn [7].

E-waste recycling activities, including dismantling and open burning, have generated a large amount of heavy metals and persistent toxic substances on existing abandoned farm soils in Hong Kong [9].

Other very noxious substances that can be found in soil are polycyclic aromatic hydrocarbons (PAHs). They are very toxic to many living organisms and with reluctance to degradation and high lipophilicity; they form a class of very dangerous compounds [8].

Polycyclic aromatic hydrocarbons (PAHs) are usually produced via the incomplete combustion of organic substances and are comprised of a diverse group of organic compounds [9].

Among hydrocarbon pollutants, diesel oil is a complex mixture of alkanes and aromatic compounds that are frequently reported as soil contaminants leaking from storage tanks and pipelines or released in accidental spills.

4. Remediation Techniques Used in Hong Kong

According to the Environmental Protection Department [10] various remediation methods have been historically applied to contaminated soils in Hong Kong.

Biopiles involve heaping contaminated soils into piles and stimulating aerobic microbial activity within the soils through aeration and/or the addition of minerals, nutrients and moisture. This technique may not be effective for high levels of contaminants.

Soil Vapour Extraction removes vapours from soil above the water table by applying a vacuum to pull out the contaminated rich vapours. It is very effective in treating volatile organic compounds (VOC). However, elevated organic content or extremely dry soils have a high absorption of VOCs, which result in a reduced removal rate.

Stabilisation/solidification traps or immobilises contaminants within the host soil. It is the only method of remediation used to adequately treat heavy metal contamination. However it does not trap organics and has depth limitations. Its long-term effectiveness has also not been demonstrated.

Thermal desorption heats the contaminants within the soil until they vaporise into gas that is subsequently collected and treated. The results have different degrees of effectiveness.

5. Commonly-Used Bioremediation

One of the best approaches to restoring contaminated soil is the use of microorganisms capable of degrading toxic compounds in a bioremediation process [11]. This system is used to clean up petroleum hydrocarbons. Diesel oil contamination in soils can also be promoted by stimulating the indigenous microorganisms; by introducing nutrients and oxygen into the soils in a process called biostimulation or through the inoculation of an enriched microbial consortium into said soils (bioaugmentation). In Hong Kong, microorganisms are currently used in biopiles.

Microbial remediation can also be used successfully as some microorganisms absorb, precipitate, oxidise and reduce heavy metals in soils [3].

Furthermore, maggots and earthworms living underground can take heavy metals in the soil, though they may only absorb a select amount of contaminants.

Microbial bioremediation has been successful for the degradation of certain organic contaminants, but is ineffective in addressing the challenge of toxic metal contamination [12]. Toxic metals can only be remediated by their removal from soil.

A new cost-effective "green" technology known as phytoremediation involves the use of particular types of plants capable of hyper-accumulating contaminants in the ground.

There are various types of phytoremediation:

1) Phytoextraction known as phytoaccumulation or phytoabsorption. This process came from the discovery of a variety of wild plants that concentrate high amounts of essential and non-essential heavy metals in their foliage. The degree of accumulation of metals such as Zn, Ni, and possibly Cu often reaches 1% - 5% of the dry weight [12]. Pb is extremely insoluble and not generally available for plants to uptake in the normal range of soil pH. To acquire these soil-bound metals, phytoextracting plants have to mobilise them into the soil solution. This can be accomplished in different ways, from metal chelating molecules to acidifying the soil with protons extruded from the roots. However, this is also very dangerous as these chelating molecules increase the solubility of metals within the soil [13].

2) Phytodegradation, where plants take up and break down contaminants through the release of enzyme and metabolic processes such as photosynthetic oxidation and reduction. In this process organic pollutants are degraded and incorporated into the plant or broken down in the soil.

3) Phytovolatilisation, in which plants take up volatile contaminants and following transpiration, release non-toxic substances into the atmosphere.

4) Phytostabilisation, in which some plants can sequester or immobilise contaminants. This method limits the movements of contaminants through erosion, leaching and wind or soil dispersal.

5) Rhizodegradation, where the roots of some plants assist in the microbial degradation of contaminants.

6. The Use of Plants and Wildflowers as Bioremediation

An abundance of plants and flowers can be used to clean up contaminated soil. In one particular family of plants (Brassicaceae), there are several that work well as accumulators and hyper-accumulators. Mustard (*Brassica juncea Linn.*) soaks up heavy metals such as Cr, Ni, Pb, U and Zn and it also acts as a hyper-accumulator for Cu [14].

A group of plants from different families share the characteristics of being accumulators for Cr, Pb and Hg. These are the rape seed plant (*Brassica napus Linn.*), the water hyacinth (*Eichhoria crassipes*) and hydrilla (*Hydrilla verticillata*).

Coconut palm (*Cocos nucifera L.*), corn (*Zea mays*) and sunflower (*Helianthus annus L.*) can store up radioactive elements such as Caesium and Uranium.

Common osier (*Salix viminalis Linn.*) and sunflower have also shown promise in their ability to amass hydrocarbons, as well as degrade PAHs in soils [15].

Sunflowers are also one of several flowers capable of accumulating heavy metals. Among the wildflowers there are a few which can trap multiple metals. Within the Brassicaceae family there are three such flowers: Arabidopsis bundles Cd, Fe and Zn, Thlapsi (Pb, Zn, Cd and Ni) and Brassica Rapa Sylvestris (Cd and Cr) [16].

Other flowers efficient in accumulating heavy metals, especially Pb, Cd and Zn, are *Cistus salvifolius* (Cistaceae), Aster (Asteraceae), *Hypericum perforatum* (Hypericaceae), Yarrow (*Achillea millefolium*) and Chives (*Allium schoenoprasum*). The last two are very effective accumulator for Cadmium.

Plants and wildflowers are not alone in their ability to potential uptake contaminants. Vetiver grass (*Vetiveriazizanioides L. Nash*) is also very effective (**Figure 2**). Although it has been used in Hong Kong for land protection and to mitigate soil erosion, it has potentials within the bioremediation field with a high tolerance to a range of trace elements such as As, Cu and Cd [17]. Other grasses worth of mention are Colonial Bentgrass (*Agrostis castellana*) which accumulates As, Pb, Zn, Mn and Al, for hydrocarbons rhizodegradation/accumulation the Buffalo grass (*Buchloe dactyloides*) and Bermuda grass, or lawn grass (*Cynodon dactylon*) have shown remarkable results in treating PAH in soil and they are also cost effective considering the low maintenance required [15].

(a)   (b)   (c)

(d)   (e)   (f)

Figure 2. (a) *Hypercum perforatum*, (b) Sunflower, (c) Rape seed plant, (d) Chives, (e) Vetiver grass and (f) Coconut palms.

7. Conclusions

PAHs and heavy metals are pervasive in Hong Kong soils. Although the worse contamination is found in urban and orchard soils, other areas such as rural and forest soils are also affected [6].

The reason for soil contamination is without doubt linked to anthropogenic activities that have been increasingly sharp during recent years. The misuse of urban and rural soil, combined with the lenient enforcement of environmental policies, has generated PAH and heavy metal contamination in cities like Hong Kong, posing a serious threat to their inhabitants.

A few techniques commonly used to clear contaminated soil can be so aggressive that they can also damage the organic material present within it. Moreover, some of them are fruitless when faced with certain types of contaminants.

The use of plants and wildflowers instead can be seen as a greener and gentler way to treat this issue. However, there are multiple factors that might influence the final result, such as the extent of the soil contamination, the availability and accessibility of contaminants and the ability of the plant and its associated microorganisms to intercept, absorb, accumulate and/or degrade the contaminants [18].

As this field expands with time, it is important to understand whether or not it can be used effectively. If we consider urban soil, wildflowers reduce the amount of contaminants and at the same time, provide a pleasing visual to the public. This method, therefore, can be deemed suitable to clean up recreational areas, parks and other such urban spaces.

More aggressive methods would probably be more suitable for e-waste and industrial zones. Phytoremediation requires time and may not be appropriate if areas affected by heavy contamination need to be cleaned up.

The depth at which the contaminants are in the ground also poses a challenge. If the depth is so great that roots cannot reach the heavy metals and PAHs, they will not extract from the soil. Another concern is how to safely dispose the plants, which after phytoaccumulation, are holding all the pollutants. Burning them would release heavy metals back into the ecosystem and if they are to be stored, it must be done properly.

In orchard and agricultural fields, the use of certain varieties of plants and vegetables can reduce the amount of contaminants in the soil. Many of them are edible but after accumulating pollutants, they can no longer be sold in the market. This problem can adversely affect the economy of local farmers, as they will temporarily lose part of their income.

Future research will probably lead to a more effective use of phytoremediation over large-scale areas and be more feasible, especially for agricultural lands.

Recent studies are also focusing their attention on harvesting only roots as a method to maximise the removal of soil pollutants as the roots sequester the majority of the contaminants taken up in most plants [19].

Despite its constraints, phytoremediation has potential and must be considered as a green option for the clean-up of urban and other contaminated soils. It is a non-invasive technique which has proved to be also cheaper than other conventional methods. It is worth exploring and implementing in cities such as Hong Kong and in many other environments.

Acknowledgements

The author wishes to thank Judy Wu and Cristina Pandolfo for their contributions to this paper.

References

[1]   Li, X., Lee, S., Wong, S., Shi, W.Z. and Thornton, I. (2004) The Study of Metal Contamination in Urban Soils of Hong Kong Using a GIS-Based Approach. *Environmental Pollution*, **129**, 113-124. http://dx.doi.org/10.1016/j.envpol.2003.09.030

[2]   Girard, J.E. (2005) Principles of Environmental Chemistry. Jones and Bartlett Publishers, Inc., Burlington, 47-53.

[3]   Su, C., Jiang, L. and Zhang, W. (2014) A Review on Heavy Metal Contamination in the Soil Worldwide: Situation, Impact and Remediation Techniques. *Environmental Skeptics and Critics*, **3**, 24-38.

[4]   Lai, K.W. (2011) Geotechnical Properties of Colluvial and Alluvial Deposits in Hong Kong. *The 5th Cross-Trait Conference on Structural and Geotechnical Engineering* (*SGE*-5), Hong Kong, 13-15 July 2011, 735-744.

[5]   Jim, C.Y. (1998) Urban Soil Characteristics and Limitations for Landscape Planting in Hong Kong. *Landscape and Urban Planning*, **40**, 235-249. http://dx.doi.org/10.1016/S0169-2046(97)00117-5

[6]   Chen, T.B., Wong, J.W.C., Zhou, H.Y. and Wong, M.H. (1997) Assessment of Trace Metal Distribution and Contamination in Surface Soils of Hong Kong. *Environmental Pollution*, **96**, 61-68. http://dx.doi.org/10.1016/S0269-7491(97)00003-1

[7]   Wong, M.H. (1987) A Review on Lead Contamination of Hong Kong's Environment. In: Hutchinson, T.C. and Meema, K.M., Eds., *Lead, Mercury, Cadmium and Arsenic in the Environment*, Chapter 14, 217-223.

[8]   Chung, M.K., Hu, R., Cheung, K.C. and Wong, M.H. (2007) Pollutants in Hong Kong Soils: Polycyclic Aromatic Hydrocarbons. *Chemosphere*, **67**, 464-473. http://dx.doi.org/10.1016/j.chemosphere.2006.09.062

[9]   Man, Y.B., Kang, Y., Wang, H.S., Lau, W., Li, H., Sun, X.L., Giesy, J.P., Chow, K.L. and Wong, M.H. (2013) Cancer risk Assessments of Hong Kong Soils Contaminated by Polycyclic Aromatic Hydrocarbons. *Journal of Hazardous Materials*, **261**, 770-776. http://dx.doi.org/10.1016/j.jhazmat.2012.11.067

[10]   Environmental Protection Department (2011) Practice Guide for Investigation and Remediation of Contaminated Land.

[11] Bento, F.M., Camargo, F.A.O., Okeke, B.C. and Frankenberger, W.T. (2005) Comparative Bioremediation of Soils Contaminated with Diesel Oil by Natural Attenuation, Biostimulation and Bioaugmentation. *Bioresource Technology*, **96**, 1049-1055. http://dx.doi.org/10.1016/j.biortech.2004.09.008

[12] Raskin, I., Smith, R.D. and Salt, D.E. (1997) Phytoremediation of Metals: Using Plants to Remove Pollutants from the Environment. *Current Opinion in Biotechnology*, **8**, 221-226. http://dx.doi.org/10.1016/S0958-1669(97)80106-1

[13] Prasad, M.N.V. and de Oliveira Freitas, H.M. (2003) Metal Hyperaccumulator in Plants-Biodiversity Prospecting for Phytoremediation Technology. *Electronic Journal of Biotechnology*, **6**, 285-321. http://dx.doi.org/10.2225/vol6-issue3-fulltext-6

[14] Khokhar, A.L., Rajput, M.T., Ahmed, B. and Tahir, S.S. (2012) Checklist of Flowering Plants Used in Phytoremediation Found in Sindh, Pakistan. *Sindh University Research Journal* (*Science Series*), **44**, 497-500.

[15] McCutcheon, S.C. and Schnoor, J.L. (2003) Phytoremediation, Transformation and Control of Contaminants. Wiley, New York.

[16] Pandolfo, C. (2012) La fitorimediazione e i fiori spontanei. Università degli Studi di Catania, Facoltà di Agraria. http://dryades.altervista.org/La_fitorimediazione_e_i_fiori_spontanei_-_Dott.ssa_Cristina_Pandolfo.pdf

[17] Khan, A.G. (2003) Vetiver Grass as an Ideal Phytosymbiont for Glomalian Fungi for Ecological Restoration of Heavy Metal Contaminated Derelict Land. *Proceedings of the 3rd International Conference on Vetiver and Exhibition*, Guangzhou, October 2003, 466-474.

[18] Vangronsled, J., Herzig, R., Weyens, N., Boulet, J., Adriaensen, K., Ruttens, A., Thewys, T., Vassilev, A., Meers, E., Nehnevajova, E., van der Lelie, D. and Mench, M. (2009) Phytoremediation of Contaminated Soils and Groundwater: Lessons from the Field. *Environmental Science and Pollution Research*, **16**, 765-794. http://dx.doi.org/10.1007/s11356-009-0213-6

[19] Negri, M.C., Hinchman, R.R. and Gatliff, E.G. (1996) Phytoremediation: Using Green Plants to Clean up Contaminated Soil, Groundwater and Wastewater. *Proceedings of the International Topical Meeting on Nuclear and Hazardous Waste Management, Spectrum* 96, Seattle, 18-23 August 1996, 1-10.

12

Understanding of Traditional Knowledge and Indigenous Institutions on Sustainable Land Management in Kilimanjaro Region, Tanzania

Richard Y. M. Kangalawe[1], Christine Noe[2], Felician S. K. Tungaraza[3], Godwin Naimani[4], Martin Mlele[5]

[1]Institute of Resource Assessment, University of Dar es Salaam, Dar es Salaam, Tanzania
[2]Department of Geography, University of Dar es Salaam, Dar es Salaam, Tanzania
[3]Department of Sociology and Anthropology, University of Dar es Salaam, Dar es Salaam, Tanzania
[4]Department of Statistics, University of Dar es Salaam, Dar es salaam, Tanzania
[5]Alpha and Omega Consulting Group Limited, Dar es Salaam, Tanzania
Email: a_ocg@yahoo.com

Abstract

The paper is based on a study whose objective is to provide an understanding of the extent to which traditional knowledge and indigenous institutions for natural resource governance remain relevant to solving current land degradation issues and how they are integrated in formal policy process in Kilimanjaro Region. Data collection for this study combined qualitative and quantitative methods. A total of 221 individuals from households were interviewed using a structured questionnaire; 41 in-depth interviews and 24 focus group discussions were held. Findings indicate that the community acknowledges that there is traditional knowledge and indigenous institutions regarding sustainable land management. However, awareness of the traditional knowledge and practices varied between districts. Rural-based districts were found to be more aware and therefore practiced more of traditional knowledge than urban based districts. Variations in landscape features such as proneness to drought, landslides and soil erosion have also attracted variable responses among the communities regarding traditional knowledge and indigenous practices of sustainable land management. In addition, men were found to have more keen interest in conserving the land than women as well as involvement in other traditional practices of sustainable land management. This is due to the fact that, customarily, it is men who inherit and own land. This, among other factors, could have limited the integration of traditional knowledge and indigenous institutions in village by-laws and overall policy process. The paper concludes by recommending that traditional knowledge and indigenous institutions for sustainable land management should be promoted among the younger generations so as to capture their interest, and ensure that successful practices are effectively integrated into the national policies and strategies.

Keywords

Indigenous Institutions, Natural Resources Management (NRM), Sustainable Land Management (SLM), Traditional Knowledge, Kilimanjaro Region, Tanzania

1. Introduction

The importance of the Kilimanjaro landscape in providing ecosystem services such as catchment for a range of water uses, local climate modification through the impacts of forest cover, tourism and support for local livelihoods cannot be overstated. However, an extensive process of degradation has seriously threatened the ability of the landscape in providing these services. Land degradation is a major problem in many parts of Tanzania. This problem is largely manifested in the form of severe soil erosion, silting, deforestation and decrease of land productivity. These environmental problems are mainly due to unsustainable agricultural activities, uncontrolled felling of trees for firewood and charcoal, frequent and uncontrolled burning of forests, unsustainable mining activities, overstocking, insecure land tenure and limited community participation in environmental activities. A set of complex and interrelated factors apply, including for example, rapid increase of human population and subsequent reliance on natural resources, land use change, poor land management practices and climate change. The Kilimanjaro Region has thus continued to face a number of environmental challenges including increased pressure on natural resources and subsequent land degradation. Although Tanzania has a comprehensive legal and policy framework for land management, there is a major weakness in sectoral linkages in the policy development and implementation process as well as little participation or integration of traditional rules and regulations for managing the unique features of the ecosystem.

While community groups have traditionally lacked representation in policy development and implementation, there is also little knowledge of the implications of these policies among community groups and how they contradict customary laws and systems of land and natural resource management. This paper presents findings from a study on indigenous institutions and knowledge in natural resources management in project pilot districts in Kilimanjaro Region conducted in November 2013. The paper first reviews relevant literature on indigenous institutions and knowledge in natural resources management. Second, the paper discusses the methodology used in the study in Section 3. This is followed by a presentation of results together with a discussion of these findings. Finally the paper draws conclusions.

2. Literature Review

The literature on traditional knowledge has gained rapid currency as researchers and natural resource managers have increasingly considered it as a valuable contributor to natural resource management and biodiversity conservation. Whereas some scholars have questioned what is agreed as indigenous knowledge, key actors and their borders [1] [2], others have debated about the viability of indigenous knowledge as a standalone regime capable of sustaining natural resource management [3] [4]. Although there is currently a general agreement that successful development strategies must incorporate traditional knowledge, controversies still exist particularly in the academic arenas [1]. As a result the terms "indigenous knowledge" and "local knowledge" have largely been used synonymously [5].

While most scholars and practitioners agree on basic issues such as the usefulness of traditional knowledge in the local cultural and environmental settings, what should be explored is the role that the concept of traditional knowledge plays in facilitating or discouraging cross-situational collaboration among actors working for indigenous and non-indigenous institutions of environmental governance such as local natural resources regimes, state agencies working with these regimes and co-management boards [6]. This view fits well in the widely utilized construct of indigenous knowledge which state that:

"It is a cumulative body of knowledge, practice and belief evolving by adaptive processes and handed down

through the generations by cultural transmission about the relationship of living beings (including humans) with one another and with their environment" [7].

Indigenous institutions have been defined to include norms and procedures that shape people's actions. These procedures define practices, assign roles and guide interactions. Examples of these traditional institutions include traditional leadership, traditional healers, ritual forests, traditional midwives and various taboos and sacred sites and practices. These institutions play key role in the management of natural resources through different form of indigenous technical knowledge [8] [9]. Both local and other literatures identify three key features that characterize the indigenous resources management: First, the indigenous social organization that controls access to natural resources within the community. Second, the customary norms and procedures for control, acquisition, maintenance and transfer for natural resources and finally, the indigenous utilization techniques for conserving and preserving resources [9]-[11].

Whereas institutions have generally included codes of conduct that define practices, assign roles and guide interactions, these institutions are made up of formal constraints (rules, laws and constitutions), informal constraints (norms, behaviour, conventions and self-imposed codes of conduct) and their various enforcement characteristics [12] [13]. Local institutions differ based on their functions and objectives. They encompass many different types of indigenous organizations and functions such as village-level governance, acceptable methods of community resource mobilization, security arrangements, conflict resolution, and asset management and lineage organizations [14]. It is against this background that the World Bank's framework considers indigenous knowledge as the basis for local decision-making in all aspects of life—food production, education, health, natural resource management and relationships [15].

In comparing the effectiveness of informal and formal institutions in sustainable Common Pool Resources (CPR) management in Sub-Saharan Africa, some scholars argue that informal institutions have contributed to sustainability by creating a suitable environment for joint decision-making, enabling exclusion at low cost for resource users and by using locally agreed sanctions [4]. As informal institutions are embedded in communal structures, they allow the incorporation of the communities' mechanisms and knowledge about the sustainable management and utilization of Common Pool Resources into the Common Pool Resources management [16]. A common pattern in all these cases points to the fact that informal institutions have evolved internally from the society and acted in the interest of the community, which has created a sense of commitment, ownership and responsiveness among the Common Pool Resources users. This in turn contributes to the achievement of sustainability outcomes, particularly prevention of Common Pool Resources degradation and improvement of the Common Pool Resources conditions, in terms of quantity and quality.

The importance of combining indigenous and non-indigenous institutions for land and natural resource management is further reflected in the widespread adoption of international strategies that establish a link between poverty alleviation, sustainable development and biodiversity conservation. The international strategies and initiatives that exemplify this link include but not limited to the Convention on Biological Diversity on Traditional Knowledge, Innovations and Practices, the Ramsar Convention on Wetlands (Resolution IX.21 of 2005—Taking into account the cultural values of wetlands) and the Millennium Ecosystem Assessment (Linking Local Knowledge and Global Science in Multi-Scale Assessments). As such, the recent Global Strategic Plan for Biodiversity (2011-2020) declared in the COP 10 (Strategic Goal E, target 18) that by 2020:

Traditional knowledge, innovations and practices of indigenous and local communities relevant for the conservation and sustainable use of biodiversity, and their customary use of their biological resources should be respected, subject to national legislation and relevant international obligations, and fully integrated and reflected in the implementation of the Convention with the full and effective participation of the indigenous and local communities, at all relevant levels [17].

It is important to note that indigenous knowledge and institutions have their limitations. In some cases these limitations have hindered their integration into the formal system. In fact, even the World Bank's indigenous knowledge framework identifies some of the indigenous practices that are not beneficial to sustainable development (giving examples of slash and burn agriculture to make its point) [2]. Accordingly, it is suggested that despite the strengths, some traditional practices and institutions have weaknesses that may limit their use in management plans that favour sustainability [14]. This calls for the critical examination of the roles, issues and challenges of indigenous knowledge and their institutions in natural resource management governance.

In the context of Tanzania, literature on traditional knowledge and institutions indicate that prior to independence in 1961, fundamental principles of land access and management were closely linked to chieftainship. Traditional institutions had chiefs as the most respectable law enforcers and the British colonial government recognized the strength of these institutions hence their adoption as native authorities and chiefs as local tax administrators [18]-[20]. However, during the early years of national building, Mwalimu Nyerere considered tribal identities through chiefs as inherently negative and challenging basic principles and objectives of building the national unity [21]. In 1963, just two years after independence, native authorities were abolished and chiefs were officially retired [18]-[22]. Conceptually, the decision to abolish chiefdoms declared their institutions officially inactive across the country.

After the abolition of chieftaincy in 1963 other forms of traditional leadership prevailed and are still effective in some parts of the country. A few communities continued to transfer their traditional knowledge to the new generations due to other reasons including religion and modernity [23] [24]. As the society advances through formal education systems and religious doctrine, children are strictly taught science and God's creations in place of traditions that invoke evil spirits. The raising population and disaggregation of land, rising demands for food and cash, the globalization that commodity land and its resources coupled with the ongoing climate change effects have progressively dissolved traditionally developed values while also strengthening the role of modern institutions over those of the indigenous people [25]. Yet, the informal and formal institutions have their remarkable differences but remain distinctively influential on human behaviour towards sustainable land management [4]. It is suggested that, comparatively, formal institutions have contributed less to sustainable resource management than the informal institutions although under the prevailing situation they are considered important for the implementation of globally agreed strategies for resource management. Based on this narrative, it is argued that formal institutions may indeed have a crucial role to play in sustainable resource management but only if they are equipped with the appropriate power and legitimacy.

The decentralization process that took place throughout most parts of the Third World since the 1990s was a remarkable change that aimed to strengthens formal institutions. In most situations these institutions have contributed less to sustainable resource management due to several factors, including unclear responsibility and power sharing in the decentralization process and their low endurance to change with political conditions. Thus, decentralization has had less effective in achieving the sustainability outcomes in resource management than the informal institutions. Again it is suggested that decentralization could make important contributions in the implementation of strategies and technologies to sustainable resource management if local levels of resource management were equipped with appropriate power and legitimacy [4].

Literature on traditional knowledge and institutions in Kilimanjaro Region and the neighbouring areas of Usambara show that there existed various land-based practices that implemented different traditional measures of land management in different times and for different purposes. Some studies focus on specific uses of land such as for agriculture, water and forest [26]-[31], while few have examined the use of indigenous knowledge in natural resource management generally but for limited geographical locations within the region and the neighbouring areas [8]-[14] and [32] [33]. In these studies, indigenous knowledge and their institutions are generally differentiated from one part of the region to another but a clearer difference is observed when the region is divided into two distinct zones: the Moshi zone in the north (which includes the Municipality, Hai, Rombo and Siha districts) and the Pare zone in the south (which includes Mwanga and Same districts). A good example is noted for the famous practice of intensive land use by intercropping coffee and banana and the combination of this farming practice with forestry and livestock in a sophisticated homestead farm, which is fairly documented but mainly for the Moshi zone [26]. Overall, there is neither study that has comprehensively documented indigenous knowledge and their institutions across the region nor have there been specific focus on Sustainable Land Management. This section is therefore intended to summarize issues that arise from the literature as the basis for assessing and documenting current indigenous knowledge and institutions in the region.

Early expeditions to Kilimanjaro back in the 1880's left Sir Harry Johnston who lead the expedition "*amazed by the skill with which the Chagga used tiny channels to irrigate the terraced hillsides and the time spent in turning the soil, manuring it with ashes and raking it with wooden hoes*" [27]. It is argued that the system flourished during the colonial period despite a temporary prohibition of furrow construction in 1923. The ban did not last for long as the law was quickly reversed when the British colonial government realized the essential role of irrigated agriculture in the economy [27]. A study which was conducted in 1952 described the Chagga as the most progressive African communities that have been able to produce coffee for the Western world without destroying their African social structure and system of land tenure [34]. The traditional land management referred to above is connected to the use of water for irrigation as well as intensive land use by intercropping coffee and banana, forestry as well as livestock in a sophisticated homestead farm called "*kihamba*," which qualifies to be a "complex agrosilvipastoral system". Basically, the *kihamba* belongs to the clan and is distributed according to the number of members in the extended family including the paternal kin [26].

Studies that have further investigated the role of *Kihamba* system in land management suggest that in normal conditions, trees planted in *Kihamba* of one clan could amount to over 40 families including many kinds of herbaceous and woody palms that are either grown or semi-domesticated in the small plots [25] [26]. The mixed cropping of many plants is possible by planting crops of different heights. The description of this mixture is captured well in the passage:

"... as the bottom layer on the land surface is covered with beans, cocoyam, potatoes and a variety of leaf vegetables such as 'loshuci' and 'nyafu', the lower layers between 1 and 2.5 meters would be filled with arabica coffee, traditional medicine trees, such as the 'iwonu' and shrubs such as 'kweme'. The middle layer is mainly banana trees, yams called 'nduu' or 'iko', fruit trees, such as avocado, and wood for fuel such as 'mduka'. The top layer consists of rather large trees for fuel wood and the highest are timber wood plants. Within this mixture of plants is another component plant 'isale' seen everywhere, is a very important plant for the Chagga as a boundary marker, as a symbol to identify lineage and to warn non-members not to trespass on their land. No one will pass through the places where isale is grown without permission of a lineage member. Furthermore, isale is used as a medicine to care stomach aches" [26].

It is implied in the foregoing passage that although it is generally believed that the use of manure is rare in Sub-Saharan countries (due to shifting cultivation and fire, in which soil fertility is mainly preserved from wood ashes and recovered after fallow), the Chagga have used manure traditionally for centuries in the kihamba system. The leaves of bananas and other trees as well as remains of some crops are spread over the soil surface, which has mulching effects that enhance soil conservation, retain soil moisture and control weeds on the fields.

In addition to kihamba, the Chagga would have a farmland (*rema*) on the dry plains where maize for the family food, beans for adding variety to family diet and an acre of millet is grown for making *mbege* to drink and for sale [34]. It is on the basis of this that the Chagga people have developed one of Africa's most impressive traditional systems of water management, which convey water from streams and springs along the steep slopes of river valleys to densely populated interfluves where farmlands are located [28]. This type of water use practice is called the "*mfongo*" irrigation system [35]. Other studies describe how the water system functioned well prior to "villagisation" when the furrows were still owned and managed by localized lineages [27]. It is also implied that the local control over the furrows ensured the survival of the central traditions governing distribution and maintenance thereby enabling farmers to use the mountain's permanent water resources throughout the year [25]-[28] [35].

Land practices and institutions in Pare (Mwanga and Same districts) did not differ significantly especially in the highlands where farming practices were more or less the same with the Chagga communities. For example, reference to banana grove "*nkonde*" and irrigation system "*nvongo*" are similar to the Chagga "*kihamba*" and the "*mfongo*" respectively [36]. The hillside groves were managed by mulching with cut banana leaves and intercropping with pumpkins, yams, and beans. However, these techniques could not make up for the gradual loss of fertility and moisture from the sandy red loam soils so each banana grove also received regular inputs of manure and water. As cattle were kept inside the houses, the animals' urine was directed towards the closest banana grove through a special hole at the base of the wall. These irrigation practices are not different from those called "*marombo*" [37].The tenure and management rights of irrigation water intake were inherited within the lineage of the man who constructed it in the first place, and until recently orders from this manager (the *mgawamaji*) was respected by the community members. Every farmer who used the irrigation water participated in cleaning and repairing the furrows, and in return the mgawamaji gave them permanent water rights (*kiaze*) (but no management rights) [36].

Likewise, the land practices that ensured allocation to every male member of family lineage in Chagga applies in Pare. Although this system gave men relatively secure tenure in the past it is also accountable for the rapid fragmentation of farms as the area's population grew throughout the 20th Century [24] [36]. Establishing the rate at which these fragmentations have reduced land area coverage from one land use to another is beyond the scope of this study. However, existing literature support generally that land in the highlands was under the control of an elderly male land owner (known locally as *mwenyekisaka*). Male initiation in sacred forests was a particularly important institution for social control because it provided military training, conferred marriageability, and bound local youths and immigrants into a moral community that had sworn allegiance to the chief. This traditional knowledge transfer system ensured sustainable land and other resource management but it is currently threatened by the rate of fragmentation.

The reviewed literature suggests, on one hand, that informal institutions in Africa did not in themselves offer a long-lasting solution to problems of sustainable land management, particularly in rapidly and dramatically changing environments. On the other hand, however, formal institutions have not managed to address the challenges that have come with modernity hence the need to borrow from and/or revert to the traditional value and rules of resource management. Published evidence from previous studies supports the argument that informal institutions have existed for centuries and have contributed to sustainable land management by mobilizing social capital, solving collective action problems and serving as entry points for interventions in sustainable land management. However, these institutions, particularly for Kilimanjaro, are poorly documented in the existing literature; this has being identified as a gap which called for this study. Although poorly or not documented, traditional practices and institutions have undergone changes that threaten their use for the current challenges of sustainable land management. Yet, there are suggestions that, when compared to formal institutions, informal institutions have a higher potential to survive regardless of the changing socioeconomic and political conditions [31]. Hence, conditions that influence the effectiveness of these institutions need to be identified. These have included, for example, the high rate of population growth and its associated impacts on scarcity of resources, the increasing land use change and the lack of human and financial capacities.

3. Methodology

3.1. The Study Area

This study was undertaken in Kilimanjaro Region. Kilimanjaro Region is located in the north eastern part of Tanzania. It lies between latitudes 2°25'S and 4°15'S, and between 36°25'30"E and 38°10'45"E. The region shares a common border with Kenya in the north, to the southeast it borders Tanga Region; to the south and west the region borders Arusha Region. Kilimanjaro Region covers an area of 13,209 km^2 equivalent to about 1.4% of the area of the entire Tanzania mainland. Administratively the region is divided into seven districts, namely, Rombo, Mwanga, Same, Hai, Siha, Moshi Rural and Moshi Municipality. Seasonal rainfall distribution greatly influences agricultural practices. In the Kilimanjaro Region, there are two rainy seasons—a major one in April-May and a minor one in September-November, and two dry seasons, a major one in December-January and a minor one in July-August. There is marked variation in the amount of rainfall according to altitude and the direction of the slope in the mountainous areas. The mean annual rainfall varies from 500 mm in the lowlands to over 2000 mm in the mountainous areas (over 1600 meters above sea level). Temperatures are closely related to altitude. During the rainy season, extra cloud cover and evaporative cooling tend to reduce maximum temperatures. Cloud cover also tends to raise minimum temperatures. The hot season lasts from October-March with high humidity; temperatures going up as far as 40°C in the lowlands. In the mountainous areas temperature ranges from about 15°C - 30°C.

According to the 2012 National Population and Housing Census, the region had a population of 1,640,087 [40]. For 2002-2012, the region's average annual population growth rate was 1.8 percent, which ranked 24th highest in the country. It was also the eighth most densely populated region with 124 people per square kilometer [38]. Although the region could be described as one of the densely populated regions in the country it is worth noting that the population is unevenly distributed. The highlands are the most populated areas with an average of 600 persons/km^2. Likewise the intermediate zone, which lies between 900 - 1100 meters above sea level, has a high population density of 250 persons/km^2. At district level, Rombo and Moshi Rural districts are the most populated with 160 and 256 people persons/km^2 respectively [38]. The population distribution pattern in the region is by and large influenced by soil fertility. This explains the reason for concentration of the population in the highland areas. In addition, environmental degradation is increasingly taking place due to poor farm management system such as non-use of soil erosion control methods. In this regard, the region needs to intensify land management practices in order to improve land productivity per unit area. Generally the experience of land limitation is the factor which mostly contributes to the movement of people out of the region.

3.2. Sample Size and Sampling Procedures

The selection and determination of sample size was based on the information obtained from the project office, particularly in relation to the number of villages in the project area. The District Focal Persons provided some useful information on villages located in high, middle and lower strata. In order to attain a 90% confidence the required minimum sample size for the entire region was estimated to be 164 households. This gave us an estimation error of 5% (the difference between the actual and estimated parameter value). The sample size was obtained using the following expression [39].

$$n_o = \frac{z_{\alpha/2}p(1-p)}{d^2}$$

If we take 90% confidence interval, the $z_{\alpha/2}$ is 1.28, if we assume the difference between the actual value and estimated value of the parameter is 5 percent and if we assume the proportion of households involved in the natural resources management in Kilimanjaro Region is 50 percent, then the required sample size will be approximately

$$n_o = \frac{1.28^2 \times 0.5 \times 0.5}{(0.05)^2} = 164.$$

In each selected ward or village a total of 30 households were supposed to be selected and interviewed, however, there were variations in the number of households interviewed per ward or village in each district due to availability of respondents in the selected villages. A total of 221 households were interviewed for the household questionnaire as follows: There were 27 households interviewed in Rombo, 29 in Hai, 29 in Siha, 30 in Same, 30 in Mwanga, 31 in Moshi Rural and 45 in Moshi Municipal.

The heads of the households were the ones who were targeted to be interviewed but if they were not available during the survey other members in the households were interviewed. Heads of households were interviewed because it is believed that they have information regarding the households' composition. The data set shows that most of the respondents were heads of households (76%) followed by spouse (20.8%), children (2.7%) and other relatives (0.5%).

On the sampling procedures, there were two levels of sampling; one for the villages and another for the households in the selected villages. Sampling for quantitative part of the study was guided by the existing stratification of villages in the Sustainable Land Management project area. Since there was already an established stratum that the project uses in other studies, a Stratified Random Sampling design was therefore used on the basis of the existing strata, which are *Upper*, *Middle* and *Lowlands* for Rombo, Hai, Siha, Mwanga, Same and Moshi Rural District; *Eastern*, *Central* and *Western* for Moshi Municipality. A total of 25 villages (approximately 3 from each district) were randomly selected, using computer-generated random numbers, for the study. One village was randomly selected from each stratum in the catchment area. Respondents were selected from all the districts in Kilimanjaro Region according to whether they are in the upper, middle or low land areas. According to stratification criteria used one village or ward was selected in each stratum per district.

Qualitative data was particularly important in this study due to the nature of the study. There were different sources and procedures to acquire qualitative information. It is important to note that not everybody possessed, understood and practiced the knowledge that we sought to assess and document. Therefore purposive sampling design was used to select people who could provide the required information on traditional knowledge and practices of natural resources management in Kilimanjaro Region. These people were elderly people, religious and traditional leaders, healers, representatives of development partners, NGOs, government officials etc. In addition, a snowball sampling techniques was used to locate other members who possessed traditional knowledge on natural resources management. Through these two methods, the research teams managed to ask those individuals to provide the needed information on traditional knowledge and practices in natural resources management. In this regard, the number of people contacted was determined by the information saturation.

3.3. Methods

Both qualitative and quantitative methods were used to collect data from different sources. While some of these sources were primary sources, other sources were secondary, that is, government policies and laws, project documents and published research materials on the subject. Different techniques were used to collect qualitative information. These included 41 in-depth interviews held with key informants. In addition, a total of 24 FGDs were conducted with groups of traditional farmers, pastoralists, water users, bee-keepers, forest and wildlife user groups and village councils. Group composition ensured gender representation (women, men, youths and special groups). A questionnaire was designed to capture the information related to general understanding of villagers on indigenous knowledge, the status and use of indigenous knowledge, how the existing policies and laws integrate (or do not integrate) and important issues of the knowledge that could facilitate Sustainable Land Management. The questionnaire had both structured questions and open-ended questions

Data analysis was focused on understanding of traditional knowledge and indigenous institutions on sustainable land management. Therefore qualitative information from focus group discussions and in-depth interviews were transcribed and analyzed to see what themes would emerge. The qualitative data contributed to the overall developing picture on natural resources management in Kilimanjaro Region. Quantitative data from the household questionnaire was computer processed using the Statistical Package for Social Science (SPSS) programme version 21. Frequency distribution and cross-tabulations were used to compare different variables within and across the study sites. Where relevant the Pearson Chi-square statistics was used to test the significance relationships between comparable variables.

4. Results and Discussions

4.1. State of Traditional Knowledge on Sustainable Land Management

This section assesses the state of local knowledge on sustainable land management in various districts in Kilimanjaro Region, by addressing issues related to awareness of traditional knowledge and practices, state of traditional knowledge and practices, custodians of traditional knowledge, ways of acquiring traditional knowledge and practices, role of gender relations and changes on traditional knowledge and practices.

4.1.1. Awareness of Traditional Knowledge and Practices

In order to establish levels of understanding of traditional knowledge on sustainable land management, respondents in the study areas were asked to state whether they were aware of various traditional knowledge and practices related to sustainable land management. Majority of respondents (93%) indicated to be aware of those practices, while the remaining 7% were not aware (**Table 1**).

As shown in the table, respondents in Same, Siha, Mwanga, Hai and Rombo districts were more aware of the traditional methods and practices of sustainable land management than those in Moshi Urban District. Respondents in Same, Siha, Hai, Mwanga and Rombo districts were more aware of the traditional methods and practices than those in other districts due to fact that these districts are rural based and therefore still practice some of these traditional methods and practices. Secondly, these districts often experience droughts during dry seasons and landslides and soil erosion during rainy seasons, conditions that have impelled communities in those districts (Same, Siha, Mwanga, Hai and Rombo) to adopt more sustainable land management practices.

The level of awareness of traditional knowledge and practices of sustainable land management was not significantly different between respondents of different education levels. However, there were considerable

Table 1. Percent response on awareness of traditional methods and practices of sustainable land management in Kilimanjaro Region by district.

District	Number of respondents	Aware of traditional methods of SLM	
		Yes	No
Same	30	100	0
Siha	29	100	0
Mwanga	30	96.7	3.7
Hai	29	96.6	3.4
Rombo	27	96.3	3.7
Moshi rural	31	90.3	11.9
Moshi urban	45	80.0	20.0
Total	221	93.2	6.8

Source: Compiled from Field Work Data, November 2013; ($X^2 = 22.731$, df =12, p < 0.030).

differences between gender groups. It was noted that more male (59%) respondents were aware of the traditional methods and practices of sustainable land management compared to 41% among female respondents. The difference in awareness of the traditional methods and practices of sustainable land management between male and female respondents was statistically significant ($X^2 = 6.56$, df = 2, p < 0.038). The main reason for the observed trend was that customarily, men are the ones who inherit land in Kilimanjaro Region, and thus have more keen interest in conserving the land than women. It was noted for instance that "*Women are strictly restricted from going to water catchment areas with their clay water pots so as not to pollute water in catchment areas*" (in-depth interview). This indicates the use of threats and taboos so as to conserve water catchments.

While there has never been any evidence of pollution, this has customarily been practiced with the aim of protecting the respective water sources. The inadequate engagement of women in resource management is also echoed by other studies in Kilimanjaro Region. For instance, Magigi and Sathiel [40] reported that despite progress on various initiatives to mainstream gender in development initiatives, equality, ownership and use of resources have been minimal in developing countries, resulting into decreasing local economic development, food security, ecological integrity and sustainable livelihoods.

Further analysis indicated that there are variations on the awareness of traditional methods and practices of sustainable land management by age groups. Respondents aged 46 years and above appeared to be more aware of traditional methods and practices of sustainable land management than other age groups. These differences in awareness between respondents of different age groups could be explained by the fact that more people (77.7%) aged 46 years and above have more experience with traditional methods and practices, and often considered as the custodians of culture of the various ethnic groups. On the other hand, people aged 17 - 35 years and those aged 36 - 45 years are mainly youths who have spent most of their lives in schools and therefore least attached to land management issues and have the least developed interest and commitment to land management matters.

Figure 1 presents the sources of traditional knowledge in Kilimanjaro Region. While a considerable proportion (54.8%) did not mention to use traditional knowledge/practices, it is evident that those who use it have learnt about it from elders and other villagers (52% and 29% respectively). It is clear from **Figure 1** that traditional practices are rarely learnt from government officials and/or conservation projects and organizations as indicated by the small proportions of respondents, but rather from peers and elders. For enhanced sustainability in land management therefore there is a need for more integration of the relevant traditional knowledge systems and practices with land related policies and governance systems.

4.1.2. Existing Traditional Knowledge and Practices

Table 2 summarizes various traditional knowledge and practices in various study sites as explained during in-depth interviews and focus group discussion. **Table 2** reveals that in all the study sites traditional knowledge and practices are applied in sustainable land management and people have considerable understanding of these methods as exemplified by different respondents during in-depth interviews.

1) Cultivation ridges and traditional irrigation canals

Indeed, the study revealed the existence of many traditional practices of sustainable land management in the Region. Among these included the use of cultivation ridges (*Matuta*) along the slopes (**Figure 2**). **Figure 2** also shows the water canal constructed for irrigation as well as for reducing soil erosion. These old practices of controlling soil erosion were enforced by traditional leaders, *or* Chief. Historically, everybody was required to do so in slopes. However, it was claimed by farmers during this study that this practice is currently used on a limited scale due to lack of manpower and inadequate knowledge and interest among the youths to construct terraces. These reasons may be associated with modernity which has kept the young generation in schools and much away from farm activities.

2) Tree planting

Tree planting in homesteads and farms is another long tradition which is still practiced today (**Figure 3**). There are trees that have traditionally been identified for their significance in land protection, provision of shades, water retention, prevention of soil erosion, border demarcation and some have symbolic importance for various cultural practices. Through traditional knowledge of identifying areas that were threatened by problems like landslides, the community would protect such lands by planting trees. As captured in **Figure 3**, these areas are mostly those along water ways (locally known as *mkondowamaji*) that happen to be in farms.

Overtime, however, changes have occurred in terms of the number and type of these trees. In terms of numbers, the growing human population and subsequent increase in settlements have reduced the areas available for

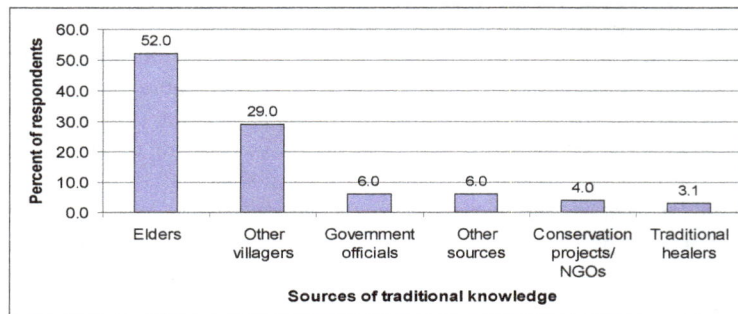

Figure 1. Percent response on where respondents learn about the traditional method/practices.

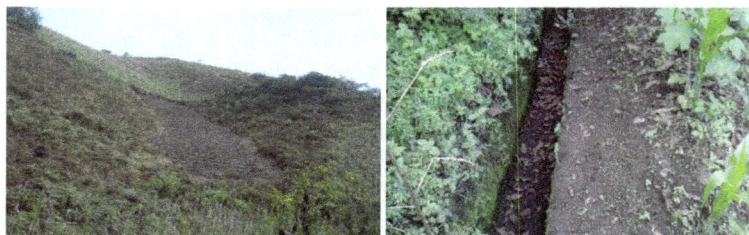

Figure 2. Ridges along the slope of the mountain in Lokiri Village in Siha District (left), and traditional irrigation canals (right).

Table 2. Various traditional knowledge and practices in the study sites.

Name of traditional knowledge and practices	Local terminology	Type of resource managed	Area of operation
Construction of terraces by using stones	"Maporomoko" in Pare language or land slide	Soil	Mwanga and Same study sites
Allocating forests for ritual ceremonies	"Mitaso" in Pare language or places for performing rituals to appease the dead	Forest and water	Rombo, Same and Mwanga study sites
Fines in kinds (cow, goats or sheep with honey beer)	"Mahaani" in Pare language or fines in kind	Water, soil, forests and other resources	All study sites (Rombo. Hai, Same, Mwanga, Moshi rural)
Reforestation by planting local trees (Mifumo and Mikuyu)	"Panda miti" or reforestation	Soil, water and forest	All study sites (Rombo. Siha Hai, Same, Mwanga, Moshi rural & Moshi urban)
Both fallowing by leaving the land idle to regain fertility	"Kulazashamba" or fallow	Soil	All study sites (Rombo. Siha Hai, Same, Mwanga, Moshi rural & Moshi urban)
Ant poaching	"Ngoloshoi" in Chagga language or prohibition of poaching	Wildlife	Siha, Hai, Rombo and Same study site
Beekeeping		Forest	Siha, Mwanga, Rombo and Same
Threats and taboos		Water and forest	All study sites (Rombo. Siha Hai, Same, Mwanga, Moshi rural & Moshi urban)
Strip terraces and planting guatemala graces		Soil	All study sites (Rombo. Siha Hai, Same, Mwanga, Moshi rural & Moshi urban)
Construction of ridges	"Matuta or Mfuloi" for Chagga language or ridges	Soil	Moshi rural, Siha, Hai and Rombo study sites
Growing grasses and Trees	"Ihaninyika" in Pare language or indigenous forest	Soil	Moshi rural, Siha, Hai and Rombo study sites

Source: Compiled from Field Work Data, November 2013.

tree planting hence fewer trees can be grown around the homesteads. An elderly respondent commented, for example, that on average 4 - 5 households have established settlements in only one acre compared to one household in the past years. In this case, land for tree planting is limited across much of the region especially in

Figure 3. A homegarden at Ibukoni Village, Rombo District.

the highlands and midlands. Among other challenges associated with tree planting was the complaint from farmers following the ban by the government through the regional administration to cut trees without permit from the District authorities. Because of the above situation, people are currently reluctant to plant trees as they are arguing that there is no need to plant a tree which you cannot harvest when you need to do so.

It is important to note also that trees are part of family assets and an important source of cash incomes. It is in this connection that there is high monetary value for forest products, which continue to be an impetus for deforestation. Although the government regulates harvesting of some tree species and that people are required to obtain permits from village offices for cutting trees, illegal activities continue due to corruption and the inadequate law enforcement. Notable timber products of threatened species still have readily available markets.

3) Use of farmyard manure

The use of farmyard manure (*samadi* or *Boru*) is common in Kilimanjaro Region. Manure comes from livestock wastes and is one of the important traditional methods of agricultural land management particularly in Rombo District. It is a most common practice especially where stall feeding of livestock is practiced. Although this is still practiced to date, respondents acknowledged that there are growing challenges of stall feeding as more open areas that were a source of fodder are either turned to settlements or access is restricted by the government (especially in forests). Following these restrictions, animal rearing specifically cattle is increasingly becoming expensive because famers have to buy the fodder. Some respondents suggested also that there has been reluctance in using *boru* generally due to the availability of industrial fertilizers which are easy to handle compared to farmyard manure. *Boru* comes from livestock wastes. Narratives indicated also that other kinds of livestock wastes (chicken and goats) are mixed with ashes and filled in bags and kept for some 20 - 30 days to decompose after which it is distributed to the farm. This was reported to be not only very effective in increasing soil fertility but also the ashes have repelling effect for some insects which cause damage to understorey plants in the home gardens.

4) Mulching

Mulching (locally known *asmasoro*) is another traditional and common practice in Kilimanjaro Region. It is usually done using crop residues and remains from livestock feeds, farm, tree leaves from different kinds of mixed plants farmland, which together make an excellent mulch (**Figure 4**). This method has been used over generations. The importance of this practice is that the mulch covers the top of the soil hence preventing it from direct sunshine and runoff. This tradition helps to retain soil moisture while increasing fertility after decomposition of the mulch, and protects the soil from erosion by reducing surface runoff.

5) Mixed cropping

Mixed cropping is another old tradition which is based on kihamba farming system where many kinds of crops are grown in same plots. In the past the diversity of plants in the *kihamba* was very high and could amount to 42 families [41]. However, currently there are very few farmers with the kind of mixed cropping recorded in the past. In the present study, only a few such plots with dense mixed cropping were observed which may be interpreted to represent a general failure of *kihamba* farming system.

6) Rotational cropping/fallowing

Rotational cropping/fallowing is another traditional practice of conserving the land in many parts of

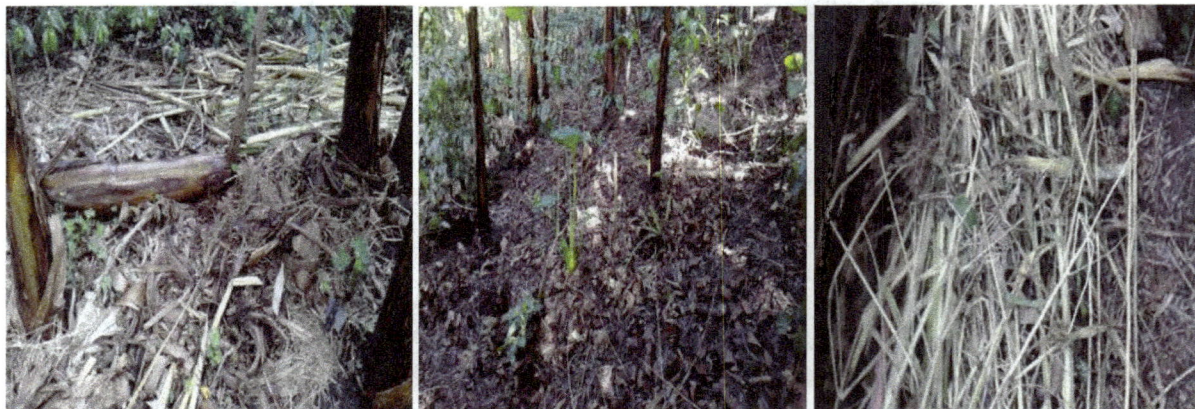

Figure 4. Mulching (*masoro*) in Rombo District.

Kilimanjaro Region. However, the increasing pressure on land associated with population growth has in many places made rotational cropping and/or fallowing generally not feasible. Consequently same fields are continuously being cultivated resulting in soil degradation and decreased productivity [42]-[45].

The role of mixed cropping has generally been to increase organic manure in the soil, creating shades that assist in soil water retention and preventing soil erosion among other things. These are all basic principles of sustainable land management that communities have attached to their culture and traditions for generations across much of the slopes of Mt. Kilimanjaro. These practices have been effective in protecting the land while also ensuring the availability of varieties of food crops. As changes in culture and traditions continue due to various reasons, including modernity, these practices have also changed over time.

7) Protection of some species by the chiefs

As it is with the government today, whereby certain species are protected and may not be utilized without permission, even in the past some species were ownership and protected by the local chief. Traditional leaders such as chiefs claimed ownership of certain species of plants and animals hence other members of the community would not be able to use these species. By their royal status these species were therefore protected wherever they were found in the chieftaincy. **Table 3** presents some of the species and others that have benefited from traditional knowledge of their significance. Notwithstanding their importance to the community and the environment, most of these tree species are locally perceived as currently extinct or threatened with extinction.

8) Use of intimidations, threats and taboos

Intimidations, threats and taboos have traditionally been used to conserve various resources. The uses of intimidations and taboos were based on the prior knowledge of the importance of particular resources. Specifically, traditional leaders introduced taboos to protect water catchments and plants around these catchments. Most of these taboos were meant to intimidate people so that they do not haphazardly harvest these species. Among the taboos recorded in this study were narratives about the presence of big snakes and/or demons in certain trees; women not allowed to reach certain areas of catchments; some trees crying if they were cut which was interpreted to signify bad lack for whoever did it. These taboos created fear and respect to certain places and species by community members. In the case of women, they are the firewood collectors hence preventing them from going to water catchments ensured that these catchments remained intact.

It was claimed, for instance, that in conserving forests among the Pare it is common to find threats like "Any person who cuts firewood or trees in a forest conserved for rituals and thanksgiving purposes or in water catchment areas will be fined a cow, goat or sheep with honey beer depending on the intensity of destruction he/she had made". This shows that various sanctions are inbuilt in traditional knowledge and practices and have for generation been effectively used to conserve the local resources in various communities.

In an interview with an elderly man (95 years) in Rau Ward, Moshi Municipality, as part of the present study revealed how traditional knowledge can be used in the interpretations of current environmental changes such as changes in climate and shortages of water and low soil fertility, associating them with the failed traditional resource management systems.

"... *this mountain that you see was not climbed by everyone like it is now. It had very special use and*

Table 3. Traditionally protected species and their significance.

Local name	Botanic name	Importance	Current status (local perception)
Mvule	Milicia	Chief's tree (Mtiwamila)—with total protection. Only Mangi could use it. Used for sacrifices	Threatened with extinction
Mgangafumu		Chief's tree (Mtiwamila)—with total protection	Locally extinct
Misesewe	Albiziaglaberrima	Shade, water retention	Threatened with extinction—they are replaced by timber species. Also, they are currently targeted for white timber
Masale	Trigonellafoenum	Prevent soil erosion, reduce conflicts (is a respected border mark)	Still maintains its traditional values but are few
Mfumu	Vitexdoniana	Water retention	Threatened with extinction
Mruka	Alfzeliaquanzensis	Shade, water retention and prevent soil erosion.	Threatened with extinction
Mapala	Acer pensylvanicum	Prevent soil erosion and are evergreen throughout the seasons	Threatened with extinction
Mapasa	Mcobotini	Prevent soil erosion. Are planted as fences across steep slopes	Threatened with extinction
Mkuyu	Ficussycomorus	Water retention	Threatened with extinction
Mringaringa	Acacia albida	Water retention	Threatened with extinction
Mihanzi	Albus	Water retention	Threatened with extinction
Masteria, Majaniyatembo, makatimara and makengera	Diplurida/ arachnida	Control soil erosion but also nutritious animal feeds	Threatened with extinction
Mnengu	Hymenaeaverrucosa	Moisture retention—hosts some insect species (called firondo) that emit water during dry season	Threatened with extinction

Source: Compiled from Field Work Data, November 2013.

therefore only chosen people from the community, mainly men, could climb to practice offerings for propitiate the spirits of the dead (matambiko). This was mainly accompanied with slaughtering of goats and cows, pouring of milk and honey. An exception could only be made when there was a need for special sacrifice and a young woman could be taken [for sacrifice]. As these men descended after offerings, they would clean water furrows all the way down. Indeed, by the time they arrive home it would have rained and the water would have a clear path to follow" [46].

Although the importance of sacrifices in causing rainfall and water availability is contestable, the narrative above suggests that some of the traditional practices were useful in conserving the mountain forests, clearing and maintaining water furrows. This ensured catchment protection and reduced land degradation that could result from the lack of clear paths for water. Two kinds of challenges exist today. One, there is no longer a well-organized system of water distribution to farms by elderly people (the "*Mfongo*" institution). The interviewed elderly men who used to distribute water to people's farms indicated passion and trust in the institution yet they confessed that the system broke in their hands for failing to transmit the knowledge to the young people.

The second challenge relates to the fact that some of the government-led resource management strategies that seek to upgrade traditional practices contribute to their destruction. Examples were given, for example, the use of cement in concretizing some of water sources and furrows that are threatened by degradation. Participants in FGDs in Rau Ward were of the view that modern methods of sustainable land management do not necessarily produce good results. In Rau, for example, a once-off sponsorship that was secured through local politicians made concrete walls for the main furrow and replaced elephant grasses that had been planted for long along the banks. In a span of only four years, the wall broke and since then there was no any renovation plan, the water force has increased hence causing land degradation than it was before.

4.2. Factors for Changes in Traditional Knowledge and Practices on Natural Resources Management

This study examined factors for changes in traditional knowledge and practices on land, forest, water and wildlife management. Results from focus group discussions and in-depth interviews in the study sites indicated that community members have been observing the changes on knowledge and practices on natural resources (land, forest, water and wildlife) management over time. The main reasons explained by respondents revolved around the spread of modern oriental religions (Christian and Muslim), which all are against traditional beliefs, population growth due to demographic processes of increase in fertility and net immigrations which have increased high demand of resources leading to reduction of arable land and grazing land, protected forests, and forests, and water catchment areas which were all formally protected.

The increase in family size has led to population pressure on the existing land thus the existence of a miss-match between number of people and the available resources. Another reason mentioned for these changes was the climatic changes which have caused many water sources to dry. In addition is the eradication of the traditional leaders which based on chiefdoms and their replacement by government leadership, such as ten cell leaders, sub village chairpersons, and village chairperson, have contributed to the decline of many traditional practices.

The other factor regards the methods of acquiring traditional knowledge and practices. Various methods of acquiring traditional knowledge and practices were mentioned by respondents in the study areas. Through focus group discussions and in-depth interviews participants mentioned field training by members of families/clans, traditional meetings, ritual ceremonies, Government expertise like extension officers and forest agencies as among the different methods used to acquire traditional knowledge and practices. It was noted during a focus group discussions and in-depth interviews *"When young boys are passed from boyhood to manhood in ritual ceremonies they are taught how to conserve the environment, how to handle wives and children; they have to learn from the previous knowledge"*. This indicates that young boys are taught to handle the land resources during ritual ceremonies. However, as noted earlier in the paper, most of the you generation spent most of the young age at school with very limited touch to livelihoods in the rural setting, which to a large extent make them to miss such training.

Furthermore due to globalization and interactions between different communities traditional customs are no longer practiced all the times, for example, ritual ceremonies in some societies especially in urban centers do not exist any longer. According to elderly people interviewed during focus group discussions and in-depth interviews, reinstatement traditional knowledge specifically through ritual ceremonies could be the only way to ensure that traditional knowledge on sustainable land management is passed from one generation to the other.

4.3. Indigenous Institutions for Natural Resource Management

An inquiry was made regarding the respondents' awareness of the existence of indigenous institutions for sustainable land resource management in their areas. Responses indicated that 62.0% were aware, while the remaining 38.0% of respondents were not aware (**Figure 5**). Findings in **Figure 5** indicate responses differed from one district to another with Same and Siha having largest proportions of respondents reporting of awareness. The level of awareness of these institutions is, in this case, closely related to two things; first is how the community is culturally organized and, second, the kind of land management challenges that communities have faced for generations. Accordingly, it is possible that Same and Siha districts have more land management challenges hence the high levels of dependence on traditional practices and institutions.

Quantitative data suggest that there is a clear understanding of the difference between traditional and modern institutions and that the percentage of respondents indicating awareness of the existence and importance of indigenous institutions is relatively high (64%), as compared to 36% reporting to be unaware. Nevertheless, this level of awareness is not directly related to the current levels of reliance on indigenous institutions of land management. Practically, most respondents made reference to indigenous institutions that have existed in the past and those that are only marginally used at the moment. Reference to the past institutions is clearer when subjected to content analysis. Qualitative data suggested, for example, that the use of indigenous institutions has been interfered by modernity in its different forms including through modern religions and schools, reliance on market economy which has encouraged importation of industrial products such as fertilizers and the democratic governance that allow the majority decision to rule even in matters that have traditionally been managed through cultural norms and beliefs. As a result, the young generation is less aware of both the practices and indigenous institutions. For example, **Table 4** demonstrates that youths between the ages of 17 - 35 years who participated in the study were 26 and majority of them (65.3%) were unaware of any indigenous institutions. As such, this

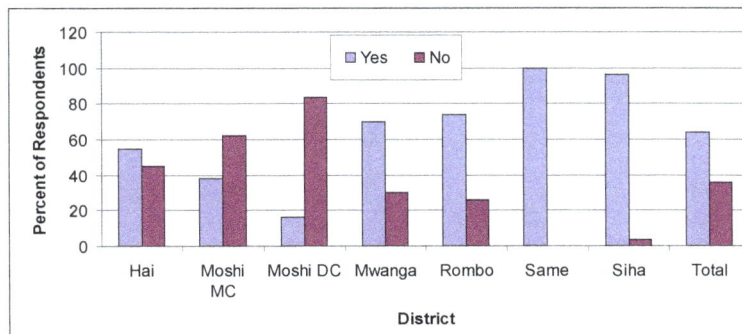

Figure 5. Percent responses on awareness of indigenous resources management institutions.

Table 4. Awareness of indigenous institutions by age group.

Age group	Yes	No	Total
17 - 35	34.7	65.3	100
36 - 45	65.6	34.4	100
46+	65.7	34.3	100
Total	62.0	38.0	100

Source: Compiled from Field Work Data, November 2013.

knowledge is mostly held by the age of 36 years and above as demonstrated by more than 65% of respondents in each age group who reported to be aware of indigenous institutions. Pearson Chi-Square test also confirmed that there is a significant relationship ($X^2 = 17.221$, df = 4, p < 0.002) between the level of awareness of indigenous institutions and age of respondents. This means that as age increases, awareness of indigenous institutions also increases. According to this data, modern institutions are currently better known by the younger generation, which suggest that the transfer of this traditional knowledge has been constrained by various factors as discussed elsewhere in the paper.

Table 5 shows the responses on the types of land management institutions. It is clear from the **Table 5** that there has been a mix between traditional and modern structures of these institutions. However, according to the respondents, family/clan organizations, traditional leaders, followed by the legacy of Mangi chieftainship, are the main indigenous institutions dealing with land management issues in the area. Other indigenous institutions are the resource user groups and community-organized security (known locally as *Sungusungu*). It should be noted however that the existence of these institutions cannot be generalized for all the districts. In Same district, for instance all the identified indigenous institutions exist, suggesting that there may be little or no use of modern institutions. This trend is also noted in the neighbouring Mwanga District. The impression created by the two districts is that reliance on indigenous knowledge and institutions is motivated by the type and intensity of problems that exist in the respective society. A similar argument was reported by [40]-[45]. Indeed, the two districts share similar landscape characteristics that might explain challenges in land management hence their continued dependence on these indigenous institutions.

Although **Table 5** presents both indigenous and modern institutions of land management as understood by respondents, the focus of this section is on indigenous institutions. In the following we present brief accounts of the current status and relevance of these indigenous institutions based on the collected quantitative and qualitative data.

4.3.1. Families/Clan Organization

This type of institutions was found across the districts with Same and Mwanga having the highest number of respondents who recognized family/clan organization as an important institution in sustainable resource management at the local level. Whereas in these districts these organizations are known by clan names (for example, Wamwanga and Wafinanga), family organizations in the rest of the region have mainly made reference to the kihamba system. Although there is a slight difference across the districts, the point of convergence for these

Table 5. Percent response on indigenous institutions by district.

Indigenous institution		Modern institution	
Name of institution	Percent of respondents	Name of institution	Percent of respondents
Family/clan organization	74.9	Cooperative societies (KNCU, Kibong'oto Wandri)	5.6
Traditional leadership	64.9	Village environmental committee	37.5
Mangi chieftainship	49.6	Government/Extension officers	30.2
Social and resource user groups	27.9	Village land council	18.0
Community-organised security (Sungusungu)	2.2	Don't know	39.6

Source: Compiled from Field Work Data, November 2013.

organizations is the family ownership of land and forests. This was particularly echoed in Mwanga and Same districts where families/clans own sacred forests that are used for various ritual functions. Conservation of these sacred forests is therefore the responsibility of these families/clans. Similarly, Kihamba belongs to the clan and is distributed according to the number of members in the extended family including the paternal kin [26] [41]-[45]. However, the Kimamba organization has changed substantially with the abolition of chieftaincy and formalization of land management practices. In this study, respondents indicated that they recognize Kihamba as one of the strongest indigenous land management institution that has nevertheless lost its importance due to many reasons including high population growth and the subsequent division of land into small individual units that are mainly used for settlements.

4.3.2. Traditional Leadership

Traditional leaders are normally wise men and women who have earned respect due to their involvement in solving existing problems in the society. Reference to these leaders in Kilimanjaro Region is made particularly in relation to the possession of traditional knowledge and the ability to transfer the same across generations. This was a concern of 64.9% of respondents in the study (**Table 5**). Comparatively, Siha district indicated the highest level of recognition of the role of traditional leaders followed by Same and Mwanga districts, as reported by 20.6%, 16.7% and 11.1% of respondents, respectively. Traditional leaders are also currently useful due to their integration in the village government through Village Land Councils, which deal particularly with land-use conflict resolution and awareness raising about tree planting for soil conservation and border demarcation. Although village land councils are not run in traditional ways per se, traditional leaders share their experiences while the rest rely on trainings that are *offered* by the government and civil society organizations. Focus group discussions revealed that land councils do not have the kind of "powers of knowledge and command" that previous traditional systems of leadership had. However, it is so far the only way that the government has worked with traditional leaders. Strengthening them could therefore work to bridge the gap between conventional and traditional land management systems.

4.3.3. Mangi Chieftainship

Chieftaincy was abolished by the government in the 1960s. However, respondents still speak highly of the role of chiefs in keeping order in the society. Although practically Mangi chieftaincy has long been dormant following the government decision, reference to Mangi chieftaincy is made with pride and confidence particularly in relation to its legacy in the protection and management of land and other natural resources. During this study this was reiterated more in Rombo district. Respondents described Mangi chieftaincy to have been a well-organized structure of council members and messengers (commonly known in Kiswahili as *wajumbe* and *matarishi* respectively). At different levels, these members and messengers ensured that traditional laws were enforced and order maintained. As indicated **Table 5** about 49.6% of respondents could still recall how this institution operated. Understanding how indigenous institutions lost power of managing resources is an important starting point for the search of ways to harmonize indigenous institutions and modern policies and laws that govern land and other natural resources in the country.

4.3.4. Social and Resource User Groups

Villagers are organized in various social groups most of which are for women and youths who participate in different activities related to protection of water catchments, tree planting as well as traditional dances. Women groups are particularly more numerous in number. The most common social and resource user groups evident in Kilimanjaro Region are the traditional dances and the irrigation water institution. Traditional dances (e.g. "Mbasa" in Machame) have their roles in knowledge transfer. Although there were no specific dances for particular resources, the message about traditional values and their importance was reported to be carried through songs and sharing of drinks and meat. These dances no longer exist but they are still recalled by elderly respondents as reported by 27.9% of respondents (**Table 5**).

The Irrigation water institution (the *Mfongo* institution), is a very unique and strong water management traditional institution. This institution is still existing but far smaller in scale than it used to be in the past. The institution has also remained with old people who cannot maintain it because the young generation has not shown much interest to take it over. In Rau ward, for example, the *Mfongo* institution died when its leaders who lived in Kisarika village died. Traditional knowledge on irrigation furrow management included predictions of the amount and timing of rains which helped in the management of these traditional furrows. The leaders (known locally as *Wamekuwa Mfungo*) had accumulated knowledge that helped in interpreting the position of the moon. Certain positions of the moon were traditionally used to indicate the amount of rainfalls hence preparations for the furrows were done accordingly. Recently, the modernization of traditional furrows involved concretization of irrigation channels hence changing their management from mainly traditional leaders to mainly the village government. The changes have had mixed impacts. Whereas water is better managed in the modern furrows, there have been increased conflicts due to the overlap of responsibilities between traditional leaders and village government that finance the maintenance of the modern furrows. Also local political elites have intervened by, for instance, funding the modernization of the furrows with attached political interests. As the case of Rau suggest, modern interventions have in some cases been very unsustainable due to the lack of ownership by water users and funds for renovations.

4.3.5. Shared Characteristics of Resource Management Institutions

Survey results suggest that although there are some differences in what the existing local institutions focus on, most of them (whether indigenous or modern) share one main characteristic; the responsibility of protecting and managing natural resources. **Table 6** captures activities of different indigenous and modern institutions. There are those that focus on land and forest protection, water source protection, irrigation water management, tree planting, food crop production, and special ritual performances, among others. However, these institutions differ in their conception, management structures, rules and methods of law enforcement. Success of these institutions is therefore dependent on whether they are formal (where legal actions can be taken) or informal (with informal rules and regulations). On the latter, which is the focus of this study, means of rule enforcement depend on beliefs and specific traditional knowledge which creates taboos and fear of disgrace among community members.

As other data suggests, Same and Mwanga demonstrate that there are many local institutions that deal with land and forest protection. Local narrative indicated that traditional means of rule enforcement have had superior results in protection of these resources. This is confirmed by local perceptions of the existence of some tree species whose ownership by the chief family helped to protect such species for generations (see **Table 3**).

Respondents suggested also that there is a clear distinction between modern and indigenous institutions although there is currently a trend towards blending core values of the two. When villagers were asked to describe indigenous institutions, the following characteristics were distinct (**Table 7**). A comparison was also made regarding corresponding modern resource management institutions. The reported blending of these institutions seem to be an important entry point towards integration of indigenous and modern institutions for enhanced sustainability of resource management.

4.4. Relevance and Effectiveness of Local Knowledge in Addressing Land Degradation

Usefulness and Effectiveness of Local Knowledge in Addressing Land Degradation

An inquiry was made on the usefulness and effectiveness of various traditional practices in addressing land degradation. Findings have indicated that a large proportion of respondents implemented traditional practices with aim of preventing soil erosion and improving soil fertility, as indicated by 42.4% and 26.4% of respondents respectively (**Table 8**). Other reported usefulness of the mentioned traditional practices in sustainable land management included soil moisture conservation (7.7%), soil conservation (6.1%) and improvement of crop

Table 6. Percentage response on activities undertaken by the indigenous institution.

Activities	Hai	Moshi MC	Moshi DC	Mwanga	Rombo	Same	Siha	Total
Land and forest protection	10.2	15.3	3.2	39.9	14.8	79.8	37.8	28.7
Land protection	27.5	8.6	0	6.6	55.5	10.0	20.6	18.4
Land and irrigation furrows (mfungo)	0	0	3.2	0	0	0	13.6	2.4
Land and coffee	0	0	2.2	0	0	0	6.8	1.3
Water source protection	6.8	8.8	6.4	3.3	0	3.3	17.1	6.5
Livestock	0	0	0	6.6	0	0	0	0.9

Source: Compiled from Field Work Data, November 2013.

Table 7. Characteristics of resource management institutions as perceived by villagers.

Indigenous institutions	Modern institutions	Mixed institutions
Not legally registered and recognized by the government hence operate on their own.	Known and used by all people.	Made up of both government and traditional leaders.
Membership based on beliefs and specific traditional knowledge (e.g. ritual performance). Few people subscribe to these institutions because not everybody has enough knowledge of traditions and beliefs associated with them.	Passed through democratic procedures—discussed and agreed upon by all members of the village. Also its leaders are elected based on their acceptance by the villagers.	Both government laws and traditions apply.
Leadership based on family lineage and inheritance, loyalty and command of respect.	Overseen by the village government.	Some depend on voluntary membership (e.g. social groups and resource user groups).
Command of respect based on society's knowledge of customs and traditions.	Have modern leadership structure—Chairperson, secretary, treasurer and members of council.	
Dependence on informal transfers of knowledge.	Recognized and report to the ward and district council.	
No cash payments—traditional leaders are paid through respect from members of the community.	Supported by the government and other stakeholders (through salaries/ cash incentives, trainings).	
Through the command of respect, the word of traditional leaders works as a law.	Has legal powers to litigate hence respected.	

Source: Compiled from Field Work Data, November 2013.

Table 8. Percentage responses on the effectiveness of the traditional practices by districts.

Effectiveness of the traditional method/practice	Hai	Moshi MC	Moshi DC	Mwanga	Rombo	Same	Siha	Total
Preventing soil erosion	48.3	26.1	51.5	40	25.9	56.7	48.3	42.4
Improved soil fertility	31	32.6	18.2	13.3	51.9	6.7	31	26.4
Soil moisture conservation	6.9	4.3	18.1	10	11.1	0	3.4	7.7
Soil conservation	6.9	2.2	3.0	0	3.7	26.7	0	6.1
Improve crop productivity	3.4	10.9	0	6.7	0	0	6.9	4.0
Forest conservation (e.g. preventing tree cutting)	0	0	0	10	3.7	0	3.4	2.4
Attract rain	0	2.2	0	10	0	3.3	0	2.2
Irrigation	0	0	0	0	0	0	3.4	0.5
Marking farm boundaries	0	0	0	3.3	0	0	0	0.5
Wind break	0	0	0	0	0	0	3.4	0.5
Not applicable	3.4	21.7	9.1	6.7	3.7	6.7	0	7.3
Total	100	100	100	100	100	100	100	100

Source: Compiled from Field Work Data, November 2013.

productivity (4.0%).

As shown in **Table 8** there are variations in responses between districts. For instance, Same, Moshi Rural, Hai, Siha and Mwanga districts, in that order, were more concerned about the need for preventing soil erosion, which could be attributed to hilly landscape features with more risks of erosion compared to other districts. Information from key informants confirmed that in some of these districts, such as Same and Mwanga they experience gully erosion in various place, hence also confirming the concern by the respondents in this study on the need to address soil erosion. However, Siha has not yet experienced serious erosion. The concern in this district on soil erosion control may perhaps be attributed to higher environmental awareness.

Although other practices were mentioned by much fewer respondents, if well organised and promoted may enhance the sustainability of natural resource management, and minimizing land degradation, for instance through traditional forest conservation practices that, for example, prevent forest tree cutting.

The perceived usefulness of the traditional practices of land management in the area is also demonstrated by the duration of their existence and utilization. **Table 9** presents the responses on the length of time that respondent households have been involved in the various traditional land management practices. A larger proportion (64.9%) of male respondents reported that these practices have, for generations, been practiced and have been effective, followed by 27.3% who reported to have been involved since childhood. However, as with other practices, there were considerable variations between districts. For instance, in Rombo, Mwanga and Moshi Rural Districts interviewed people indicated to have experienced or practiced such traditional land management practices for generations as reported by 93.8%, 83.3% and 72.2% of respondents respectively. Some of the respondents have experienced such practices only recently, as reported by 57.9% and 52.6% of respondents in Hai and Same districts respectively (**Table 9**). The variations in responses could be explained by the earlier reported landscape features that have influenced the need for particularly traditional land management practices/methods.

Comparing such experiences by gender, it appears that there are no significant differences between those reporting traditional/indigenous practices to have been practiced for generations or since childhood (**Table 9**). However, there was a larger proportion of male respondents (18%) reporting that such practices have been in place only recently, compared to only 5.6% among female respondents. This may be influenced by the relatively shorter duration of stay in the area among the female respondents. As such, this group of respondents may have lesser experience on traditional/indigenous or sustainable land management practices of the area. This demonstrates that traditional knowledge is acquired through experience to the local environment. Presence of respondents indicating that they are unaware or having limited knowledge of traditional practices of Kilimanjaro Region could be attributed to issues of marriage and migration where some people migrate and/or get married into the region from other places with different traditional practices. The men in this category mainly attributed their lack of such knowledge to migration, as they moved into the region for various reasons such as business and/or employment. However, these findings indicate that for more sustainable resource management, gender concerns have to be considered especially because in many parts of Tanzania and elsewhere in the world women play a crucial role in resource manage. Gender mainstreaming may need recognition and documentation of the good practices and having gender mainstreaming strategy [40].

4.5. Integration of Indigenous Knowledge in Policy Process and Governance

4.5.1. Strengths and/or Weakness of Traditional Institutions in NRM

The traditional institutions under play in Kilimanjaro Region and there characteristics have been presented in

Table 9. Percentage response on perceived existence and length of use of traditional practice.

Sex	Length of time	Hai	Moshi MC	Moshi DC	Mwanga	Rombo	Same	Siha	Total
Male	Existed for generations	57.9	47.1	72.2	83.3	93.8	52.6	47.4	64.9
	Since childhood	31.6	47.1	16.7	0	6.3	42.1	47.4	27.3
	Recently	19.0	17.0	18	18	16.0	19.0	19.0	18.0
Female	Existed for generations	40	31.0	69.2	66.7	90.9	54.5	60.0	58.9
	Since childhood	30.0	37.9	15.4	25.0	0	36.4	40	26.4
	Recently	20.0	3.4	7.7	8.3	0	0	0	5.6
	Total	100	100	100	100	100	100	100	100

Source: Compiled from Field Work Data, November 2013.

Section 4.3. Historically, local institutions have ensured that local resources are managed sustainably. This indicates that they have been strong and effective throughout the historical times. This is confirmed by the fact that these institutions have managed to sustain the local resources for generations, as indicated by majority of people in the region. However, as indicated there are people in Kilimanjaro Region that seems to be unaware of the existence of traditional or indigenous institutions for sustainable land management. This raises some questions as to the sustainability of the institutions themselves. It also raises concerns on the future land resource management in the area. Such experience may indicate that the traditional institutions have not adequately promoted themselves to capture the interest, attention and practice of all generations, possibly in favour of more modern and conventional practices as discussed earlier. This may be viewed to be among the weaknesses facing traditional institutions in the region, and possibly elsewhere in the country and globally.

An alternative argument may be that the inadequate recognition of indigenous institutions indicates that national level institutions have become of greater influence at the local level, thereby superseding the indigenous institutions. To ensure that succession of traditional practices related to sustainable land management is achieved across generations, such indigenous and local experiences have to be integrated into the national policies, strategies and plans so that they are as well recognized and enforced at the higher level.

4.5.2. Potential for Integration of Indigenous Institutions in Formal Policy Processes

Although the Tanzanian government has not established any specific policy which deals with traditional, local, or indigenous knowledge, considerable attempts have been made to integrate such knowledge systems by including policy issues in various sector policies and strategies. **Table 10** presents examples of the policy statements and issues that have been integrated in selected sector policies, strategies and/or regulations. As can be seen in this table there are several policies and policy issues that support the integration of traditional, local, or indigenous knowledge in respective sector activities, which have relevance to sustainable land management. These may be interpreted to form an important platform for ensuring that local natural resource management institutions are integrated in sector policies and strategies. A few examples are further elaborated in the following below.

National Environmental Policy (1997): The Government promulgated the National Environmental Policy in 1997, which among other things reflects on the outcome of the United Nations Conference on Environment and Development. It recognizes the Rio Declaration and Agenda 21. The policy reiterates that:

> "*In Agenda 21 the need to move from a development model in which sectors act independent of each other, to a model in which there is integration across sectors, where decisions take into account inter-sectoral effects, to improve inter-sectoral coordination. This involves the integration of policies, plans and programmes of interacting sectors and interest groups to balance long-term and short-term needs in environment and development.*"

Although the National Environmental Policy contains no specific statements on local, indigenous or traditional knowledge these issues are implied by the fact that there are relevant parts of the National Environmental Policy that indirectly deal with local, indigenous or traditional knowledge, especially those dealing with technology, biodiversity and public participation [47]-[55]. This is one of the reasons why it may be advisable in Tanzania to discuss the issue of local, traditional/indigenous knowledge in tandem with access to resource management issues.

Sustainable Industrial Development Policy 1996-2020: The Government in 1996 formulated the Sustainable Industrial Development Policy, 1996-2020. The Policy does not expressly local, indigenous or traditional knowledge, but is relevant in the discussion of these issues as it deals also with issues of intellectual property rights. The Policy states that:

> "*Tanzania has adequate intellectual property laws to regulate intellectual property. There are legislations to regulate of acquisition of patent rights in new inventions and innovations, and assurance of effective protection of all such patent rights, as well as to protect the right to use trade and service marks, the right to sue for infringement and pass-offs as well as legislations for copyrights and neighbouring rights. Tanzania being a signatory to the World Trade Organization will abide by the trend of protection within the Trade Related Aspects of Intellectual Property Rights*".

The mention of copyright and neighbouring rights, patents, trademarks and services covers knowledge as well

Table 10. Examples of the policy statements and issues integrating indigenous knowledge.

Policy/strategy	Statements/issues relevant to indigenous knowledge
National Environmental Policy of 1997	Reiterates the need for the integration of policies, plans and programmes of interacting sectors and interest groups to balance long-term and short-term needs in environment and development, including the need to improve inter-sectoral coordination.
Agriculture and Livestock Policy 2013	Section 3.1 Research and Development (p. 11) The policy recognized the importance of indigenous/traditional knowledge in relation to agricultural research and it states that: Indigenous knowledge shall be integrated into scientific research; Section 3.5 (Agricultural Extension Services, page 14) states that "The transformation of agricultural extension services is important in order to impart the right tools, knowledge and skills as well as ensuring farmers adhere to good agricultural practices", which implicitly addresses good indigenous practices. Section 3.26 Gender states that: 1) The Government shall facilitate equal access to land to both men and women; 2) Participation of men and women in decision making processes to improve their access to productive resources shall be enhanced; and 3) Awareness creation and sensitization of communities on negative cultural attitudes and practices shall be promoted in collaboration with the ministry responsible for gender.
The National Forest Policy of 1998	The National Forest Policy formulated in 1998 makes extensive reference to biological diversity issues. It mainly deals with flora and it has an express provision on local, indigenous and traditional knowledge in relation to sustainable forest management. The Policy states in Section 2.4 (p. 13) that: "That there have been inadequate consultations to encourage grassroots participation in forestry planning and the potential of indigenous knowledge has not been fully utilized. This is partly due to limited resources for participatory consultations." Such participatory consultations may in this case integrate local/indigenous institutions relevant to sustainable land management.
National Beekeeping Policy of 1998	The National Beekeeping Policy was issued at the same time with the National Forest Policy in 1998. However, the National Beekeeping Policy has no provisions recognizing the potential of indigenous knowledge in beekeeping management and the use of honey and its products, such as the use of honey in traditional medicine, knowledge that needs to be shared and protected [54].
Forest Regulations of 2004	The provisional Rule 52(23) stated issues relevant to indigenous knowledge on community intellectual property right, that: "The community intellectual property rights of the local communities, including traditional professional groups, particularly traditional practitioners, shall at all times remain inalienable, and shall be further protected under the mechanism established by the relevant law relating to intellectual property rights".
The National Wildlife Policy of 1998	The Wildlife Policy of Tanzania that was adopted in March 1998 emphasizes among other things the need for ensuring the participation of the people through community based natural resource management arrangements. Section 3.3.8 recognizing the intrinsic value of rural people, states that "enhancing the use of indigenous knowledge in the conservation and management of natural resources". page 18 (viii) Section 3.2.1 on wildlife protection "to transfer management of WMA to local community thus taking care of corridors, migratory routes and buffer zones and ensure that local community obtain substantial tangible benefits from wildlife conservation. Has a link with Article 8J on Convention on Biodiversity conservation. Under the part dealing with recognition of intrinsic value of wildlife to the rural people the policy states under Section 3.3.8.9 (viii) the need for: "enhancing the use of indigenous knowledge in the conservation and management of natural resources".
Cultural Policy of 1997	Section 3.5.1 States that: Traditional knowledge, skills and technology which are environmentally friendly shall be identified and their use encouraged. (pp. 9-10)
National Strategy for Growth and Reduction of Poverty (MKUKUTA), 2005, Cluster 1; Growth and Reduction of Income Poverty	Cluster strategy Section 2.4.5 (for operational target stated increased Agricultural growth from 5% in 2003 to 10% in 2010) (p. 6) states that: "Increase training and awareness creation on safe utilization and storage of agro-chemicals (including agriculture and livestock inputs, e.g. cattle dips), and the use of integrated pest control, eco-agricultural techniques, and use of traditional knowledge." (p. 6) Goal 4: cluster strategy item 4.7.1 states that: "Develop programmes for increasing local control and earnings in wildlife management areas, and establish locally managed natural resources funds, taping on local traditional knowledge." (p. 11) Goal 6. cluster strategy item 6.2.1 states that: "Develop and promote utilization of indigenous energy resources and diversification of energy sources." (p. 13)
National Fisheries Sector Policy and Strategy Framework of 1997	The National Fisheries Sector Policy and Strategy Statement recognize local/traditional knowledge in the part dealing with improved knowledge of the fisheries resources base. Section 3.3.2 (p. 9) of the policy shows that the fisheries sector is constrained with scanty information on traditional/local knowledge of fisheries resources (p. 3). Therefore the policy develops an objective (Section 3.3.2) to enhance knowledge of the fisheries resource base. One of the strategies to achieve this is to "Facilitate and promote acquisition and documentation of traditional fisheries knowledge." (p. 9).

Continued

The National Health Policy of 2003	Stated issues relevant to traditional knowledge, for instance: In Section 4.2 dealing with traditional medicine and alternative healing systems, the policy recognizes that "The role of traditional and alternative health care to Tanzanian people is significant. It is estimated that about 60% of the population use traditional and alternative care system for their day-to-day health care. Traditional and alternative healing services and conventional health services are complementary to each other". Policy objective 2.4.9 "Promote traditional medicine and alternative healing system and regulate the practice"). On research, Section 3.9.3 states: "Research in traditional medicines will focus on the identification of traditional remedies, screening of traditional herbal and medicinal materials and assessing the efficacy and safety of the products". Although the policy does not specifically mention traditional/indigenous knowledge, it states issues relevant to it, for instance by having statement supporting working together with the traditional healers, traditional nurses and recognising the importance of both traditional and alternative medicines

Source: Compiled from Field Work Data, November 2013 [47]-[54].

as its protection and application. However, taking into account that traditional intellectual property right systems have been problematic when dealing with local, traditional/indigenous knowledge, innovations and technologies under the Convention on Biological Diversity and trade related aspects of intellectual property rights [55].

The National Science and Technology Policy of Tanzania, 1996: The general objective of the National Science and Technology Policy of 1996 include the need to "Establish appropriate legal framework for the development and transfer of technology including intellectual property rights, monitoring and controlling of the choice and transfer of technology, as well as biosafety". Page 5 of this policy states that: "Thus one primary function of a National Science and Technology Policy is to establish relative priorities of programmes for generating new knowledge and to determine strategies for the application of science and technology for development." On environment—the major sectoral objectives include preservation of the biological diversity, cultural richness and natural beauty of Tanzania. Like many other policies this National Policy on Science and Technology does not expressly provide for local, traditional or indigenous knowledge, which may contribute immensely to development of science and technology at the local level.

As much as the policies recognize the usefulness and the potential of local, traditional or indigenous knowledge that is not followed up with sufficient strategies and activities to effectively share and protect it. There is a need to review the relevant policies and come out with strategies and action plans on the development and protection of local, traditional or indigenous knowledge. As noted in Section 4.3, some of the respondents in Kilimanjaro Region seem to be unaware of indigenous institutions for sustainable land management. While this raises some questions as to the sustainability of the institutions themselves, it also raises concerns on the future land resource management in the area. Inadequate recognition of indigenous institutions may mean that the national level institutions have become of greater influence at the local level, though they may lack the influence of traditional knowledge and practices such as taboos and local beliefs which were very effectively in the management of natural resources in Kilimanjaro as well as other parts of the country. As such the local experiences have to be integrated into the relevant national policies for enhanced sustainability of resource management.

5. Conclusions

Indigenous institutions have been very strong and effective throughout the historical times and they for generations managed to sustain the local resources in the Kilimanjaro Region. However, while it is acknowledged that traditional knowledge and indigenous institutions still exist, they are mainly restricted at the individual level, mainly because indigenous institutions that were instrumental in pursuing traditional knowledge are no longer in place, or not as strong as they used to be in the past. This has caused a decline in the use of traditional practices for sustainable land management. In addition, increasing modernization of agriculture and other land use and resource management systems have continued to weaken the role of the intergenerational experiences regarding traditional practices of land management. This is supported by the increasing number of people in the contemporary times that are not involved and/or are not aware of the existence of these traditional practices of sustainable land management. In particular, such experience may indicate that the traditional institutions have not adequately promoted themselves to capture the interest, attention and practice of all generations, possibly in favour of more modern and conventional practices. Regardless of all these constrains and the low rating of the current use of indigenous practices and institutions, field observation proved that there were methods of land management that were still trusted and specifically useful in land protection from various types of soil erosion, landslides as well as loss of fertility. For instance, the use of terraces in hilly areas of Mwanga and Same districts, strip terraces, ridges and growing of grasses and trees across slopes has remained useful practices in Moshi District Council, Siha, Hai and Rombo districts.

Although indigenous institutions have remained something of the past they are locally rated as a superior means of dealing with land and other natural resource management problems, especially in areas with greater land management challenges. Based on the dominance of modern institutions different stakeholders agree that there is a need to mainstream traditional practices into the existing structures and policies. There is currently little documentation of the integration of indigenous institutions in village by-laws but there has been a successful but rudimentary attempt, namely, the village land councils which combine both traditional and government leaders. Although there is little in terms of documented evidences of this unity, respondents suggest that formal and informal institutions can work in ways that complement each other by integrating the systems based on cultural values and beliefs with current modernity. This can be achieved by having in places well established policy frameworks that address traditional/indigenous knowledge and institutions. The presence of various sectoral policy statements in support of indigenous practices provides an eminent stepping stone towards preserving indigenous knowledge but the gap exists because there is no platform for a comprehensive promotion of these practices.

Acknowledgements

This paper is based on a study commissioned by Kilimanjaro Regional Office and UNDP/GEF. We are grateful to them for funding this study. We are specifically thankful to the Kilimanjaro Regional Administrative Secretary's office and the Sustainable Land Management project office in Moshi for all the logistical support throughout the study. Special appreciations are due to Eng. Alfred I. Shayo, Acting Regional administrative Secretary, Mr. Paulo Shayo, the chairman of the Regional Sustainable Land Management project Technical Team, Dr. Francis Mkanda, the Project Technical Advisor and Mr. Damas Masologo, the Acting National Sustainable Land Management Project Coordinator for facilitating the implementation of this study.

References

[1]   Agrawal, A. (1995) Dismantling the Divide between Indigenous and Scientific Knowledge. *Development and Change*, **26**, 413-439. http://dx.doi.org/10.1111/j.1467-7660.1995.tb00560.x

[2]   Briggs, J. and Sharp, J. (2004) Indigenous Knowledge and Development: A Postcolonial Caution. *Third World Quarterly*, **25**, 661-676. http://dx.doi.org/10.1080/01436590410001678915

[3]   Fabricius, C., Folke, C., Cundill, G. and Schultz, L. (2007) Powerless Spectators, Coping Actors and Adaptive Co-Managers: A Synthesis of the Role of Communities in Ecosystem Management. *Ecology and Society*, **12**, 29. http://www.ecologyandsociety.org/vol12/iss1/art29

[4]   Yamia, M., Vogl, C. and Hausera, M. (2009) Comparing the Effectiveness of Informal and Formal Institutions in Sustainable Common Pool Resources Management in Sub-Saharan Africa. *Conservation and Society*, **7**, 153-164. http://dx.doi.org/10.4103/0972-4923.64731

[5]   Van Vlaenderen, H. (2000) Local Knowledge: What Is It and Why and How Do We Capture It? National Workshop on Gender, Biodiversity and Local Knowledge Systems Links to Strengthen Agricultural and Rural Development, Morogoro, FAO.

[6]   Whyte, P. (2003) On the Role of Traditional Ecological Knowledge as a Collaborative Concept: A Philosophical Study. *Ecological Processes*, **2**, 1-12.

[7]   Berkes, F. (2008) Sacred Ecology. Routledge, Manitoba, 7.

[8]   Kajembe, G.C., Malimbwi, R.E., Zahabu, E. and Luoga, E.J. (2002) Contribution of Charcoal Extraction to Deforestation: Experience from CHAPOSA. Research Report.

[9]   Boonto, S. (1993) Indigenous Knowledge and Sustainable Development. *International Institute of Rural Reconstruction Symposium Proceedings*, Cavite.

[10] Luoga, E.J. (1994) Indigenous Knowledge and Sustainable Management of Forest Resources in Tanzania. *Proceedings of the Workshop on Information for Sustainable Natural Resources of Eastern, Central and Southern Africa*, Arusha, 4-9 November 1994, 139-148.

[11] Luoga, E.J., Witkowski, E.T.F. and Balkwill, K. (2000) Differential Utilization and Ethnobotany of Trees in Kitulangalo Forest Reserve and Surrounding Communal Lands of Eastern Tanzania. *Economic Botany*, **54**, 328-343. http://dx.doi.org/10.1007/BF02864785

[12] Berkes, F. and Folke, C. (1998) Linking Social and Ecological Systems: Management Practices and Social Mechanisms for Building Resilience. Cambridge University Press, Cambridge.

[13] Kweka, D. (2004) The Role of Local Knowledge and Institutions in the Conservation of Forest Resources in the Eastern Usambara. UNESCO-Man and Biosphere, Dar es Salaam.

[14] Mowo, J., Adimassu, Z., Masuki, K., Lyamchai, C., Tanui, J. and Catacutan, D. (2011) The Importance of Local Traditional Institutions in the Management of Natural Resources in the Highlands of Eastern Africa. Working Paper No. 134, World Agroforestry Centre, Nairobi. http://www.dx.doi.org/10.5716/WP11085.PDF

[15] World Bank (1998) Indigenous Knowledge for Development: A Framework for Action. World Bank, Washington DC.

[16] Ashenafi, Z. and Leader-Williams, N. (2005) Indigenous Common Property Resource Management in the Central Highlands of Ethiopia. *Human Ecology*, **33**, 539-563. http://dx.doi.org/10.1007/s10745-005-5159-9

[17] Global Strategic Plan for Biodiversity 2011-2020. http://www.cbd.int/decision/cop/default

[18] Kiwanuka, S. (1970) Colonial Policies and Administration in Africa: The Myths of the Contract. *African Historical Studies*, **3**, 295-315. http://dx.doi.org/10.2307/216218

[19] Schneider, L. (2004) Freedom and Unfreedom in Rural Development: Julius Nyerere, UjamaaVijijini and Villagization. *Canadian Journal of African Studies*, **38**, 344-392. http://dx.doi.org/10.2307/4107304

[20] Kisangani, E. (2009) Development of African Administration: Pre-Colonial Times and Since. *Administration and Public Policy*, **1**, 1-7.

[21] Shivji, I. (2012) Nationalism and Pan-Africanism: Decisive Moments in Nyerere's Intellectual and Political Thought. *Review of African Political Economy*, **39**, 103-116. http://dx.doi.org/10.1080/03056244.2012.662387

[22] Liviga, A. (1992) Local Government in Tanzania: Partner in Development or Administrative Agent of the Central Government? *Local Government Studies*, **18**, 208-225. http://dx.doi.org/10.1080/03003939208433639

[23] Kajembe, G.C., Monela, G.C. and Mvena, Z.S.K. (2003) Making Community-Based Forest Management Work: A Case Study from Duru-Haitemba Village Forest Reserve, Babati, Tanzania. In: Kowero, G., Campbell, B.M. and Sumaila, U.R., Eds., *Policies and Governance Structures in Woodlands of Southern Africa*, Center for International Forestry Research, Jakarta, 16-27.

[24] Ylhäisi, J. (2004) Indigenous Forests Fragmentation and the Significance of Ethnic Forests for Conservation in the North Pare, the Eastern Arc Mountains, Tanzania. *Fennia*, **182**, 109-132.

[25] Maro, P. (1988) Agricultural Land Management under Population Pressure: The Kilimanjaro Experience, Tanzania. *Mountain Research and Development*, **8**, 273-282. http://dx.doi.org/10.2307/3673548

[26] Ikegami, K. (1994) The Traditional Agrosilvipastoral Complex System in the Kilimanjaro Region, and Its Implications for the Japanese-Assisted Lower Moshi Irrigation Project. *African Study Monographs*, **15**, 189-209.

[27] Grove, A. (1993) Water Use by the Chagga on Kilimanjaro. *African Affairs*, **92**, 431-448.

[28] Gillingham, E. (1999) Gaining Access to Water: Formal and Working Rules of Indigenous Irrigation Management on Mount Kilimanjaro, Tanzania. *Natural Resources Journal*, **39**, 419-441.

[29] Tagseth, M. (2008) Oral History and the Development of Indigenous Irrigation: Methods and Examples from Kilimanjaro, Tanzania. *Norwegian Journal of Geography*, **62**, 9-22.

[30] Adams, W., Brun, C. and Havnevik, K. (2010) Studies of the Waterscape of Kilimanjaro, Tanzania: Water Management in Hill Furrow Irrigation. *Journal of Geography*, **64**, 172-173.

[31] Ylhäisi, J. (2006) Traditionally Protected Forests and Sacred Forests of Zigua and Gweno Ethnic Groups in Tanzania. Doctoral Dissertation, Helsingin Yliopiston Maantieteen Laitoksen Julkaisuja, Helsingin.

[32] Mgumia, H. and Oba, G. (2003) Potential Role of Sacred Groves in Biodiversity Conservation in Tanzania. *Environmental Conservation*, **30**, 259-265. http://dx.doi.org/10.1017/S0376892903000250

[33] Msuya, S. and Kideghesho, J. (2009) The Role of Traditional Management Practices in Enhancing Sustainable Use and Conservation of Medicinal Plants in West Usambara Mountains, Tanzania. *Tropical Conservation Science*, **2**, 88-105.

[34] Munger, S. (1952) African Coffee on Kilimanjaro: A Chagga Kihamba. *Economic Geography*, **28**, 181-185. http://dx.doi.org/10.2307/141027

[35] Tagseth, M. (2006) The "Mfongo" Irrigation Systems on the Slopes of Mt. Kilimanjaro, Tanzania. In: Tvedt, T. and Jakonsson, E., Eds., *A History of Water Volume I: Water Control and River Biographies*, I.B. Tauris, London, 488-506.

[36] Sheridan, M. (2002) An Irrigation Intake Is Like a Uterus: Culture and Agriculture in Pre-Colonial North Pare, Tanzania. *American Anthropologist*, **104**, 79-92. http://dx.doi.org/10.1525/aa.2002.104.1.79

[37] Baumann, O. (1891) Usambara und seine Nachbargebiete. Dietrich Reimer, Berlin.

[38] United Republic of Tanzania, URT (2013) Population and Housing Census: Population Distribution by Administrative

Areas. National Bureau of Statistics, Ministry of Finance, Dar es Salaam and Office of Chief Government Statistician, President's Office, Finance, Economy and Development Planning, Zanzibar.

[39] Cochran, W.G. (1977) Sampling Techniques. 3rd Edition, John Wiley & Sons, New York.

[40] Magigi, W. and Sathiel, A. (2014) Gender Consideration in Sustainable Land Management Project Activities on the Highlands of Kilimanjaro Region: Lessons and Future Outlook. *Open Journal of Soil Science*, **4**, 185-205. http://dx.doi.org/10.4236/ojss.2014.45022

[41] O'kting'ati, A., Maghembe, J.A., Fernandes, E.C.M. and Weaver, G.H. (1985) Plant Species in the Kilimanjaro Agroforestry System. *Agroforestry Systems*, **2**, 177-186. http://dx.doi.org/10.1007/BF00147032

[42] Kangalawe, R.Y.M. (2001) Changing Land-Use Patterns in the Irangi Hills, Central Tanzania: A Study of Soil Degradation and Adaptive Farming Strategies. Ph.D. Dissertation, Department of Physical Geography and Quaternary Geology, Stockholm University, Stockholm.

[43] Kangalawe, R.Y.M. (2014) Nutrient Budget Analysis under Smallholder Farming Systems and Implications on Agricultural Sustainability in Degraded Environments of Semiarid Central Tanzania. *Journal of Soil Science and Environmental Management*, **6**, 44-60. http://dx.doi.org/10.5897/JSSEM13.0390

[44] Kajembe, C., Mwaipopo, S., Mvena, K. and Monela, G. (2002) The Role of Traditional Leadership, Institutions and Ecological Knowledge in Sustainable Management of Miombo Woodlands in Handeni District, Tanzania. Policies, Governance and Harvesting Miombo Woodlands, Harare.

[45] Kajembe, G.C., Mwaipopo, S. and Kijazi, M. (2003) The Role of Traditional Institutions in the Conservation of Forest Resources in East Usambara, Tanzania. *International Journal of Sustainable Development and World Ecology*, **10**, 101-107. http://dx.doi.org/10.1080/13504500309469789

[46] Data Collected from In-Depth Interviews with Elderly Person in Rau Ward. Moshi Municipality, Kilimanjaro Region, Tanzania.

[47] United Republic of Tanzania, URT (2013) Agriculture Policy. Ministry of Agriculture, Food Security and Cooperatives, Dar es Salaam.

[48] United Republic of Tanzania, URT (1997) Cultural Policy: Policy Statement of 1997. Dar es Salaam.

[49] United Republic of Tanzania, URT (1997) National Fisheries Sector Policy and Strategy Framework. Ministry of Natural Resources and Tourism, Dar es Salaam.

[50] United Republic of Tanzania, URT (2003) National Health Policy. Ministry of Health and Social Welfare, Dar es Salaam.

[51] United Republic of Tanzania, URT (1998) The National Forest Policy. Ministry of Natural Resources and Tourism, Dar es Salaam.

[52] United Republic of Tanzania, URT (1998) National Beekeeping Policy. Ministry of Natural Resources and Tourism, Dar es Salaam.

[53] United Republic of Tanzania, URT (2004) Forest Regulations of 2004, the Government Notice (GN) No. 153. Ministry of Natural Resources and Tourism, Dar es Salaam.

[54] United Republic of Tanzania, URT (2005) National Strategy for Growth and Reduction of Poverty (MKUKUTA). Government Printer, Dar es Salaam.

[55] Kabudi, P.J. (2003) Benefits and Risks of Sharing Local and Indigenous Knowledge in Tanzania: The Legal Aspects and Challenges. Report No. 5, Links Project: Gender, Biodiversity and Local Knowledge Systems for Food Security, FAO, Rome.

13

Soil Nutrients, Landscape Age, and *Sphagno-Eriophoretum vaginati* Plant Communities in Arctic Moist-Acidic Tundra Landscapes

Joel A. Mercado-Díaz*, William A. Gould, Grizelle González

USDA Forest Service, International Institute of Tropical Forestry, Jardín Botánico Sur, 1201 Calle Ceiba, Río Piedras, Puerto Rico
Email: *joel_pr19@hotmail.com

Abstract

Most research exploring the relationship between soil chemistry and vegetation in Alaskan Arctic tundra landscapes has focused on describing differences in soil elemental concentrations (e.g. C, N and P) of areas with contrasting vegetation types or landscape age. In this work we assess the effect of landscape age on physico-chemical parameters in organic and mineral soils from two long-term research sites in northern Alaska, the Toolik Lake and Imnavait grids. These two sites have contrasting landscape age but similar vegetation composition. We also used correlation analysis to evaluate if differences in any of these parameters were linked with between-site variation in the abundance of growth forms. Our analysis was narrowed to soils in *Sphagno-Eriophoretum vaginati* plant communities. We found no significant differences between these sites for most parameters evaluated, except for total Ca which was significantly higher in organic soils from Imnavait vs. Toolik and total Na which was significantly higher in mineral horizons from Toolik compared to Imnavait. Moreover, the abundance of non-*Sphagnum* mosses was positively correlated with total Ca in organic soils, whereas the abundance of forbs, non-*Sphagnum* mosses and bryophytes was negatively correlated with total Na in mineral soils. We suggest that differences in the concentration of these two elements are most likely tied to landscape age differences between these sites. However, since observed dissimilarity in terms of total Ca in organic soils and total Na in mineral soils is concordant with correlation patterns observed between these elements and the aforementioned growth forms, it is likely that existing differences in vegetation composition between these sites are also influencing the concentration of these elements in soils, particularly that of Ca, since non-*Sphagnum* mosses are dominant above organic soils and are therefore expected to significantly influence biogeochemical processes at this horizon. Thus, we conclude that

---

*Corresponding author.

except for organic Ca and mineral Na, there is little difference between these sites in terms of their soil physico-chemical properties. We suggest that most of the influence of landscape age on evaluated parameters is masked by factors such as moderate cryoturbation and similarities in terms of vegetation properties and climate. These observations are relevant as they suggest a linkage between soil chemistry and vegetation composition in this tundra region.

Keywords

Arctic Alaska, Soil Nutrients, Moist-Acidic Tussock Tundra, Vegetation

---

1. Introduction

Studies have shown that landscape age (time since deglaciation) is an important factor influencing essential biogeochemical processes in Arctic tundra soils [1]-[6]. For instance, Hobbie and Gough [7] compared foliar and soil nutrients between landscapes that were deglaciated > 50,000 (moist-acidic tundra [MAT]) and less than 11,500 years ago (moist non-acidic tundra [MNT]) and found higher rates of net N mineralization, cation exchange capacity and exchangeable base cations in soils at the geologically older site. Contrasting landscape age has also been proposed as a potential factor explaining variation in litter decomposition rates [8] and the rate of processes like C and N cycling in tundra soils [9].

In the Alaskan Arctic, differences in landscape age are also intimately related to differences in soil acidity [2] [3]. A number of important biological attributes are influenced by gradients of soil pH in this region. For example, some studies have documented the relationship between contrasting vegetation types with a distinct pH boundary that separates MAT and MNT [7] [9]. Considerable variation in species and growth form dominance within some tussock tundra communities has been associated with variation in soil pH [10] [11]. Differences in soil pH can also affect specific plant community attributes like vascular-plant species richness [12]. More recently, Eskelinen *et al.* [13] proposed an indirect effect of soil pH on vegetation via the evolution of bacteria-based microbial communities in alkaline soils where the properties of forb-produced organic matter were possibly sustaining the prevalence of soil bacteria.

Vegetation can also influence important soil processes like nutrient cycling which regulate nutrient concentrations and the size of C and N pools in tundra soils [14]. Since vegetation effects on soils are thought to be primarily related to the accumulation of organic material and nutrients [15], variation in specific traits influenced by plant communities, such as litter chemistry, is believed to play an important role in some of these processes. For instance, altered litter quality resulting from changes in species composition is known to affect processes like soil N mineralization [16]. However, some evidence appears to indicate that other landscape-scale soil processes are less likely to be significantly affected by vegetation. For example, Hobbie and Gough [8] demonstrated that variation in plant species composition did not account for differences in litter decomposition between MAT and MNT.

The linkage between patterns in vegetation composition and soils in tundra landscapes has not been characterized thoroughly, although several studies have contributed significantly in this direction. Chu and Grogan [17] found that variation in total C and N, Dissolved Organic Carbon (DOC) and N (DON), mineral N and N mineralization potential in organic soils in Daring Lake, Canada has been directly related to differences in vegetation types. Likewise, Eskelinen *et al.* [13] observed that forb-rich non-acidic heaths were associated with low C:N and low soluble N:phenolics ratios in soils, whereas shrub-dominated acidic heaths were associated with high values of these ratios. Understanding the reciprocal relationship between vegetation and soil dynamics in Arctic tundra ecosystems is challenging mostly because vegetation is ultimately controlled by meso-topographic relationships (slope position and soil moisture), micro-scale disturbances and factors related to long-term landscape evolution [4] [18]. This suggests that plant communities exhibit considerable spatial variability along soil chemical gradients.

In this study we compare a number of soil physico-chemical parameters from organic and mineral horizons of two study sites in northern Alaska that differ in landscape age, *i.e.*, time since deglaciation 55 k and 125 k years, but have similar vegetation composition [18]. These are long-term vegetation monitoring areas which have been intensively used since the late 1980s to study the effects of climate change on tundra vegetation [19]-[21]. Our analysis is focused on soils derived from areas dominated by *Sphagno-Eriophoretum vaginati* plant communities' sensu Walker *et al.* [18], the most ubiquitous plant community within these sites. Little information on soil chemistry has been published from soils within these long-term monitoring sites. Only recently the works of Whittinghill and Hobbie [5] [6] and Keller *et al.* [23] have shown that the two landscape ages represented by these two study sites are similar in terms of several soil chemical parameters. The present work aims to advance the understanding of soil processes in this region by including analyses on physico-chemical properties of both mineral and organic horizons and other aspects of soil chemistry that are not necessarily evaluated in detail by those studies.

We also assess between-site differences in the abundance of a number of growth forms and subsequently evaluate the relationship between those differences and the variation between sites in terms of particular physico-chemical parameters to make inferences regarding how vegetation composition may be affecting soil physico-chemical properties at these two sites, and vice versa. These observations will contribute to a better understanding on how vegetation composition, landscape age and soil chemistry are reciprocally linked in this region.

2. Materials and Methods

2.1. Study Area

This study was carried out in two 1 km^2 research grids, Toolik Lake and Imnavait Creek, in the vicinity of the Toolik Lake Field Station, Alaska, located north of the Brooks Range in the Southern Foothills Physiographic Province of the Alaskan North Slope [24] [25] (**Figure 1**). The region is underlain by continuous permafrost which is 250 - 300 m thick [26]. Both grids are dominated by MAT vegetation; however their landscapes are slightly different due primarily to differences in glacial age [18] [27] [28]. The Imnavait grid is in the headwaters of the Imnavait Creek, a small beaded tributary of the Kuparuk River basin [18]. This site lies on the Sagavanirktok (Middle Pleistocene) glacial drift which deglaciated about 125,000 years ago [27] [28]. Topography is

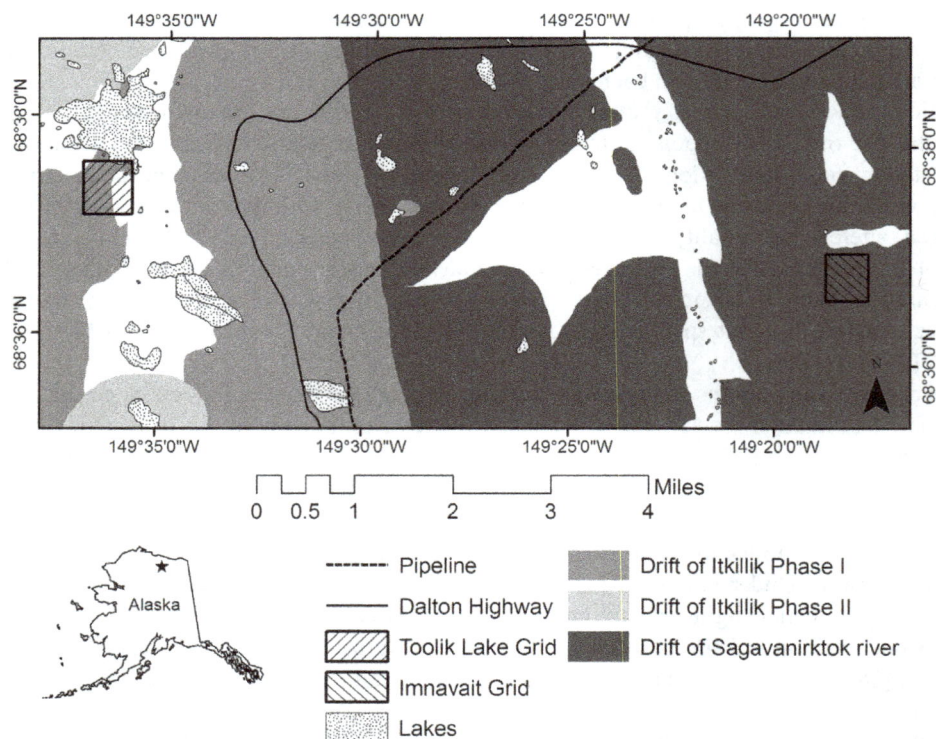

Figure 1. Landscape ages and the location of the Toolik Lake and Imnavait Creek 1 km^2 grids in the Upper Kuparuk River region in Northern Alaska.

dominated by gently rolling hills and elevation within the watershed varies from about 770 to 980 m [22]. Most of the Toolik Lake grid lies in a younger substrate (Itkillik I glacial drift) which deglaciated during the late Pleistocene (ca. 55,000 years), but includes areas of Itkillik II outwash towards the east [27] [28]. The landscape at this site is more heterogeneous than at Imnavait and is dominated by small glacial lakes, kames and moraines. Elevations range from 670 to 850 m [18].

Both grids share similar meteorological conditions due to their close proximity (<12 km). Mean annual surface atmospheric temperature (SAT) from 1989-2008 have been −8.5°C; whereas average annual precipitation during the same period was 312 mm [29]. Linear trend analysis of mean annual SAT and mean annual precipitation revealed no trends in these parameters suggesting that they have remained stable over the last two decades [29].

2.2. Study Design

There are 157 permanent 1m^2 vegetation plots within these grids, 72 at Imnavait and 85 at Toolik. Plots are located equidistantly at 100 meters from each other within the grids [19]. These are non-manipulative plots that are currently being studied for analyzing long term changes in vegetation composition and structure in this region [30]. We limited our analysis to a subset of these plots that are established in areas with *Sphagno-Eriophoretum vaginati* plant communities. This plant community has a broad spatial extent which allows us to extrapolate results to a landscape-scale level. Likewise, confounding effects that may result from grid-scale variation in plant communities could be reduced by focusing on a single plant community. The *Sphagno-Eriophoretum vaginati* plant community is the most dominant plant community within the grids and is considered the zonal vegetation of mesic slopes throughout the Arctic Foothills [18]. It typically occurs on ice-rich sediments with shallow active layers and low soil pH [31]. The most conspicuous plant species is the tussock-forming sedge *Eriophorum vaginatum* which dominates particularly in stable hillslope shoulders and upper backslopes; whereas shrubs like *Betula nana* and *Salix pulchra* tend to become dominant on footslopes and associate with deep *Sphagnum* spp. mats, other mosses like *Aulacomnium turgidum*, *Hylocomnium splendens* and lichens like *Peltigera aphtosa* and *Cladonia* spp. [18] [22] [32]. *Sphagno-Eriophoretum vaginati* plant communities' sensu Walker *et al.* [18] are contained within the *Moist-tussock sedge, dwarf shrub, moss tundra* physiognomic unit [24] and mostly coincide with *Eriophorum vaginatum-Sphagnum* spp. plant communities' sensu Walker *et al.* [22].

Plant communities within the Toolik and Imnavait grids were identified using geographical layers prepared by Walker [33] and published maps of the region [24]. There are 33 vegetation types represented in the Upper Kuparuk river region [22] [31]. Walker *et al.* [18] classified vegetation of both Toolik and Imnavait grids into five associations and 15 community types. We performed an overlay analysis using a geographic information system (GIS) and selected 24 plots (12 at Toolik and 12 at Imnavait) representing *Sphagno-Eriophoretum vaginati* plant communities. We extracted growth form abundance data from selected plots from a long-term plant community dataset. This data corresponded to vegetation sampling realized in 2007 (Imnavait) and 2008 (Toolik) [34].

2.3. Sampling and Nutrient Analysis

Soil samples were collected 1 - 2 meters apart from selected plots in areas with visually similar vegetation composition. Samples of organic and mineral soils were collected near each plot following the procedures described below. Mineral soils near two of the 24 plots (both at Toolik) were frozen at the time of collection and not sampled. Soil sampling was realized during August, 2008.

A shovel was used to create a 25 × 25 cm pit to collect soil samples in each selected plot. Pits reached an approximate depth of 20 cm, or deeper until the upper 5 - 7 cm portion of the mineral horizon was exposed. Three samples of mineral soils (n = 3) were taken at each pit by pushing a stainless steel soil core with plastic core inserts (aprox. 98 cc.) horizontally into the soil profile. These samples were used individually to calculate soil moisture, bulk density and elemental concentrations in this horizon. Mineral soil moisture was calculated after oven drying samples at 105°C for 48 hours in laboratory facilities at Toolik Lake Field Station. Depending on soil conditions, each organic soil sample (n = 1) was extracted horizontally at the soil profile or vertically using a small bread knife and were then placed into labeled cloth bags. When samples were collected vertically, we removed the upper layer of soil where live material was evident and the lower layer where the organic layer diffuses into the mineral soil layer. Each organic soil sample had approximately twice the volume of mineral soil samples. All samples (except those used for estimating mineral soil moisture) were sent to the International Institute of Tropical Forestry Soils Laboratory (USDA-USFS) in San Juan, Puerto Rico and were processed within 8 days. Mineral soil samples for elemental concentration and bulk density determinations were oven dried at 40°C for two weeks, whereas those from organic soils were air dried over the same period. We used a Foss Tecator Cyclotec (model 1093) mill to ground the samples and then passed them through a 1 mm stainless steel sieve. Roots of considerable size and other live material were excluded from the samples.

Total Nitrogen (N) and Total Carbon (C) were analyzed using the dry combustion method by means of a LECO TruSpec CN Analyzer [35]. The procedure used is a modified version of the Organic Application Note titled "Carbon and Nitrogen in Soil and Sediment" obtained from LECO Corp. [36]. The dry combustion method was also used to determine Total Sulfur (S), utilizing the LECO TruSpec (Add-On Module) S Analyzer [37]. The procedure used is from LECO [38] and is titled "Sulfur in Cement, Fly Ash, Limestone, Soil, and Ore". In the dry combustion method a small weighted sample is combusted in a high temperature furnace (950°C for the LECO TruSpec CN Analyzer and 1450°C for the LECO TruSpec S Analyzer) and in a stream of purified oxygen. Total Carbon is measured as CO_2 by an infrared detector and Total Nitrogen is determined as N_2 by a thermal conductivity cell detector. Total Sulfur as SO_2 is also detected by an infrared.

Ground material was analyzed for elemental concentrations of Phosphorus (P), Aluminum (Al), Potassium (K), Calcium (Ca), Magnesium (Mg), Manganese (Mn), Sodium (Na) and Iron (Fe). Soil samples were digested with concentrated HNO_3, 30% H_2O_2 and concentrated HCl using a modified version of the method recommended by Luh Huang and Schulte [39] and analyzed by means of a Spectro Plasma Emissions Spectrometer (Spectro Ciros ICP).

Subsamples were oven dried at 105°C for 24 hrs. A moisture factor was calculated and applied to each analysis [40]. These same subsamples were later ignited at 490°C in a muffle furnace (for at least 8 hrs.) to determine Loss-on-Ignition (LOI). Soil pH was measured in a 1:1 (soil:water) solution using an Orion Ionanalyzer Model 901 with a combination pH electrode [41].

Soil moisture of mineral and organic soil samples was determined by calculating the percentage weight loss after drying. Organic horizon thickness was averaged for each plot and was calculated using three different measurements of the organic horizon width at randomly selected areas of each pit.

2.4. Statistical Analysis

We analyzed and treated as independent variables each physico-chemical parameter and the abundance of particular growth forms. We used Exploratory Data Analysis (EDA) to corroborate if these variables conformed to parametric testing assumptions. We assessed normality using the Shapiro-Wilk test along with Normal Q-Q plots. Except for "% Moisture" which was analyzed using non-parametric Mann-Whitney U-test, results from EDA supported the use of Independent Samples T-Test for most variables. Variables that initially failed EDA tests were mostly affected by few extreme outliers and behave normally after their exclusion.

We used Pearson's correlation coefficient (r) to evaluate how physico-chemical parameters that were significantly different between sites relate to growth forms that were also found to have significantly different abundances at each site. We combined data from both sites that corresponded to these two variables. This resulted in 22 data points for correlation analysis. Correlation tests were performed after absolute abundances values in the dataset were converted to percent cover of growth forms at each plot.

3. Results

We found no significant differences between Imnavait and Toolik in terms of the thickness of the organic horizon and the bulk density of the mineral horizon (**Table 1**). Soils at both sites were similarly acidic whereas comparable LOI values are indicative of similar organic matter content. Only percent moisture was significantly higher in mineral soils from Imnavait vs. Toolik. Nonetheless, higher moisture recorded at Imnavait is possibly linked to higher precipitation activity that was observed at this site during sampling.

There were no differences in the concentration of C, N and S between both sites for both organic and mineral horizons (**Figure 2(a)**, **Figure 2(b)** and **Table 1**). There were also no significant differences between sites for the concentration of other elements, except for Ca which was significantly higher in organic horizons of the older Imnavait site (M = 5.56, SD = 4.21) compared to the younger Toolik site (M = 2.74, SD = 1.97), t (21) = 2.09, p < 0.05 (**Figure 3(a)**) and Na which was significantly higher in mineral soils of Toolik (M = 0.27, SD = 0.07)

Table 1. Physical and chemical properties of mineral and organic soils from the *Sphagno-Eriophoretum vaginati* plant communities in Toolik Lake and Imnavait Creek grids. Between-site comparisons were assessed with Independent Samples T-Test unless otherwise noted. Sample size within parentheses unless otherwise indicated with superscripts. SD = Standard deviation. Statistical significance achieved at p < 0.05.

| Variables | Mineral | | | | | Organic | | | | |
| | Imnavait (n = 12) | | Toolik (n = 10) | | p-values | Imnavait (n = 12) | | Toolik (n = 12) | | p-values |
	Mean	SD	Mean	SD		Mean	SD	Mean	SD	
Total elements										
C (%)	3.41[a]	2.13	4.01[c]	1.84	ns	34.85	7.36	29.15	10.74	ns
N (%)	0.14[a]	0.12	0.15[c]	0.10	ns	1.55	0.34	1.24	0.47	ns
S (%)	0.02[a]	0.01	0.02[c]	0.01	ns	0.19	0.05	0.17	0.08	ns
Al (mg/g)	12.18[a]	3.55	11.88	2.75	ns	8.57	4.78	8.70	3.09	ns
P (mg/g)	0.45	0.19	0.45[c]	0.18	ns	1.51	0.44	1.37	0.67	ns
Na (mg/g)	0.17[b]	0.10	0.27	0.07	p < 0.05	1.49	0.87	1.26	0.55	ns
Mn (mg/g)	0.30[a]	0.20	0.26	0.20	ns	6.70	7.80	4.81[a]	7.48	ns
Ca (mg/g)	0.79[a]	0.30	0.86	0.37	ns	5.56[a]	4.21	2.74	1.97	p < 0.05
Fe (mg/g)	27.20	7.82	27.51	8.26	ns	25.36[a]	12.64	22.32	12.27	ns
Mg (mg/g)	2.48[a]	0.84	2.14	0.28	ns	1.38[a]	0.33	1.14	0.53	ns
K (mg/g)	0.73[a]	0.27	0.79	0.20	ns	0.84[a]	0.22	0.78	0.26	ns
Organic horizon thickness (cm; n = 3)	-	-	-	-	-	13.04	1.22	10.63[c]	1.88	ns
Bulk density (g/cc)	1.25	0.14	1.46	0.12	ns	-	-	-	-	-
% Moisture	30.34	19.02	12.65	15.84	p < 0.05[*]	-	-	-	-	-
LOI (%)	10.52	6.18	9.88[c]	3.48	ns	68.64	12.95	58.23	19.74	ns
pH (H$_2$O)	4.53	0.12	4.41	0.19	ns	4.80	0.72	4.34	0.53	ns

[a] n = 11; [b] n = 10; [c] n = 9; [*] Mann-Whitney U-test.

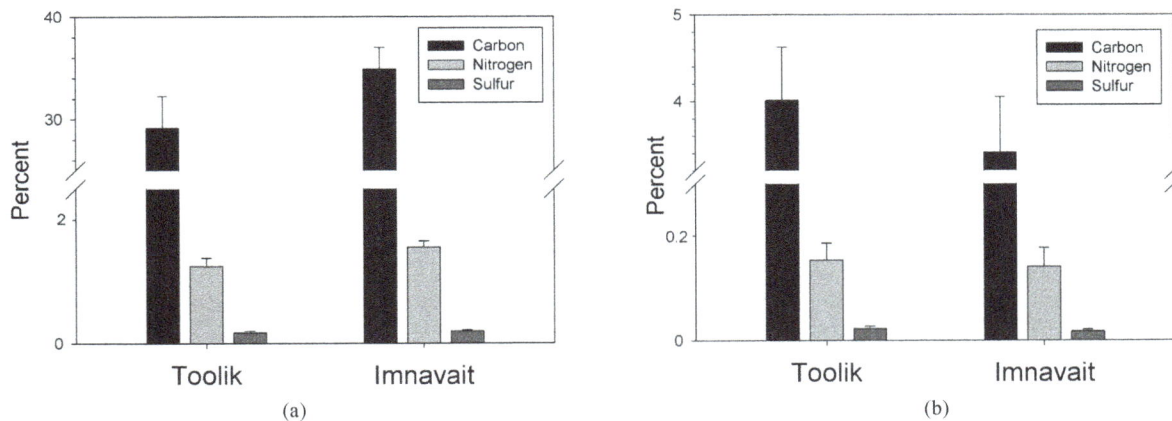

Figure 2. Differences between Toolik Lake and Imnavait Creek grids in terms of mean total concentration (%) of C, N and S in (a) organic horizons and (b) mineral horizons.

compared to Imnavait (M = 0.17, SD = 0.10), t (20) = 2.75, p < 0.05 (**Figure 3(b)** and **Table 1**).

We found significant between-site differences in the abundances of several growth forms (**Table 2** and **Figure 4**). The abundance of shrubs was significantly higher at Toolik (M = 32.25, SD = 5.36) vs. Imnavait (M = 26.51, SD = 7.04); t (22) = −2.25, p < 0.05. In contrast, forbs were significantly more abundant at Imnavait (M = 5.99, SD = 5.52) vs. Toolik (M = 1.60, SD = 1.69); t (22) = 2.64, p < 0.05. Similarly, non-*Sphagnum* mosses at Imnavait (M = 27.42, SD = 4.92) were more abundant than at Toolik (M = 10.60, SD = 5.31); t (22) = 8.05, p < 0.00,

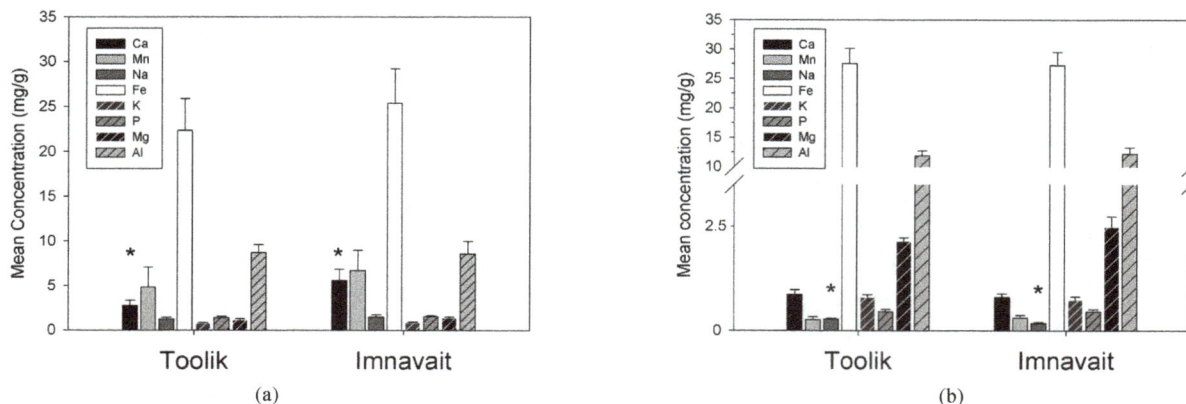

(a)   (b)

Figure 3. Differences between Toolik Lake and Imnavait Creek grids in terms of mean total concentration (mg/g) of Ca, Mn, Na, Fe, K, P, Mg and Al in (a) organic horizons and (b) mineral horizons. Asterisks indicate significant differences ($p < 0.05$).

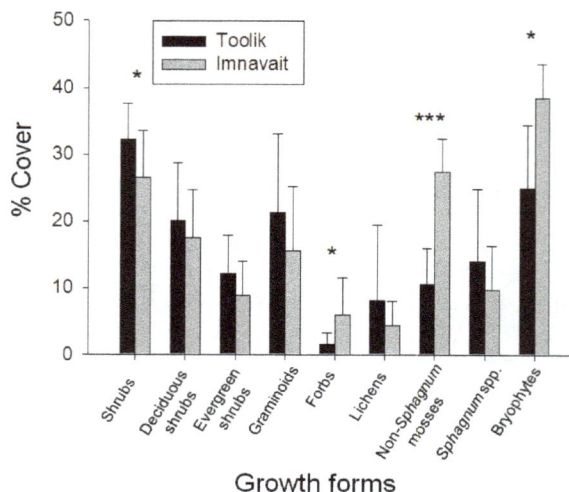

Growth forms

Figure 4. Differences between Toolik Lake and Imnavait Creek grids in terms of mean abundance of growth forms (% Cover). Abundance data correspond to years 2007-2008. Asterisks indicate significant differences ($p < 0.05$).

Table 2. Abundances (% Cover) of main growth forms in 24 one meter squared plots with vegetation classified as *Sphagno-Eriophore- tum vaginati* plant community. Abundance data correspond to years 2007-2008. Between-site comparisons were assessed with Independent Samples T-Test. Sample size within parentheses unless otherwise indicated with superscripts. SD = Standard deviation. Statistical significance achieved at $p < 0.05$.

Type	Growth form	Toolik		Imnavait		p-values
		Mean (n = 12)	SD	Mean (n = 12)	SD	
Vascular	Shrubs	32.25	5.36	26.51	7.04	<0.05
	Deciduous shrubs	20.12	8.61	17.56	7.21	ns
	Evergreen shrubs	12.13	5.77	8.94	5.06	ns
	Graminoids	21.31	11.77	15.63	9.52	ns
	Forbs	1.60	1.69	5.99	5.52	<0.05
Non-vascular	Lichens	8.14	11.39	4.44	3.59	ns
	Non-*Sphagnum* mosses	10.60	5.31	27.42	4.92	<0.001
	Sphagnum spp.	13.96	10.93	9.68	6.65	ns
	Bryophytes	24.99	9.45	38.50	5.07	<0.05

and bryophytes more abundant at Imnavait (M = 38.50, SD = 5.07) than at Toolik (M = 24.99, SD = 9.45); t (22) = 4.36, p < 0.05.

Because Ca in organic soils and Na in mineral soils were the only elements exhibiting a statistically significant difference between sites, correlation analysis was focused on evaluating the relationship of these two elements with the abundance those growth forms found to have contrasting abundances between sites. In this sense, we found the abundance of non-*Sphagnum* mosses to be positively correlated with the total concentration of Ca in organic soils (r = 0.48, n = 24, p < 0.05). We also found that the total concentration of Na in mineral soils was negatively correlated with the abundances of forbs (r = −0.486, n = 22, p < 0.05), non-*Sphagnum* mosses (r = −0.470, n = 22, p < 0.05) and bryophytes (r = −578, n = 22, p < 0.01).

4. Discussion

Determinations of several physico-chemical parameters, including total C, N and S and other elemental concentrations of soils occurring in *Sphagno-Eriophoretum vaginati* plant communities at Toolik Lake and Imnavait Creek grids, indicate that for most elements evaluated, and in spite of a 70 k year difference in landscape age, there are no significant differences between sites. In general, main patterns observed for many of the variables measured are not distant to what has been reported previously for this area [3] [5]-[7] [9] [23] [42]-[45]. Our observations confirm that both sites have similar soil acidity [3] [5] [43] and mineral soils bulk density values comparable to those reported in other studies [7] [42]. As expected, and in accordance with most of these studies, total C for both Toolik and Imnavait was 7 to 10 times larger in organic soils vs. mineral soils. Total N exhibited a similar pattern. There were no significant differences between Toolik and Imnavait in terms of the total concentration of both C and N in organic and mineral soils, hence confirming observations made by Whittinghill and Hobbie [5]. No between-site differences were found in terms of the amount of S in both organic and mineral soils, still, the concentration of this element in these soils is almost negligible.

Although some of our findings serve mostly as a confirmation of previously reported similarities in elemental concentrations between these two areas [5] [6] [24], others represent intriguing aspects of the soil chemistry of this region that are worth highlighting. For instance, similar to findings of Keller *et al.* [23], Fe had the largest concentration in both mineral and organic soils at both sites. Studies have demonstrated that Fe (as well as Al, the second highest) tends to accumulate in soils from tundra and other ecosystems [46]-[49]. Moreover, Fe is more soluble in acidic soils; therefore its availability tends to increase under these conditions [46] [50]. High concentrations of Al and Fe have been linked to substantial increases in the sorption capacity of organic soils. This situation may lead to reductions in P availability, causing plants to become P limited [47] [51] [52]. These observations suggest that due to its effect on P availability, changes in the concentration of Fe or Al in these soils could affect significantly the vegetation in this region. This is particularly relevant for wet sedge and some MAT areas where P availability is known to limit primary productivity [53] [54].

We consider atypical the higher concentration of Ca in organic soils from Imnavait. Organic soils from both sites have similar pH. Because soil pH is known to affect the availability of Ca [12] [46], contrasting Ca concentrations are expected to be more commonly be found between soils with significantly different acidity, as is the case for MAT vs. MNT tundra [3] [7], and less frequently between areas with the same type of vegetation (MAT), such as Toolik and Imnavait. Accordingly, Whittinghill and Hobbie [5] found that exchangeable Ca concentrations were not significantly different between soils derived from the landscape ages represented by these sites. It is likely that higher total Ca in organic soils from Imnavait is linked to differences in landscape age between these sites. This observation would agree with findings from Keller *et al.* [23] which found that organic and mineral horizons from Sagavanirktok glacial surfaces had higher Ca concentration than soils from Itkillik I surfaces (not statistically corroborated). Perhaps other processes that were not evaluated in this study, such as those related to biological and/or chemical immobilization, are variable in soils from these two landscape ages, and may account for some of these differences. However, as evidenced in this study, it is likely that the total Ca in organic soils is being influenced by between-site differences in vegetation composition, specifically, the abundance of non-*Sphagnum* mosses. In general, bryophytes tend to be in direct contact with underlying organic soils in these ecosystems and usually exert an influence on a number of soil processes [55]. Compared to Toolik, Imnavait had a significantly higher abundance of non-*Sphagnum* mosses and significantly higher concentration of Ca in organic soils, which agrees with the positive correlation found between these two parameters. Considering that both sites are influenced by similar climate and that plots are located in terrain of similar relief, dominated by MAT tundra; it seems plausible that besides landscape age, subtle variation in the abundance of non-*Sphagnum* mosses is contributing towards the differentiation of *Sphagno-Eriophoretum vaginati* plant community areas in terms of total Ca in organic soils. The work of van der Welle [56] partly support this hypothesis and revealed a significant positive correlation between non-*Sphagnum* moss cover and soil Ca content in several tussock tundra sites. Nevertheless, it remains to be tested if the effect of contrasting abundances on organic soil Ca concentration occurs via specific plant traits, like between-species variation in tissue concentration of this element, or via other plant community factors, like differences in species richness of non-*Sphagnum* mosses.

In terms of its ecosystemic effects, a higher total Ca in Imnavait soils imply that there may be less available substrate for microbial activity at that site. This is expected to diminish microbial respiration due to stabilization of organic matter by cation bridging with Ca ions [6]. However, rates of microbial activity are higher on older, more acidic landscapes [5] [8] [9] which may partly compensate for the effects of higher Ca on microbial respiration in these soils.

A significantly higher concentration of Na in minerals soils of Toolik was an unexpected finding considering that total Na have been found to be higher in mineral soils from Sagavarnirktok vs. Itkillik I glacial surfaces [23] (not statistically corroborated). However, because the dynamics of Na in mineral soils has not been characterized thoroughly in this region, we believe that between-site differences in the chemical composition of the parent material, perhaps the presence of Na-rich bedrock in sampled plots at Toolik, would be the most logical explanation for these observations. Although less likely, part of this dissimilarity might also be tied to differences in vegetation composition. Contrary to patterns observed between the abundance of non-*Sphagnum* mosses and total Ca in organic soils, significantly lower abundances of forbs, non-*Sphagnum* mosses and bryophytes in Toolik are responsible for the observed negative correlation between these growth forms and total Na in mineral soils. Assuming that forbs and bryophytes have low Na tissue concentration compared to other growth forms, it could be argued that in areas where they flourish (like Toolik), soils could reflect low Na concentration. However, in first instance, this assumption is undermined by the general scarcity of information regarding the concentration of Na in tissues of tundra plant species. Additionally, the effect of these growth forms on deeper mineral soil biogeochemical processes is expected to be negligible considering that their vertical distribution is limited to upper soil layers. Lastly, higher Na in mineral soils at Toolik might also be linked to other factors not evaluated in this study, for example, between-site variation in rates of chemical weathering of parent materials.

Soil Na has not been properly characterized in studies of soil chemistry in tundra ecosystems, except for few notable studies [3] [7] [23]. As an element, Na is present in small quantities in soils and is not an essential element for plant growth. Nevertheless, Na concentration in soils or water has been shown to promote maximal biomass yield and perform other important functional roles in plants [57]. High levels of Na could be detrimental to soil structure, soil permeability and plant growth, and in spite of its overall low levels, projected climate warming in tundra ecosystems may result in reduced soil moisture conditions [58] which may in turn affect Na dynamics in soils. Under this scenario, it becomes important to increase efforts towards a better understanding of Na dynamics in tundra soils as they are likely to change under a warming climate.

Based on our results and those of others [5] [6] [23] [42]-[44], it is evident that landscape age differences between our sites do not result in a robust chemical imprint that might help discriminate these sites in terms of their soil physico-chemical properties. While we were able to detect significant differences in two (Ca in organic soils and Na in mineral soils) of the 11 chemical parameters evaluated, these differences are subtle compared to the conspicuous differences in concentration of other elements, like Ca, between MAT and MNT [5] [7]. It has been recognized that variation in elemental concentrations in tundra soils of the Toolik-Imnavait region are tied to larger landscape-scale factors like changes in vegetation physiognomy and pH which show considerable variation at different spatial and temporal scales in this region [1] [3] [5] [7] [9] [18] [23]. Additionally, other processes like cryoturbation and frost boil formation could cause the continuous movement and mixing of soil material [3] [15] [59], thus preventing long term soil stability and causing continuous homogenization of nutrients throughout soil horizons [42] [60]. Freeze-thaw processes like cryoturbation are more frequent on MNT [3] [61], though discontinuous surface organic horizons resulting from cryoturbation are common throughout the southern foothills [62]. Therefore, we suggest that while possibly occurring in a lower frequency and magnitude, these processes are at least partly responsible for the little differentiation that organic and mineral soils of these two sites reflect in terms of most of the elemental concentrations and other physico-chemical properties evaluated in this study.

Most importantly, after detecting the differences between Toolik and Imnavait in terms of the concentration of Ca in organic soils and Na in mineral soils, we were able to discern that abundance patterns of several growth forms at these sites were concordant with correlations found between these growth forms and the concentration of these elements in organic and mineral soils respectively. These observations are relevant as they demonstrate that minimal differences in soil chemistry, possibly linked to contrasting landscape ages between these two sites, are influencing aboveground patterns in vegetation composition. In turn, by directly influencing the biophysical environment through variation in moisture or temperature regimes, through differences in root mycorrhizal processes [33], or by indirectly influencing the rates of litter input and decomposability [8], this variation in vegetation composition could be reciprocally influencing soil chemistry, therefore perpetuating the differentiation between these two sites in terms of these two elements.

5. Conclusion

Investigations analyzing the relationship between soil chemistry and vegetation in tundra ecosystems in this region have shown that conspicuous soil chemical differences (e.g. pH and Ca concentration) between some different aged landscapes (e.g. Itkillik II vs. Itkillik I) are accompanied by notable differences in vegetation types (MAT vs. MNT) [3] [7] [9]. In this respect, our study indicates that soil evolution may be governed by vegetation cover as much as by time. Landscapes of different age (Itkillik I vs. Sagavariktok) and very similar vegetation type (*i.e.*, MAT) are not distinguishable in terms of most of their soil physical and chemical properties. However, total concentration of elements like Ca and Na may vary between them. This variation, although likely linked to differences in the chemical composition of the geologic substrates underlying these soils, appears to reflect some level of agreement with abundance patterns of specific growth forms. These observations are relevant as they testify in favor of the reciprocal relationship between soil chemistry and vegetation composition in moist acidic tundra landscapes in Arctic Alaska.

Acknowledgements

We thank Amy Breen Carroll and Sayuri Ito for their work during the collection of soil samples. We are profoundly grateful to the personnel of the IITF Chemistry Laboratory for helping in the analysis and interpretation of soil chemical results. Thanks to the staff of the Toolik Lake Field Station (UAF) for their hospitality and for coordination of laboratory space. Ariel E. Lugo reviewed the original version of the manuscript and provided helpful comments. This work is based on support by the United States National Science Foundation awards OPP-0632277 and OPP-0856710. This research was conducted in cooperation with the University of Puerto Rico.

References

[1] Marion, G.M., Hastings, S.J., Oberbauer, S.F. and Oechel, W.C. (1989) Soil-Plant Element Relationships in a Tundra Ecosystem. *Holarctic Ecology*, **12**, 296-303.

[2] Walker, D.A., Auerbach, N.A. and Shippert, M.M. (1995) NDVI, Biomass, and Landscape Evolution of Glaciated Terrain in Northern Alaska. *Polar Record*, **31**, 169-178. http://dx.doi.org/10.1017/S003224740001367X

[3] Bockheim, J.G., Walker, D.A., Everett, L.R., Nelson, F.E. and Shiklomanov, N.I. (1998) Soils and Cryoturbation in Moist Non-Acidic and Acidic Tundra in the Kuparuk River Basin, Arctic Alaska, USA. *Arctic and Alpine Research*, **30**, 166-174. http://dx.doi.org/10.2307/1552131

[4] Walker, D.A. (2000) Hierarchical Subdivision of Arctic Tundra based on Vegetation Response to Climate, Parent Material and Topography. *Global Change Biology*, **6**, 19-34. http://dx.doi.org/10.1046/j.1365-2486.2000.06010.x

[5] Whittinghill, K.A. and Hobbie, S.E. (2011) Effects of Landscape Age on Soil Organic Matter processing in Northern Alaska. *Soil Science Society of America Journal*, **75**, 907-917. http://dx.doi.org/10.2136/sssaj2010.0318

[6] Whittinghill, K.A. and Hobbie, S.E. (2011) Effects of pH and Calcium on Soil Organic Matter Dynamics in Alaskan Tundra. *Biogeochemistry*, **111**, 569-581. http://dx.doi.org/10.1007/s10533-011-9688-6

[7] Hobbie, S.E. and Gough, L. (2002) Foliar and Soil Nutrients in Tundra on Glacial Landscapes of contrasting Ages in Northern Alaska. *Oecologia*, **131**, 453-462. http://dx.doi.org/10.1007/s00442-002-0892-x

[8] Hobbie, S.E. and Gough, L. (2004) Litter Decomposition in Moist Acidic and Non-Acidic Tundra with different Glacial Histories. *Oecologia*, **140**, 113-124. http://dx.doi.org/10.1007/s00442-004-1556-9

[9] Hobbie, S.E., Miley, T.A. and Weiss, M.S. (2002) Carbon and Nitrogen Cycling in Soils from Acidic and Non Acidic

Tundra with Different Glacial Histories in Northern Alaska. *Ecosystems*, **5**, 761-774.
http://dx.doi.org/10.1007/s10021-002-0185-6

[10] Walker, D.A. and Walker, M.D. (1996) Terrain and Vegetation of the Imnavait Creek Watershed. In: Reynolds, J.F. and Tenhunen, J.D., Eds., *Landscape Function and Disturbance in Arctic Tundra Ecological Studies*, Vol. 120, Springer-Verlag, Berlin, 73-108. http://dx.doi.org/10.1007/978-3-662-01145-4_4

[11] Walker, D.A., Epstein, H.E., Jia, G.J., Balser, A., Copass, C., Edwards, E.J., Gould, W.A., Hollingsworth, J., Knudson, J., Maier, H.A., Moody, A. and Raynolds, M.K. (2003) Phytomass, LAI, and NDVI in Northern Alaska: Relationships to Summer Warmth, Soil pH, Plant Functional Types, and Extrapolation to the Circumpolar Arctic. *Journal of Geophysical Research*, **108**, 8169-8185. http://dx.doi.org/10.1029/2001JD000986

[12] Gough, L., Shaver, G.R., Carroll, J., Royer, D.L. and Laundre, J.A. (2000) Vascular Plant Species Richness in Alaskan Arctic Tundra: The Importance of Soil pH. *Journal of Ecology*, **88**, 54-66.
http://dx.doi.org/10.1046/j.1365-2745.2000.00426.x

[13] Eskelinen, A., Stark, S. and Männistö, M. (2009) Links between Plant Community Composition, Soil Organic Matter Quality and Microbial Communities in Contrasting Tundra Habitats. *Oecologia*, **161**, 113-123.
http://dx.doi.org/10.1007/s00442-009-1362-5

[14] Hobbie, S.E. (1992) Effects of Plant Species on Nutrient Cycling. *Trends in Ecology & Evolution*, **7**, 336-339.
http://dx.doi.org/10.1016/0169-5347(92)90126-V

[15] Walker, D.A., Epstein, H.E., Gould, W.A., Kelley, A.M., Kade, A.N., Knudson, J.A., Krantz, W.B., Michaelson, G., Peterson, R.A., Ping, C.L., Raynolds, M.K., Romanovsky, V.E. and Shur, Y. (2004) Frost-Boil Ecosystems: Complex Interactions between Landforms, Soils, Vegetation and Climate. *Permafrost and Periglacial Processes*, **15**, 171-188.
http://dx.doi.org/10.1002/ppp.487

[16] Hobbie, S.E. (1996) Temperature and Plant Species Control over Litter Decomposition in Alaskan Tundra. *Ecological Monographs*, **66**, 503-522. http://dx.doi.org/10.2307/2963492

[17] Chu, H. and Grogan, P. (2010) Soil Microbial Biomass, Nutrient Availability and Nitrogen Mineralization Potential among Vegetation-Types in a Low Arctic Tundra Landscape. *Plant and Soil*, **329**, 411-420.
http://dx.doi.org/10.1007/s11104-009-0167-y

[18] Walker, M.D., Walker, D.A. and Auerbach, N.A. (1994) Plant Communities of a Tussock Tundra Landscape in the Brooks Range Foothills, Alaska. *Journal of Vegetation Science*, **5**, 843-866. http://dx.doi.org/10.2307/3236198

[19] Walker, D.A., Walker, M.D., Gould, W.A., Mercado, J., Auerbach, N.A., Maier, H.A. and Neufeld, G.P. (2010) Maps for Monitoring Long-Term Changes to Vegetation Structure and Composition, Toolik Lake, Alaska. In: Bryn, A., Dramstad, W. and Fjellstad, W., Eds., *Mapping and Monitoring of Nordic Vegetation and Landscapes*, Vol. 1, Norsk Institutt for Skog og Landskap, Ås, Norway, 121-123.

[20] Walker, D.A., Lederer, N.D. and Walker, M.D. (1987) Permanent Vegetation Plots: Site Factors, Soil Physical and Chemical Properties and Plant Species Cover. Department of Energy, R4D Program Data Report, Plant Ecology Laboratory, Institute of Arctic and Alpine Research, Boulder, National Snow and Ice Data Center. Identifier Number: ARCSS110.

[21] Walker, D.A. and Barry, N. (1991) Toolik Lake Permanent Vegetation Plots: Site Factors, Soil Physical and Chemical Properties, Plant Species Cover, Photographs, and Soil Descriptions. Data Report 48, Department of Energy R4D Program, Institute of Arctic and Alpine Research, University of Colorado, Boulder.

[22] Walker, D.A., Binnian, E., Evans, B.M., Lederer, N.D., Nordstrand, E. and Webber, P.J. (1989) Terrain, Vegetation and Landscape Evolution of the R4D Research Site, Brooks Range Foothills, Alaska. *Holarctic Ecology*, **12**, 238-261.

[23] Keller, K., Blum, J.D. and Kling, G.W. (2007) Geochemistry of Soils and Streams on Surfaces of Varying Ages in Arctic Alaska. *Arctic, Antarctic, and Alpine Research*, **39**, 84-98.
http://dx.doi.org/10.1657/1523-0430(2007)39[84:GOSASO]2.0.CO;2

[24] Walker, D.A. and Maier, H.A. (2008) Vegetation in the Vicinity of the Toolik Field Station, Alaska. Biological Papers of the University of Alaska 28, Institute of Arctic Biology, Fairbanks.

[25] Wahrhaftig, C. (1965) Physiographic Divisions of Alaska U.S. Geological Survey Professional Paper 482. US Government Printing Office, Washington DC.

[26] Osterkamp, T.E., Petersen, J.K. and Collet, T.S. (1985) Permafrost Thicknesses in the Oliktok Point, Prudhoe Bay and Mikkelsen Bay Areas of Alaska. *Cold Regions Science and Technology*, **11**, 99-105.
http://dx.doi.org/10.1016/0165-232X(85)90010-2

[27] Hamilton, T.D. (2003) Glacial Geology of Toolik Lake and the Upper Kuparuk River Region. Biological Papers of the University of Alaska No. 26, University of Alaska Printing Services, Fairbanks.

[28] Hamilton, T.D. (2003) Surficial Geology of the Dalton Highway (Itkillik-Sagavanirktok Rivers) Area, Southern Arctic Foothills, Alaska. Alaska Division of Geological & Geophysical Surveys Professional Report 121, Alaska, 32 p.

[29] Cherry, J., Déry, S.J., Cheng, Y., Stieglitz, M., Jacobs, M.S. and Pan, F. (2014) Climate and Hydrometeorology of the Toolik Lake Region and the Kuparuk River Basin: Past, Present and Future. In: Hobbie, J.E. and Kling, G.W., Eds., *Alaska's Changing Arctic: Ecological Consequences for Tundra, Streams and Lakes*, Oxford University Press, New York, 31-60.

[30] Gould, W.A. and Mercado-Díaz, J.A. (2014) Decadal-Scale Changes of Vegetation from Long-Term Plots in Alaskan Tundra. In: Hobbie, J.E. and Kling, G.W., Eds., *Alaska's Changing Arctic: Ecological Consequences for Tundra, Streams and Lakes*, Vignette 5.5, Oxford University Press, New York, 130-131.

[31] Walker, D.A., Hamilton, T.D., Maier, H.A., Munger, C.A. and Raynolds, M.K. (2014) Glacial History and Long-Term Ecology in the Toolik Lake Region. In: Hobbie, J.E. and Kling, G.W., Eds., *Alaska's Changing Arctic: Ecological Consequences for Tundra, Streams and Lakes*, Oxford University Press, New York, 61-80.

[32] Shaver, G.R., Laundre, J.A., Bret-Harte, M.S., Chapin III, F.S., Mercado-Díaz, J.A., Giblin, A.E., Gough, L., Gould, W.A., Hobbie, S.E., Kling, G.W., Mack, M.C., Moore, J.C., Nadelhoffer, K., Rastetter, E.B. and Schimel, J.P. (2014) Terrestrial Ecosystems at Toolik Lake, Alaska. In: Hobbie, J.E. and Kling, G.W., Eds., *Alaska's Changing Arctic: Ecological Consequences for Tundra, Streams and Lakes*, Oxford University Press, New York, 90-142.

[33] Walker, D.A. (1996) GIS Data from the Alaska North Slope. National Snow and Ice Data Center. http://nsidc.org/data/arcss017.html

[34] Mercado-Díaz, J.A. (2011) Plant Community Responses of the Alaskan Arctic Tundra to Environmental and Experimental Changes in Climate. M.Sc. Thesis, University of Puerto Rico, Río Piedras Campus, San Juan.

[35] LECO Corp. (2006) LECO TruSpec CN Carbon/Nitrogen Determinator Instruction Manual. St. Joseph.

[36] LECO Corp. (2005) Carbon and Nitrogen in Soil and Sediment. Organic Application Note: TruSpec CN (Form No. 203-821-275). St. Joseph.

[37] LECO Corp. (2006) LECO TruSpec Add-On Module Sulfur Analyzer Instruction Manual. St. Joseph.

[38] LECO Corp. (2008) Sulfur in Cement, Fly Ash, Limestone, Soil and Ore. Organic Application Note: TruSpec S (Form No. 203-821-345). St. Joseph.

[39] Luh Huang, C.Y. and Schulte, E.E. (1985) Digestion of Plant Tissue for Analysis by ICP Emission Spectroscopy. *Communications in Soil Science and Plant Analysis*, **16**, 943-958. http://dx.doi.org/10.1080/00103628509367657

[40] Wilde, S.A., Corey, R.B., Iyer, J.G. and Voight, G.K. (1979) Soil and Plant Analysis for Tree Culture. 5th Edition, Oxford & IBH Publishing Co., New Delhi.

[41] McLean, E.O. (1982) Soil pH and Lime Requirement. In: Page, A.L., Miller, R.H. and Keeney, D.R., Eds., *Methods of Soil Analysis, Part 2, Chemical and Microbiological Properties, Agronomy Monograph Number 9*, Soil Science Society of America, Madison, 199-224.

[42] Michaelson, G.J., Ping, C.L. and Kimble, J.M. (1996) Carbon Storage and Distribution in Tundra Soils of Arctic Alaska, U.S.A. *Arctic and Alpine Research*, **28**, 414-424. http://dx.doi.org/10.2307/1551852

[43] Ping, C.L., Michaelson, G.J., Loya, W.M., Chandler, R.J. and Malcolm, R.L. (1997) Characteristics of Soil Organic Matter in Arctic Ecosystems of Alaska. In: Lal, R., Kimble, J.M., Follet, R.F. and Stewart, B.A., Eds., *Soil Processes and the Carbon Cycle*, CRC Press LLC, Boca Raton, 157-167.

[44] Ping, C.L., Bockheim, J.G., Kimble, J.M., Michaelson, G.J. and Walker, D.A. (1998) Characteristics of Cryogenic Soils along a Latitudinal Transect in Arctic Alaska. *Journal of Geophysical Research*, **103**, 28917-28928. http://dx.doi.org/10.1029/98JD02024

[45] Giblin, A.E., Nadelhoffer, K.J., Shaver, G.R., Laundre, J.A. and McKerrow, A.J. (1991) Biogeochemical Diversity along a Riverside Toposequence in Arctic Alaska. *Ecological Monographs*, **61**, 415-435. http://dx.doi.org/10.2307/2937049

[46] Moore, P.D. (2008) Tundra. Infobase Publishing, New York.

[47] Giesler, R., Petersson, T. and Högberg, P. (2002) Phosphorus Limitation in Boreal Forests: Effects of Aluminum and Iron Accumulation in the Humus Layer. *Ecosystems*, **5**, 300-314. http://dx.doi.org/10.1007/s10021-001-0073-5

[48] Birkeland, P.W., Burke, R.M. and Benedict, J.B. (1989) Pedogenic Gradients for Iron and Aluminum Accumulation and Phosphorus Depletion in Arctic and Alpine Soils as a Function of Time and Climate. *Quaternary Research*, **32**, 193-204. http://dx.doi.org/10.1016/0033-5894(89)90075-6

[49] Ugolini, F.C., Stoner, M.G. and Marrett, D.J. (1987) Arctic Pedogenesis: 1. Evidence for Contemporary Podzolization. *Soil Science*, **144**, 90-100. http://dx.doi.org/10.1097/00010694-198708000-00002

[50] Brady, N.C. and Weil, R.R. (2008) The Nature and Properties of Soils. Prentice-Hall Inc., New Jersey.

[51] Giesler, R., Andersson, T., Lövgren, L. and Persson, P. (2005) Phopshate Sorption in Aluminum and Iron-Rich Humus Soils. *Soil Science Society of America Journal*, **69**, 77-86.

[52] Kang, J., Hesterberg, D. and Osmond, D.L. (2009) Soil Organic Matter Effects on Phosphorus Sorption: A Path Analy-

sis. *Soil Science Society of America Journal*, **73**, 360-366. http://dx.doi.org/10.2136/sssaj2008.0113

[53] Shaver, G.R. and Chapin III, F.S. (1986) Effect of Fertilizer on Production and Biomass of Tussock Tundra, Alaska, U.S.A. *Arctic and Alpine Research*, **18**, 261-268. http://dx.doi.org/10.2307/1550883

[54] Shaver, G.R. and Chapin III, F.S. (1995) Long-Term Responses to Factorial, NPK Fertilizer Treatment by Alaskan Wet and Moist Tundra Sedge Species. *Ecography*, **18**, 259-275. http://dx.doi.org/10.1111/j.1600-0587.1995.tb00129.x

[55] Jäggerbrand, A.K., Björk, R.G., Callaghan, T. and Seppelt, R.D. (2011) Effects of Climate Change on Tundra Bryophytes. In: Tuba, Z., Slack, N.G. and Stark, L.R., Eds., *Bryophyte Ecology and Climate Change*, Cambridge University Press, New York, 211-236.

[56] van der Welle, M.E.W., Vermeulen, P.J., Shaver, G.R. and Berendese, F. (2003) Factors Determining Plant Species Richness in Alaskan Arctic Tundra. *Journal of Vegetation Science*, **14**, 711-720. http://dx.doi.org/10.1111/j.1654-1103.2003.tb02203.x

[57] Subberao, G.V., Ito, O., Berry, W.L. and Wheeler, R.M. (2003) Sodium—A Functional Plant Nutrient. *Critical Reviews in Plant Sciences*, **22**, 391-416.

[58] Xu, W., Yuan, W., Dong, W., Xia, J., Liu, D. and Chen, Y. (2013) A Meta-Analysis of the Response of Soil Moisture to Experimental Warming. *Environmental Research Letters*, **8**, 1-8. http://dx.doi.org/10.1088/1748-9326/8/4/044027

[59] Munroe, J.S. and Bockheim, J.G. (2001) Soil Development in Low-Arctic Tundra of the Northern Brooks Range, Alaska, U.S.A. *Arctic, Antarctic and Alpine Research*, **33**, 78-87. http://dx.doi.org/10.2307/1552280

[60] Michaelson, G.J., Ping, C.L. and Kimble, J.M. (2001) Effects of Soil Morphological and Physical Properties on Estimation of Carbon Storage. In: Lal, R., Kimble, J.M., Follett, R.F. and Stewart, B.A., Eds., *Assessment Methods for Soil Carbon*, Lewis Publishers, Boca Raton, 339-347.

[61] Bockheim, J.G., Walker, D.A. and Everett, L.R. (1997) Soil Carbon Distribution in Non Acidic and Acidic Tundra of Arctic Alaska. In: Lal, R., Kimble, J.M., Follett, R.F. and Stewart, B.A., Eds., *Soil Processes and the Carbon Cycle*, CRC Press, Boca Raton, 143-155.

[62] Ping, C.L., Clark, M.H. and Swanson, D.K. (2004) Cryosols in Alaska. In: Kimble, J.M., Ed., *Cryosols, Permafrost-Affected Soils*, Springer-Verlag, New York, 71-94.

14

Seasonal Evolution of the Rhizosphere Effect on Major and Trace Elements in Soil Solutions of Norway Spruce (*Picea abies* Karst) and Beech (*Fagus sylvatica*) in an Acidic Forest Soil

Christophe Calvaruso[1,2,3*], Christelle Collignon[1,4], Antoine Kies[2], Marie-Pierre Turpault[1]

[1]INRA, UR1138 "Biogeochemistry of Forest Ecosystems", Centre INRA of Nancy, Champenoux, France
[2]"Radiation Physics" Laboratory, University of Luxembourg, Campus Limpersberg, Luxembourg, Luxembourg
[3]EcoSustain, Environmental Engineering Office, Research and Development, Kanfen, France
[4]INRA, UMR1136 INRA-Nancy University "Interactions Tree-Microorganisms", Centre INRA of Nancy, Champenoux, France
Email: *chriscalva@hotmail.com

Abstract

In low-nutrient ecosystems such as forests developed on acidic soil, the main limiting factor for plant growth is the availability of soil nutrients. The aim of this study was to investigate in a temperate forest: 1) the influence of the rhizosphere processes on the availability of nutrients and trace elements during one year period and 2) the seasonal evolution of this rhizosphere effect. Bulk soil and rhizosphere were collected in organo-mineral and mineral horizons of an acidic soil during autumn, winter, and spring under Norway spruce (*Picea abies* Karst) and beech (*Fagus sylvatica*). Soil solutions were extracted by soil centrifugation. Rhizosphere solutions were enriched in K, and in Ca, Mg, and Na (principally in spring) compared to those of the bulk soil. Our study reveals seasonal variations of the rhizosphere effect for Ca, Mg, and Na under both species, *i.e.*, higher enrichment of the rhizosphere solution in spring as compared with that in autumn and winter. An enrichment of the rhizosphere solutions was also observed for trace elements regardless of the season under both species in the mineral horizon, only. In contrast, seasonal variations of the rhizosphere effect for the trace elements were observed in the solutions of the organo-mineral horizon under beech, *i.e.*, enrichment in autumn and depletion in winter. This study demonstrates that rhizosphere biological activities significantly increase nutrient bioavailability during the growth period. These complex interactions between roots, microbial communities and soils are a key-process that supports tree nutrition in nutrient-poor forest soils. This research also reveals that rhizosphere processes a) occur throughout the year, even in winter, and b) influence differently the dynamics of nutrients and trace elements in the root vicinity of the organo-mineral horizon.

Keywords

Major and Trace Elements, Rhizosphere Processes, Soil Solution, Seasonal Variations, Tree Nutrition

---

1. Introduction

In low-nutrient ecosystems such as temperate forests developed on acidic soil, plant growth is often limited by the availability of soil nutrients rather than by light or water [1]. The rhizosphere, defined as the volume of soil influenced by root activity [2], constitutes the interface between the solid soil phase, soil solution, and root system and is the zone where major processes take place [3]. The characteristics of the rhizosphere may be drastically different from those of the bulk soil; that is to say the root-free soil material. Although the rhizosphere represents only 1% to 2 % of the total soil [4] [5], it plays a central role in the maintenance of the soil-plant system, strongly influencing the availability of organic and inorganic elements in the soil and therefore the nutrition of plants [6] [7]. Because rhizospheric soil solution is the direct source of nutrients for plants [1], the amount of elements in this compartment is significantly affected by plant uptake. Element accumulation in the rhizosphere occurs when the rate of nutrient supply from the soil is higher than nutrient uptake by plant roots; in the reverse situation, depletion of nutrients takes place [8]. The element input in the rhizosphere principally results from mass flow, *i.e.*, the rapid transport of elements from the bulk soil to the root vicinity induced by root water uptake [9] and from the release of elements through soil mineral weathering and from organic matter mineralization [10]. Plant roots can enhance soil mineral dissolution and organic phase mineralization either directly by releasing protons and organic substances such as organic acids and enzymes [11], or indirectly by stimulating soil microbial activity [12]. This plant strategy to exploit and colonize nutrient-limited habitats is called the rhizosphere effect [13]. Tree roots and associated microorganisms are known to promote the mineral weathering in the rhizosphere, notably in low-nutrient environments such as forest ecosystems [14]-[16] and to increase nutrient availability in the soil, thus improving plant nutrition. Additionally, carbon input by tree roots provides the essential energy for microbial activities and stimulates degradation and mineralization of organic matter in the rhizosphere [17]. In consequence, the availability of nutrients in the tree rhizosphere results from chemical and biochemical reactions caused by complex root-microorganism-soil interactions [18] [19].

Several studies showed a significant effect of tree roots on essential nutrients such as K, Mg, Ca or P [20]-[25]. The results, however, are contrasted and a general pattern cannot be drawn. In fact, the rhizosphere effect on element availability depends principally on climatic conditions, soil properties as well as plant species and soil microorganism characteristics. In particular, it is well known that root as well as root-associated microorganism activity is high during spring and low during winter, suggesting that the rhizosphere effect on element availability may vary during the year. For example, [25] demonstrated that seasonal changes in the nutrient uptake by tree can rapidly modify the chemical composition of the rhizosphere. Reference [26] studied the seasonal influence on the behavior of exchangeable nutrients (K, Ca and Mg) in acid temperate forest soils, and even demonstrated that processes resulting from interactions between trees, microorganisms and soil influenced not only the seasonal dynamics of nutrients in the root vicinity but also those in the bulk soil. In contaminated forests, previous studies have also reported an accumulation of trace elements such as Cd, Cu, Pb, Ni, Zn or Rb in the tree rhizosphere [27]-[30]. Although tree rhizosphere is known to be mineral weathering hot-spots [31]-[33], the availability of trace elements in the rhizosphere is not documented for uncontaminated forest ecosystems.

The objective of our study was to investigate, in a temperate forest ecosystem: 1) the influence of rhizosphere processes on the availability of major elements such as K, Ca, Mg, Na, Fe, Mn and Si as well as trace elements such as Ba, Cd, Ce, Co, Cr, Cs, Cu, Ga, La, Nd, Ni, Pb, Rb, Sr, Th, U, V, Y, and Zn, and 2) the seasonal evolution of this rhizosphere effect. For that purpose, samples of bulk soil and rhizosphere were collected in organo-mineral (0 - 3 cm) and mineral (3 - 10 and 10 - 23 cm) horizons of an acid and low-nutrient soil during three seasons (autumn: November, winter: February and spring: May) under two species, *i.e.*, the evergreen Norway spruce (*Picea abies* Karst) and the deciduous oak (*Quercus sessiliflora* Smith). The soil solutions were extracted by soil centrifugation and were analysed for pH, carbon, nitrogen, major and trace elements.

2. Materials and Methods

2.1. Study Site and Soil Properties

This study was conducted in the Breuil-Chenue experimental forest site located in the Morvan (47°18'N, 4°5'E, France). The forest is situated on a plateau at an altitude of 638 m. The native forest was clear-felled and replaced in 1976 by monospecific plantations distributed in plots of 0.1 ha of various species such as oak, beech (*Fagus sylvatica* L.), Norway spruce, and Douglas-fir (*Pseudotsuga menziesii* [Mirb.] Franco). The soil derives from the "Pierre qui Vire" granite that contains quartz (34.0%), albite (31.1%), K-feldspar (24.2%), muscovite (8.9%), biotite (1.2%), and chlorite (0.5%). The soil has been classified as a Typic Dystrochrept, according to [34]. The soil characteristics of the native forest are described in detail by [35]. The bulk soil has a sandy-loam texture (55% sand and less than 20% clay). The soil is well-aerated and acid (pH$_{KCl}$ 3.1 - 4.3). The soil cation exchange capacity ranges from 9.2 to 2.7 cmolc·kg^{-1} with a base saturation lower than 10%. Humus is present as a moder in the native forest and carbon concentration is 7.3% in the A1 horizon.

2.2. Soil Sampling and Preparation

Soil samples were collected under beech and Norway spruce stand in November 2007 (autumn), February 2008 (winter) and May 2008 (spring). Soil samples were also collected during summer (July) but the quantity of water extracted by centrifugation from the soils at this period was insufficient for analyses. For each season and both stands, soil samplings were carried out in four replicates in independent plots (about 10 m distance between each soil sample) from pits of 120 × 80 cm. After removing the forest floor, soil samplings were carried out systematically at three depths, *i.e.*, 0 - 3 cm (layer I), 3 - 10 cm (layer II), and 10 - 23 cm (layer III). The choice of 0 - 3, 3 - 10 and 10 - 23 cm was motivated by the fact that these depths correspond approximately to the limits of two distinct and homogeneous soil horizons (from organo-mineral to mineral soil horizons) in both Norway spruce and beech stands [4]. The separation of soil samples into bulk soil and rhizosphere fractions was conducted in the field. In each horizon, soil material was cut and extracted from the profile. At the site, living roots with diameters < 2 mm were carefully removed by hand from each soil layer. The soil without roots was collected to give the bulk soil. Soil aggregates > 1 cm adherent to the roots were removed. The rhizosphere fraction was obtained by gently shaking fresh roots. Bulk soil and rhizosphere were sieved at 200 μm to eliminate the roots and to obtain a comparable particle size distribution in both compartments, and homogenized. These soil samples were placed in air-tight bags and stored at 4°C for 24 to 48 h. The solutions of the bulk soil and rhizosphere were extracted by centrifugation (15°C, 20 min, 3000 rpm; JOUAN KR422) for the 3 months (November, February and May), the three soil horizons (0 - 3, 3 - 10, and 10 - 23 cm) and the two tree species (beech, Norway spruce). Each extract was filtered with a pre-rinsed, 0.45 μm pore diameter, cellulose nitrate filter, and stored at 4°C. The solutions collected through this extraction protocol correspond to capillary plus gravitational solutions. This method was used because the volume of solution extracted from the rhizosphere would have been insufficient for analysis if we had separated the two types of solution.

2.3. Sample Analysis

The pH of the soil solutions was determined (pHmeter SENTRON, Argus X). Total carbon and nitrogen in soil solutions were estimated using a TOC analyzer (TOC-5050, Shimadzu). The concentration of major cations (K, Ca, Mg, Na, Fe and Mn) and trace elements (Ba, Cd, Ce, Co, Cr, Cs, Cu, Ga, La, Nd, Ni, Pb, Rb, Sr, Th, U, V, Y, and Zn only in soil samples collected in layers I and III) in the solutions was determined inductively coupled plasma atomic emission spectrometer (ICP-AES; Plasma torch JY180 ULTRACE) and by ICP-MS spectrometer (ICP-MS, VG PlasmaQuad PQ2+), respectively.

For each element, rhizosphere effects were calculated as the percentage difference between paired rhizosphere and bulk soil samples for each depth, each season and each tree species. A positive rhizosphere effect indicates a greater flux in the rhizosphere, while a negative rhizosphere effect indicates a greater flux in the bulk soil.

2.4. Statistical Analysis

The mean values were calculated from our replicates (n = 4) and are given with standard errors. For each tree species, each depth and each soil compartment, a one-factor variance analysis (ANOVA) was used to assess significant differences between the different seasons, at the threshold of $p < 0.05$. The normality of distribution and the homoscedasticity of variances were tested. Average comparisons were made using the Student-Newman-Keuls test. Before analysis, all percentages were arcsine transformed. The paired t-Student test was performed to establish significant differences for pH, carbon, nitrogen, major and trace elements between the two soil compartments (not independent samples) for each tree species, each depth and each season at the threshold level of $p < 0.05$. Statistical analyses were completed with the UNISTAT software (Unistat version 5.0, 2002, England).

3. Results

3.1. Climatic Conditions

The climatic conditions (air temperature and rainfall) measured during the period of the study in the region of the Breuil-Chenue site were presented in [26]. Briefly, the mean temperatures were 4.4°C, 2.5°C, and 14.9°C in November, February, and May 2007, respectively. The mean precipitations were 58, 63, and 122 mm in November, February, and May 2007, respectively. The mean annual temperature and precipitations were about 9°C and 1300 mm, respectively.

3.2. The pH, Carbon and Nitrogen

The pH of the rhizosphere solution was significantly inferior to that of the bulk soil solution in winter whatever the soil horizon under beech and only in layer II under Norway spruce (**Table 1**). This difference is mainly due to a strong increase of the pH in the bulk soil between the autumn and the winter, 4.35 vs. 3.79, 4.36 vs. 4.05, and 5.37 vs. 4.52 for layer I, layer II, and layer III, respectively. The pH of the rhizosphere solution was significantly superior to that of the bulk soil solution in spring in layers II and III under beech, only. No significant difference between the pH of the bulk soil and rhizosphere solutions was observed in autumn.

Figure 1 represents percentages of difference for C and N between the bulk soil and rhizosphere solutions for each depth, each season, and both species. Our results showed an enrichment of the rhizosphere in C and N in layer III whatever the season and the species, with exception of the N in spring under Norway spruce (**Figure 1**).

Table 1. Bulk soil (B) and rhizosphere (R) solution pH for the different seasons (autumn, winter, and spring), species (beech and Norway spruce), and depths (layer I: 0 - 3 cm, layer II: 3 - 10 cm, and layer III: 10 - 23 cm) (adapted from [26]). Asterisk indicates significant difference between bulk soil and rhizosphere solution pH for the same season, species and depth, according to a paired t-Student analysis at the threshold of $p = 0.05$.

Species	Layer	Compartment	Season		
			Autumn	Winter	Spring
Beech	I	B	3.79	4.35	3.80
		R	3.60	3.69*	3.78
	II	B	4.05	4.36	3.68
		R	4.00	3.80*	4.12*
	III	B	4.52	5.37	4.46
		R	4.35	4.71*	4.72*
Norway spruce	I	B	3.27	3.79	3.55
		R	3.20	3.66	3.31
	II	B	3.82	4.08	3.91
		R	3.69	3.69*	3.82
	III	B	4.12	4.62	4.40
		R	4.34	4.53	4.09

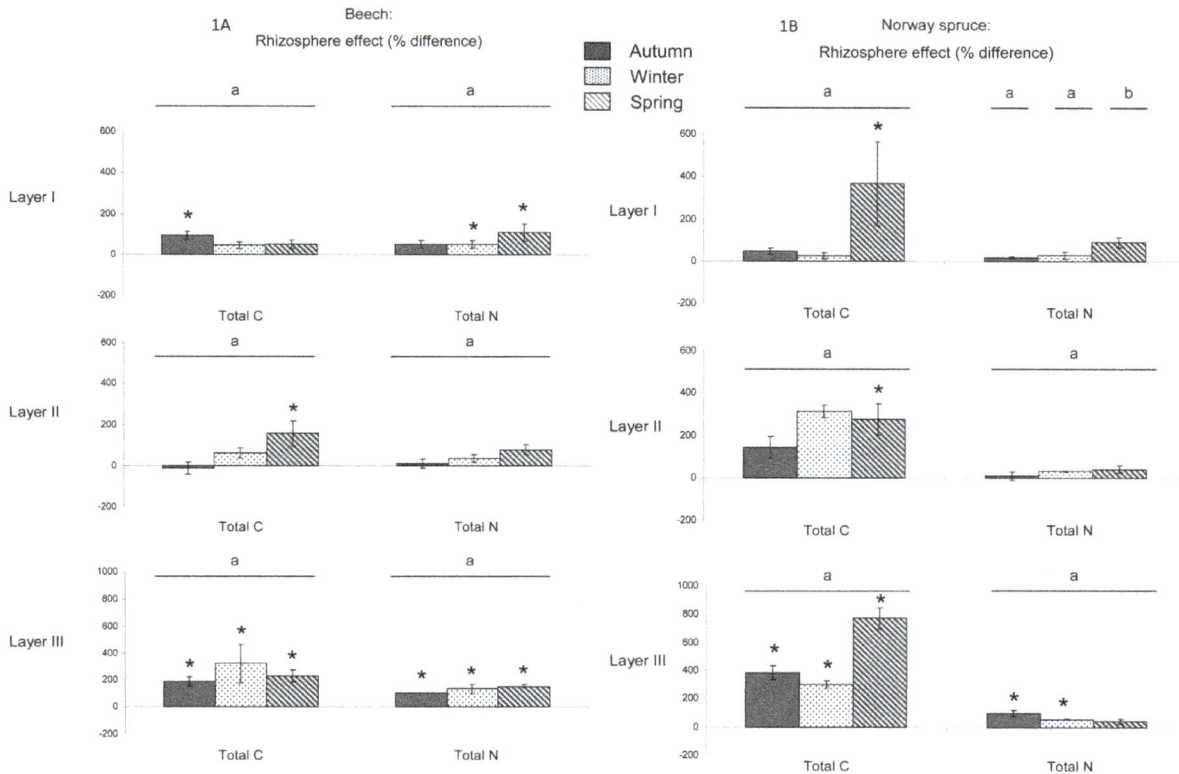

Figure 1. Rhizosphere effect for total C and total N under beech (A) and Norway spruce (B) for the three depths (layer I: 0 - 3 cm, layer II: 3 - 10 cm, and layer III: 10 - 23 cm) and the three seasons (autumn, winter, and spring). Rhizosphere effect is expressed as a percentage of difference between bulk soil solution and rhizosphere solution. Histograms represent the mean value of four replicates. Bars represent standard errors. For each depth and season, bars with an asterisk are significantly different according to a one-factor (soil compartment) ANOVA and the Student-Newman-Keuls test ($p < 0.05$). For each depth and each soil compartment, bars with the same letter (a, b, c) are not significantly different according to a one-factor (season) ANOVA and the Student-Newman-Keuls test ($p < 0.05$).

In layer I, the rhizosphere is enriched in C in autumn and in N in winter and spring for the beech as well as in C in spring for the Norway spruce. In layer II, the rhizosphere is enriched in C in spring for both species (**Figure 1**). We did not observe seasonal variations of the rhizosphere effect, except for an increase of this effect on N in spring in layer I under the beech.

3.3. Major Elements (K, Ca, Mg, Na, Fe and Mn)

Figure 2 represents percentages of difference for major elements between the bulk soil and rhizosphere solutions for each depth, season, and tree species. The rhizosphere solutions were significantly enriched in K compared to those in the bulk soil, independently of the tree species, the depth and the season. No seasonal variation of the rhizosphere effect on K was observed, except for an enrichment of the rhizosphere solution in spring in layer III under beech. For Ca, the rhizosphere solutions were enriched under beech in layers II and III in spring, and under Norway spruce in layer I in spring, and in layer III in autumn and spring. Significant seasonal variations of the rhizosphere effect on Ca were observed under beech in layers II and III, and under Norway spruce in layer I, *i.e.* higher enrichment of the rhizosphere solution in spring. For Mg, the rhizosphere solutions were enriched under beech in layer II in spring, and under Norway spruce in layers I, II, and III in spring. Significant seasonal variations of the rhizosphere effect on Mg were observed under beech in layer II and under Norway spruce in layers I and II, *i.e.* higher enrichment of the rhizosphere solution in spring. For Na, the rhizosphere solutions were enriched under beech in layer I in autumn, winter and spring, in layer II in spring, and in layer III in autumn and spring, and under Norway spruce in layers I and II in spring and in layer III in autumn, winter, and spring. Significant seasonal variation of the rhizosphere effect on Na was observed under Norway spruce in

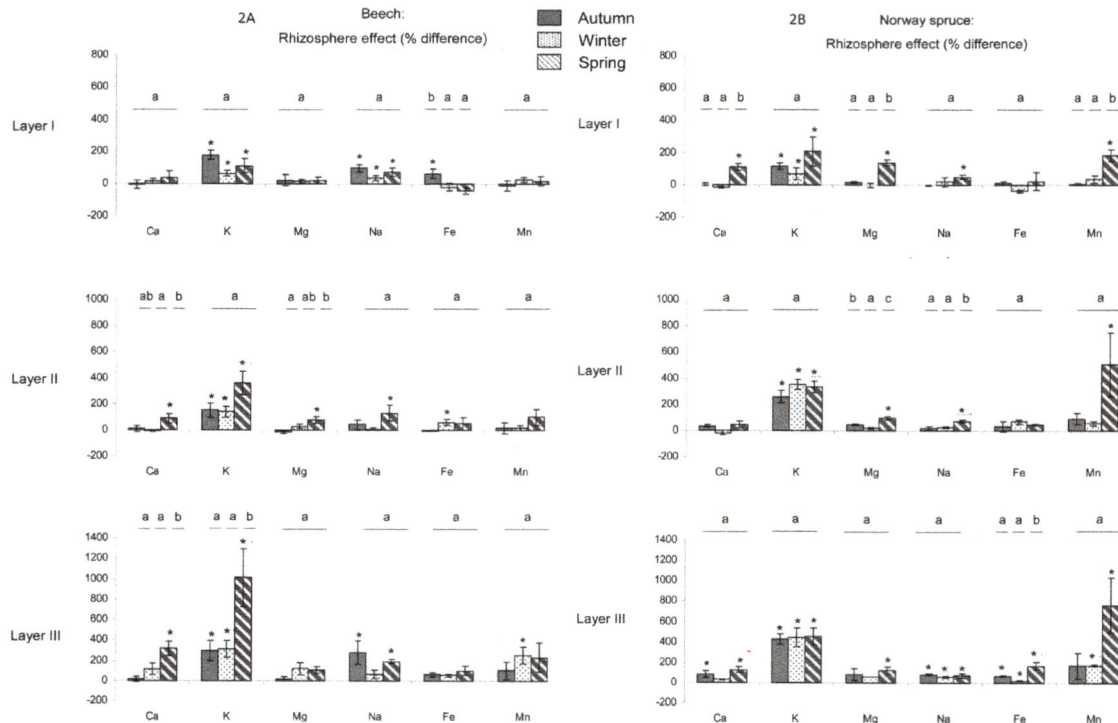

Figure 2. Rhizosphere effect for major elements under beech (A) and Norway spruce (B) for the three depths (layer I: 0 - 3 cm, layer II: 3 - 10 cm, and layer III: 10 - 23 cm) and the three seasons (autumn, winter, and spring). Rhizosphere effect is expressed as a percentage of difference between bulk soil solution and rhizosphere solution. Histograms represent the mean value of four replicates. Bars represent standard errors. For each depth and season, bars with an asterisk are significantly different according to a one-factor (soil compartment) ANOVA and the Student-Newman-Keuls test ($p < 0.05$). For each depth and each soil compartment, bars with the same letter (a, b, c) are not significantly different according to a one-factor (season) ANOVA and the Student-Newman-Keuls test ($p < 0.05$).

layer II, *i.e.* higher enrichment of the rhizosphere solution in spring. For Fe, the rhizosphere solutions were enriched under beech in layer I in autumn and in layer II in winter, and under Norway spruce in layer III in autumn, winter and spring. Significant seasonal variations of the rhizosphere effect on Fe were observed under beech in layer I, *i.e.* higher enrichment of the rhizosphere solution in autumn, and under Norway spruce in layer III, *i.e.* higher enrichment of the rhizosphere solution in spring. For Mn, the rhizosphere solutions were enriched under beech in layer III in winter, and under Norway spruce in layers I and II in spring and in layer III in winter and spring. Significant seasonal variation of the rhizosphere effect on Mn was observed under Norway spruce in layer I, *i.e.* higher enrichment of the rhizosphere solution in spring.

3.4. Trace Elements (Ba, Cd, Ce, Co, Cr, Cs, Cu, Ga, La, Nd, Ni, Pb, Rb, Sr, Th, U, V, Y and Zn)

Figure 3 represents percentages of difference for trace elements between the bulk soil and rhizosphere solutions for each depth, each season, and both species. In layer I, the general tendency was a depletion of trace elements (except for Rb) in the rhizosphere solution during the winter under both species and an accumulation of trace elements in the rhizosphere solution during the autumn under beech as compared to the bulk soil solution. In layer III, we observed an accumulation of most of the trace elements in the rhizosphere solution whatever the season and the species. Interestingly, the rhizosphere solutions were enriched in Rb whatever the depth, the species and the season.

4. Discussion

4.1. Rhizosphere Effect on pH, Carbon, and Nitrogen

Our study demonstrated that, whatever the depth (except in layer III under beech) and the species, the rhizo-

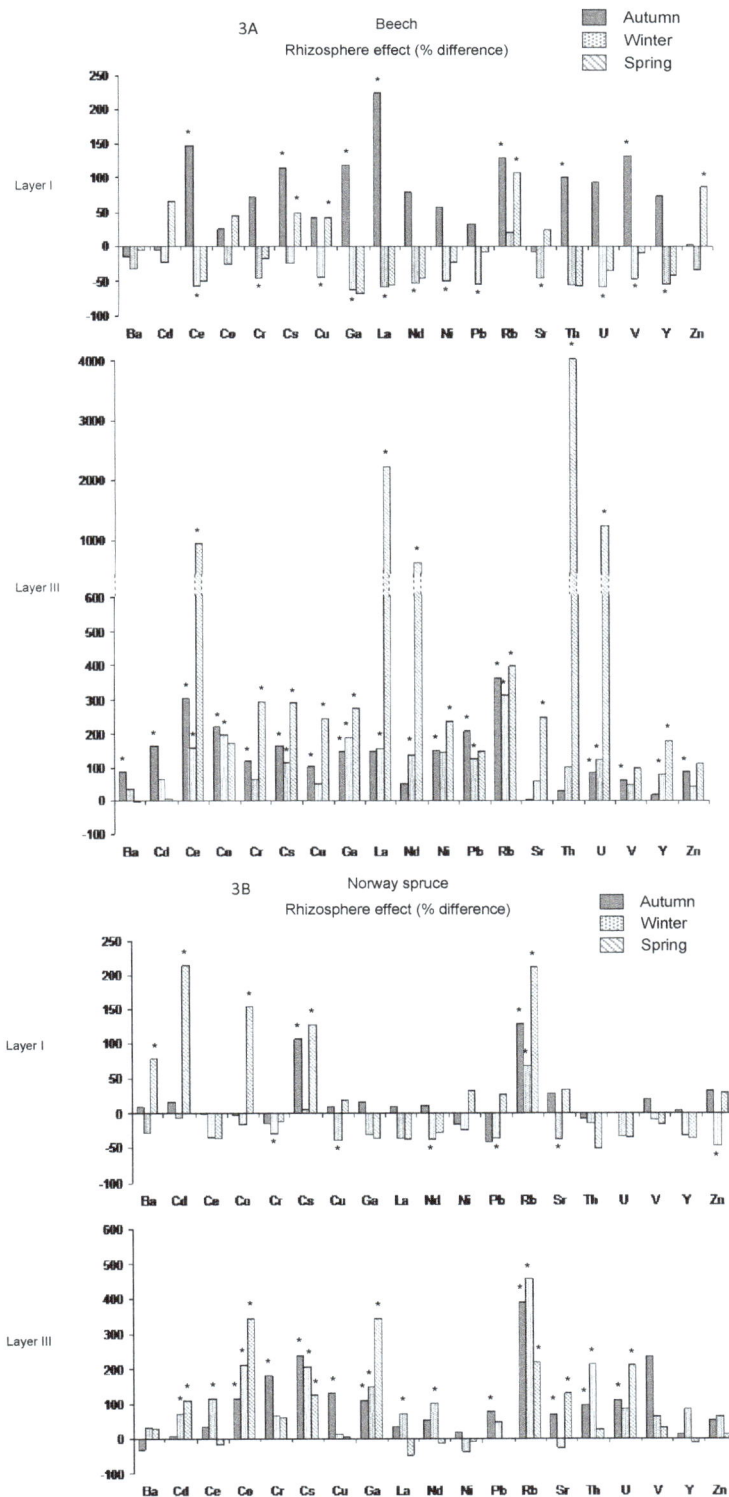

Figure 3. Rhizosphere effect for trace elements under beech (A) and Norway spruce (B) for the two depths (layer I: 0 - 3 cm and layer III: 10 - 23 cm) and the three seasons (autumn, winter, and spring). Rhizosphere effect is expressed as a percentage of difference between bulk soil solution and rhizosphere solution. Histograms represent the mean value of four replicates. Bars represent standard errors. For each depth and season, bars with an asterisk are significantly different according to a one-factor (soil compartment) ANOVA and the Student-Newman-Keuls test (p < 0.05).

sphere solution was enriched in carbon in spring as compared to that of the bulk soil. The rhizosphere solutions were also generally enriched in N. The accumulation of C in the rhizosphere solution can result from the release by roots during the growing period of an abundant amount of carbon into the surrounding soil, called rhizodeposition [36], which includes the C derived from photosynthates.

Our study also showed that pH in bulk soil solution did not significantly differ from that of rhizosphere solution in autumn and spring for both species. In contrast, the rhizosphere solution is more acidic than that of the bulk soil under beech (layers I, II, and III) and Norway spruce (layer II, only) in winter. The acidification of the rhizosphere solution can be attributed to carbonic and organic acids produced by roots and rhizosphere microflora through respiration and exudation as well as to proton release by roots to compensate for an unbalanced cation-anion uptake at the soil-root interface [37]. Because nutrient uptake is very low or inexistent in winter, it is likely that the decrease of pH in the rhizosphere mainly results from the activity of soil microorganisms. Our results also revealed an enrichment of the rhizosphere in carbon and nitrogen in winter for both species in the 10 - 23 cm horizon. The decrease of pH and the high production of acid compounds, carbon, and nitrogen in the rhizosphere can be explained by the fine-root decomposition but also by root and root-associated microorganism activity that can occur even in winter [38]-[40]. While roots tend to freeze and die at soil temperatures below −6°C, minimum temperatures for root growth are thought to be between 0°C and 5°C. Thus, if soil temperatures warm to or stay above this minimum, as was the case during our study period with a relatively warm winter (mean air temperature > 4°C), winter roots can break dormancy and become active thus stimulating rhizosphere microflora. In addition, [41] demonstrated that soil microorganisms can maintain both catabolic (CO_2 production) and anabolic (biomass synthesis) processes under frozen conditions and that no significant differences in carbon allocation from [^{13}C] glucose into [^{13}C] CO_2 and cell organic ^{13}C-compounds occurred between +9°C and −4°C.

The large quantities of carbon in the rhizosphere could enhance the nutrient availability through an increase of CEC (increase of exchange sites) and a stimulation of microbial growth and activities which favours organic matter mineralization and mineral weathering [19].

4.2. Rhizosphere Effect on Major Element Availability

For Norway spruce and beech, the rhizosphere solutions were enriched in K whatever the depth or the season as compared to those of the bulk soil. In addition, the rhizosphere solutions were generally enriched in Ca, Mg, and Na in spring as compared to the bulk soil ones. No significant depletion of the rhizosphere solution in major elements was observed during the year of the study. Our results are in agreement with previous studies led in the Breuil-Chenue site that showed an enrichment of the rhizosphere of Norway spruce, beech, and oak in exchangeable nutrients such as K, Ca and Mg in spring [4] [27], as well as the enrichment of the rhizosphere clay-sized fraction of Norway spruce and beech in elements such as K, Mg, and Fe [10]. These latter suggest that the formation of mica-like minerals in the rhizosphere in spring results from the presence in the rhizosphere solution of large amounts of K possessing a strong affinity for the high charge expandable phyllosilicate like vermiculite, which become fixed in the interlayer space [24] [42] [43]. Many authors have already observed an enrichment of exchangeable K in the rhizosphere soil and solution of different mature tree species such as Norway spruce, trembling aspen (*Populus tremuloides* Michx), and Douglas fir [8] [21] [24] [29] [44]. In addition, the accumulation of nutrients in the rhizosphere solution has already been observed in other forest sites under Douglas-fir [24] [45] and Norway spruce [20]. However, a depletion of cations (such as K, Mg and Ca) in the rhizosphere solutions was also observed during tree growth [21] [22] [46]. These authors suggest that nutrient uptake by plants can lead to a decrease of nutrients in the rhizosphere. In our study, the accumulation of nutrients in the rhizosphere solution can thus be attributed to positive differences between the nutrient inputs, principally by mineral weathering, organic matter mineralization and mass flow as well as nutrient output principally by root uptake. In the same site of Breuil-Chenue, a rate increase of organic matter decomposition [47] and of mineral weathering [10] was observed in the rhizosphere, revealing that the element mobilisation is exacerbated in the vicinity of roots due to root and root-associated microorganism activity. Furthermore, previous studies led in the Breuil-Chenue site showed that bacterial strains with a high mineral weathering efficiency were enriched in the ectomycorrhizosphere of beech and oak as compared to the bulk soil [48] [49]. Similarly, from soil samples collected on the same dates as those of our study, [50] demonstrated that the culturable bacterial communities of spruce and beech rhizospheres were characterized by a higher density and a higher potential of mineral weathering compared to the bulk soil ones. Ectomycorrhizal fungi are also known to significantly affect nutrient cycling in soils [51]. For example, [52] demonstrated that ectomycorrhizal fungi enhanced the mineral weathering and contributed thus to the increase of nutrient concentration (such as Ca, Mg, Fe and Mn) in the soil solutions of ectomycorrhizosphere of subalpine fine. These results suggest that high biological activities within the rhizosphere significantly improve soil nutrient availability. This hypothesis is in accordance with the conceptual model of nutrient availability in the soil-root system proposed by [5] and [53].

In addition, our study reveals that the accumulation of nutrients in the rhizosphere solution was generally higher in spring compared to that in autumn and in winter. This result is in disagreement with the study of [54] who observed a depletion of Ca and Mg in the rhizosphere solution of Douglas-fir in June compared to August, November and February. They suggested that the strong tree uptake during the spring season could cause a decrease of Ca and Mg in the vicinity of roots. Our results, however, corroborate the increase of exchangeable nutrients (especially K) observed in the rhizosphere of Norway spruce and beech in the same soil samples collected in spring compared to the other seasons [26]. These results suggest that even if the root nutrient uptake is high in spring, the intensity of biological activities favouring organic matter degradation and mineral weathering is such as the input-output nutrient budget is higher in spring than in autumn and winter when the root nutrient uptake is low or inexistent. In the beech rhizosphere, [55] showed that enzymatic activities involved in the organic matter degradation (such as N-acetylglucosaminidase, cellobiohydrolase and β-glucosidase) were higher at the beginning of the vegetation period (in June) than in August and September. These authors suggested an effect of a high carbon input due to photosynthetic activity in June, when leaves have been fully developed. In addition, [56] observed an increase of mineral weathering in the rhizosphere of Douglas-fir between March and June. As demonstrated by [26] and [35], nutrient stocks are very low in the soil of the Breuil-Chenue forest. According to [57], Ca and Mg pools at the Breuil-Chenue site were amongst the lowest as compared to other monitored sites in France [58]-[60]. Our results thus suggest that in such nutrient-poor soils, trees can enhance nutrient availability in spring in the rhizosphere by a positive feedback loop between roots, microbial communities and soils. This complex process may thus allow maintaining a favorable environment for tree nutrition during the tree growth period.

Interestingly, the dynamics of K in soil solutions differ from those of Ca, Mg, and Na during the year. In fact, the rhizosphere solutions were enriched in K in autumn, winter and spring as compared to the bulk soil ones while the rhizosphere solutions were enriched in Ca, Mg, and Na in spring, only. These results suggest that a weak rhizosphere activity is sufficient to mobilize K to cover tree nutrient needs all throughout the year. In contrast, as soon as the rhizosphere activity slows down, the availability of Ca, Mg, and Na decreases. This highlights the main role of biological rhizosphere processes in the bioavailability of these nutrients. As demonstrated by [56], soil exchangeable Mg pools are very small and decreased between 1974 and 2001 in the Breuil-Chenue forest, thus the biological rhizosphere processes are going to play a growing role in tree nutrition in the future.

4.3. Rhizosphere Effect on Trace Element Availability

In the present study, the trace element concentrations in soil solutions were significantly different between the rhizosphere and the bulk soil for both species, especially in layer III. These results show for the first time that the tree rhizosphere influences the availability of trace elements in uncontaminated forest soils. According to [61], the impact of roots on trace element availability can result from several processes, *i.e.*, accumulation/depletion of ionic species in the rhizosphere, acidification/alkalinization of the rhizosphere, oxidation/reduction in the rhizosphere, and complexation/chelation in the rhizosphere. As observed in this study for major elements, the rhizosphere solution was generally enriched in trace elements as compared to that of the bulk soil. Because the mineral soil represents the main source of trace elements in terrestrial environments, this enrichment could result from the increase of mineral weathering and organic matter mineralization in the rhizosphere. For example, [29] demonstrated in forest soils that microorganisms contributed to Cu increase in the tree rhizosphere as compared to the bulk soil and suggested that microbial mineralization could partly supply Cu to the solution fraction of the rhizosphere.

A significant depletion of most trace elements was observed in the rhizosphere solution of both tree species in winter and in layer I. This depletion may result from the adsorption of trace elements on oxides and hydroxides of Al, Fe or Mn and on organic matter as demonstrated by [62]-[64]. Furthermore, [65] showed, in contaminated forest soils, that the tree rhizosphere was an accumulation zone of Zn and Cu associated with inorganic amorphous solid phase due to the increase of Fe and Mn oxides in the rhizosphere as compared to the bulk soil. During the winter, the amount of dead roots can equal that of living roots [66]. This organic matter may also have adsorbed trace elements thus decreasing the amount of trace elements in the rhizosphere solution. In addition, it is known that mycorrhizal fungi play a key role in the heavy metal and radionuclide circulation in the soils [67] [68]. For example, [30] demonstrated in a forest soil that Cs was largely accumulated in fungal mycelium compared to bulk soil and rhizosphere. Consequently, the trace elements uptake by mycorrhizal fungi could contribute to the decrease of these elements in the vicinity of roots during the winter. This hypothesis is in agreement with [38] [39] that showed that ectomycorrhizal fungi can be metabolically active during winter period in temperate forest soils. According to [69], the changes of pH in soils and soil solutions may play an especially significant role in the rate of availability of certain trace elements. Between autumn and winter, we observed a high increase in the pH of the bulk soil solution. These conditions may have decreased the adsorption of trace elements in the bulk soil thus favouring their solubility.

It is interesting to note that Rb was the only trace element enriched in the rhizosphere solution independently of the tree species, the depth, and the season. This result confirms the enrichment of the rhizosphere solution in K we observed. Indeed, both K and Rb show the same uptake kinetic by plants and $^{86}Rb^+$ has often been used in studies of K uptake [70] as it seems to emulate K to a high degree due to rather similar physicochemical properties (*i.e.*, valence and ion diameter).

5. Conclusion

Living plant roots and associated microorganisms have been recognized to influence the biogeochemical parameters of soil in their vicinity. Our study demonstrates that common European tree species such as beech and Norway spruce significantly impact the chemical characteristics of the rhizosphere solution, mainly during the growing period. Whatever the depth and the species are, an enrichment of the rhizosphere solutions in K, Ca, Mg, and Na is observed in spring (autumn, winter, and spring for K). This reveals positive differences between the nutrient inputs, principally by mineral weathering, organic matter mineralization and mass flow as well as nutrient output mainly by root uptake. Hence, our results confirm that the high biological activity is a key process that improves soil fertility and supports tree nutrition in acidic and nutrient-poor forest soils. In contrast, for the trace elements we observe an enrichment of the rhizosphere solution whatever the season and the species in the mineral horizon are, only, while seasonal variations of the rhizosphere effect appear in the organo-mineral horizon. This suggests that other mechanisms affect the dynamics of trace elements in the organo-mineral horizon. Further studies in the long term are required to reach clearer conclusions and generalizations.

Acknowledgements

We acknowledge Krista Bateman for review of our English, and Zornitza Tosheva, Louisette and Dominique Gelhaye, Séverine Bienaimé, Bruno Simon, Pascal Bonnaud, Claude Nys, René Boutin, Robert Wagener, Ernest Apel, Astrid Tobias, for technical assistance. This work was supported by the Research National Fund of Luxemburg and the LorLux network.

References

[1]   Marschner, H. (1995) Mineral Nutrition of Higher Plants. 2nd Edition, Academic Press, London.

[2]   Darrah, P.R. (1993) The Rhizosphere and Plant Nutrition: A Quantitative Approach. *Plant and Soil*, **155-156**, 1-20. http://dx.doi.org/10.1007/BF00024980

[3]   Hinsinger, P., Gobran, G.R., Gregory, P.J. and Wenzel W.W. (2005) Rhizosphere Geometry and Heterogeneity Arising from Root-Mediated Physical and Chemical Processes. *New Phytologist*, **168**, 293-303. http://dx.doi.org/10.1111/j.1469-8137.2005.01512.x

[4]   Calvaruso, C., N'Dira, V. and Turpault, M.-P. (2011) Impact of Common European Tree Species and Douglas-Fir (*Pseudotsuga menziesii* [Mirb.] Franco) on the Physicochemical Properties of the Rhizosphere. *Plant and Soil*, **342**, 469-480. http://dx.doi.org/10.1007/s11104-010-0710-x

[5]   Gobran, G.R. and Clegg, S. (1996) A Conceptual Model for Nutrient Availability in the Mineral Soil-Root System. *Canadian Journal of Soil Science*, **76**, 125-131. http://dx.doi.org/10.4141/cjss96-019

[6]   Dessaux, Y., Hinsinger, P. and Lemanceau, P. (2007) Rhizosphere: Achievements and Challenges. Springer Science Press, Berlin.

[7]   Hinsinger, P., Plassard, C. and Jaillard, B. (2006) Rhizosphere: A New Frontier for Soil Biogeochemistry. *Journal of Geochemical Exploration*, **88**, 210-213. http://dx.doi.org/10.1016/j.gexplo.2005.08.041

[8]   Yanai, R.D., Majdi, H. and Park., B.P. (2003) Measured and Modelled Differences in Nutrient Concentrations between

Rhizosphere and Bulk Soil in a Norway Spruce Stand. *Plant and Soil*, **257**, 133-142. http://dx.doi.org/10.1023/A:1026257508033

[9]   Tinker, P.B. and Nye, P.H. (2000) Solute Movement in the Rhizosphere. Oxford University Press, New York.

[10] Calvaruso, C., Mareschal, L., Turpault, M.-P. and Leclerc, E. (2009) Rapid Clay Weathering in the Rhizosphere of Norway Spruce and Oak in an Acid Forest Ecosystem. *Soil Science Society of America Journal*, **73**, 331-338. http://dx.doi.org/10.2136/sssaj2007.0400

[11] Drever, J.I. and Stillings, L.L. (1997) The Role of Organic Acids in Mineral Weathering. *Colloids and Surfaces A: Physicochemical and Engineering Aspects*, **120**, 167-181. http://dx.doi.org/10.1016/S0927-7757(96)03720-X

[12] Grayston, S.J., Vaughan, D. and Jones, D. (1996) Rhizosphere Carbon Flow in Trees, in Comparison with Annual Plants: The Importance of Root Exudation and Its Impact on Microbial Activity and Nutrient Availability. *Applied Soil Ecology*, **5**, 29-56. http://dx.doi.org/10.1016/S0929-1393(96)00126-6

[13] Warembourg, F.R. (1997) The "Rhizosphere Effect": A Plant Strategy for Plants to Exploit and Colonize Nutrient-Limited Habitats. *Bocconea*, **7**, 187-194.

[14] Courty, P.E., Buée, M., Diedhiou, A.G., Frey-Klett, P., Le Tacon, F., Rineau, F., Turpault, M.P., Uroz, S. and Garbaye, J. (2010) The Role of Ectomycorrhizal Communities in Forest Ecosystem Processes: New Perspectives and Emerging Concepts. *Soil Biology and Biochemistry*, **42**, 679-698. http://dx.doi.org/10.1016/j.soilbio.2009.12.006

[15] Dakora, D.F. and Phillips, D.A. (2002) Root Exudates as Mediators of Mineral Acquisition in Low-Nutrient Environments. *Plant and Soil*, **245**, 35-47. http://dx.doi.org/10.1023/A:1020809400075

[16] Uroz, S., Calvaruso, C., Turpault, M.P. and Frey-Klett, P. (2009) Mineral Weathering by Bacteria: Ecology, Actors and Mechanisms. *Trends in Microbiology*, **17**, 378-387. http://dx.doi.org/10.1016/j.tim.2009.05.004

[17] Phillips, R.P. and Fahey, T.J. (2008) The Influence of Soil Fertility on Rhizosphere Effects in Northern Hardwood Forest Soils. *Soil Science Society of America Journal*, **72**, 453-461. http://dx.doi.org/10.2136/sssaj2006.0389

[18] Hinsinger, P., Bravin, M.N., Devau, N., Gerard, F., Le Cadre, E. and Jaillard, B. (2008) Soil-Root-Microbe Interactions in the Rhizosphere—A Key to Understanding and Predicting Nutrient Bioavailability to Plants. *5th International Symposium Interactions of Soil Minerals with Organic Components and Microorganisms*, Pucon, 24-28 November 2008.

[19] Lambers, H., Mougel, C., Jaillard, B. and Hinsinger, P. (2009) Plant-Microbe-Soil Interactions in the Rhizosphere: An Evolutionary Perspective. *Plant and Soil*, **321**, 83-115. http://dx.doi.org/10.1007/s11104-009-0042-x

[20] Arocena, J.M., Göttlein, A. and Raidl, S. (2004) Spatial Changes of Soil Solution and Mineral Composition in the Rhizosphere of Norway-Spruce Seedlings Colonized by *Piloderma croceum*. *Journal of Plant Nutrition and Soil Science*, **167**, 479-486. http://dx.doi.org/10.1002/jpln.200320344

[21] Dieffenbach, A. and Matzner, E. (2000) *In Situ* Soil Solution Chemistry in the Rhizosphere of Mature Norway Spruce (*Picea abies* [L.] Karst.) Trees. *Plant and Soil*, **222**, 149-161. http://dx.doi.org/10.1023/A:1004755404412

[22] Dieffenbach, A., Göttlein, A. and Matzner, E. (1997) *In Situ* Soil Solution Chemistry in an Acid Forest Soil as Influenced by Growing Roots of Norway Spruce (*Picea abies* [L.] Karst.). *Plant and Soil*, **192**, 57-61. http://dx.doi.org/10.1023/A:1004283508101

[23] Gottlein, A., Heim, A. and Matzner, E. (1999) Mobilization of Aluminium in the Rhizosphere Soil Solution of Growing Tree Roots in an Acidic Soil. *Plant and Soil*, **211**, 41-49. http://dx.doi.org/10.1023/A:1004332916188

[24] Turpault, M.P., Uterano, C., Boudot, J.P. and Ranger, J. (2005) Influence of Mature Douglas Fir Roots on the Solid Soil Phase of the Rhizosphere and Its Solution Chemistry. *Plant and Soil*, **275**, 327-336. http://dx.doi.org/10.1007/s11104-005-2584-x

[25] Wang, Z.Y., Gottlein, A. and Bartonek, G. (2001) Effects of Growing Roots of Norway Spruce (*Picea abies* [L.] Karst.) and European Beech (*Fagus sylvatica* L.) on Rhizosphere Soil Solution Chemistry. *Journal of Plant Nutrition and Soil Science*, **164**, 35-41. http://dx.doi.org/10.1002/1522-2624(200102)164:1<35::AID-JPLN35>3.0.CO;2-M

[26] Collignon, C., Calvaruso, C. and Turpault, M.P. (2011) Temporal Dynamics of Exchangeable K, Ca and Mg in Acidic Bulk Soil and Rhizosphere under Norway Spruce (*Picea abies* Karst.) and Beech (*Fagus sylvatica* L.) Stands. *Plant and Soil*, **349**, 355-366. http://dx.doi.org/10.1007/s11104-011-0881-0

[27] Legrand, P., Turmel, M.C., Sauvé, S. and Courchesne, F. (2005) Speciation and Bioavailability of Trace Metals (Cd, Cu, Ni, Pb, Zn) in the Rhizosphere of Contaminated Soils. In: Huang, P.M. and Gobran, G.R., Eds., *Biogeochemistry of Trace Elements in the Rhizosphere*, Elsevier, Amsterdam, 261-299. http://dx.doi.org/10.1016/B978-044451997-9/50010-6

[28] Séguin, V., Gagnon, C. and Courchesne, F. (2004) Changes in Water Extractable Metals, pH and Organic Carbon Concentrations at the Soil-Root Interface of Forested Soils. *Plant and Soil*, **260**, 1-17. http://dx.doi.org/10.1023/B:PLSO.0000030170.49493.5f

[29] Cloutier-Hurteau, B., Sauvé, S. and Courchesne, F. (2008) Influence of Microorganisms on Cu Speciation in the

Rhizosphere of Forest Soils. *Soil Biology and Biochemistry*, **40**, 2441-2451.
http://dx.doi.org/10.1016/j.soilbio.2008.06.006

[30] Vinichuk, M., Taylor, A.F.S., Rosén, K. and Johanson, K.J. (2010) Accumulation of Potassium, Rubidium and Cae-sium ([133]Cs and ^{137}Cs) in Various Fractions of Soil and Fungi in a Swedish Forest. *Science of the Total Environment*, **408**, 2543-2548. http://dx.doi.org/10.1016/j.scitotenv.2010.02.024

[31] April, R. and Keller, D. (1990) Mineralogy of the Rhizosphere in Forest Soils of the Eastern United-States-Mineralogic Studies of the Rhizosphere. *Biogeochemistry*, **9**, 1-18. http://dx.doi.org/10.1007/BF00002714

[32] Bakker, M.R., George, E., Turpault, M.P., Zhang, J.L. and Zeller, B. (2004) Impact of Douglas-Fir and Scots Pine Seedlings on Plagioclase Weathering under Acidic Conditions. *Plant and Soil*, **266**, 247-259.
http://dx.doi.org/10.1007/s11104-005-1153-7

[33] Courchesne, F. and Gobran, G.R. (1997) Mineralogical Variations of Bulk and Rhizosphere Soils from a Norway Spruce Stand. *Soil Science Society of America Journal*, **61**, 1245-1249.
http://dx.doi.org/10.2136/sssaj1997.03615995006100040034x

[34] USDA (1999) Soil Taxonomy: A Basic System of Soil Classification for Making and Interpreting Soil Surveys. 2nd Edition, Agriculture Handbook Number 436, US Government Printing Office, Washington DC.

[35] Mareschal, L., Bonnaud, P., Turpault, M.P. and Ranger, J. (2010) Impact of Common European Tree Species on the Chemical and Physicochemical Properties of Fine Earth: An Unusual Pattern. *European Journal of Soil Science*, **61**, 14-23. http://dx.doi.org/10.1111/j.1365-2389.2009.01206.x

[36] Jones, D.L., Hodge, A. and Kuzyakov, Y. (2004) Plant and Mycorrhizal Regulation of Rhizodeposition. *New Phytologist*, **163**, 459-480. http://dx.doi.org/10.1111/j.1469-8137.2004.01130.x

[37] Hinsinger, P., Plassard, C., Tang, C.X. and Jaillard, B. (2003) Origins of Root-Mediated pH Changes in the Rhizo-sphere and Their Responses to Environmental Constraints: A Review. *Plant and Soil*, **248**, 43-59.
http://dx.doi.org/10.1023/A:1022371130939

[38] Buée, M., Vairelles, D. and Garbaye, J. (2005) Year-Round Monitoring of Diversity and Potential Metabolic Activity of the Ectomycorrhizal Community in a Beach (*Fagus silvatica*) Forest Subjected to Two Thinning Regimes. *Mycorrhiza*, **15**, 235-245. http://dx.doi.org/10.1007/s00572-004-0313-6

[39] Courty, P.E., Pouysegur, R., Buée, M. and Garbaye, J. (2006) Laccase and Phosphatase Activities of the Dominant Ectomycorrhizal Types in a Lowland Oak Forest. *Soil Biology and Biochemistry*, **38**, 1219-1222.
http://dx.doi.org/10.1016/j.soilbio.2005.10.005

[40] Meinen, C., Hertel, D. and Leuschner, C. (2009) Root Growth and Recovery in Temperate Broad-Leaved Forest Stands Differing in Tree Species Diversity. *Ecosystems*, **12**, 1103-1116. http://dx.doi.org/10.1007/s10021-009-9271-3

[41] Harrysson Drotz, S., Sparrman, T., Nilsson, M.B., Schleucher, J. and Öquist, M.G. (2010) Both Catabolic and Ana-bolic Heterotrophic Microbial Activity Proceed in Frozen Soils. *Proceedings of the National Academy of Sciences of the United States of America*, **107**, 21046-210515. http://dx.doi.org/10.1073/pnas.1008885107

[42] Nettleton, W.D., Nelson, R.E. and Flach, K.W. (1973) Formation of Mica in Surface Horizons of Dryland Soils. *Soil Science Society of America Journal*, **37**, 473-478. http://dx.doi.org/10.2136/sssaj1973.03615995003700030043x

[43] Tice, K.R., Graham, R.C. and Wood, H.B. (1996) Transformations of 2:1 Phyllosilicates in 41-Year-Old Soils under Oak and Pine. *Geoderma*, **70**, 49-62. http://dx.doi.org/10.1016/0016-7061(95)00070-4

[44] Clegg, S. and Gobran, G.R. (1997) Rhizosphere Chemistry in an Ammonium Sulfate and Water Manipulated Norway Spruce [*Picea abies* (L.) Karst.] Forest. *Canadian Journal of Soil Science*, **77**, 525-533.
http://dx.doi.org/10.4141/S95-069

[45] Zhang, J.L. and George, E. (2009) Rhizosphere Effects on Ion Concentrations near Different Root Zone of Norway Spruce (*Picea abies* (L.) Karst.) and Root Types of Douglas-Fir (*Pseudotsuga menziesii* L.) Seedlings. *Plant and Soil*, **322**, 209-218. http://dx.doi.org/10.1007/s11104-009-9909-0

[46] Bakker, M.R., Dieffenbach, A. and Ranger, J. (1999) Soil Solution Chemistry in the Rhizosphere of Roots of Sessile Oak (*Quercus petraea*) as Influenced by Lime. *Plant and Soil*, **209**, 209-216.
http://dx.doi.org/10.1023/A:1004511712471

[47] Colin-Belgrand, M., Dambrine, E., Bienaimé, S., Nys, C. and Turpault, M.P. (2003) Influence of Tree Roots on Nitro-gen Mineralization. *Scandinavian Journal of Forest Research*, **18**, 260-268.
http://dx.doi.org/10.1080/02827581.2003.9728296

[48] Calvaruso, C., Turpault, M.P., Leclerc, E. and Frey-Klett, P. (2007) Impact of Ectomycorrhizosphere on the Functional Diversity of Soil Bacterial and Fungal Communities from a Forest Stand in Relation to Nutrient Mobilization Processes. *Microbial Ecology*, **54**, 567-577. http://dx.doi.org/10.1007/s00248-007-9260-z

[49] Uroz, S., Calvaruso, C., Turpault, M.P., Pierrat, J.C., Mustin, C. and Frey-Klett, P. (2007) Effect of the Mycorrhizo-

sphere on the Genotypic and Metabolic Diversity of the Soil Bacterial Communities Involved in Mineral Weathering in a Forest Soil. *Applied and Environmental Microbiology*, **73**, 3019-3027. http://dx.doi.org/10.1128/AEM.00121-07

[50] Collignon, C., Uroz, S., Turpault, M.P. and Frey-Klett, P. (2011) Seasons Differently Impact the Structure of Mineral Weathering Bacterial Communities in Beech and Spruce Stands. *Soil Biology and Biochemistry*, **43**, 2012-2022. http://dx.doi.org/10.1016/j.soilbio.2011.05.008

[51] Gadd, G.M. (2007) Geomycology: Biogeochemical Transformation of Rocks, Minerals, Metals and Radionuclides by Fungi, Bioweathering and Bioremediation. *Mycological Research*, **111**, 3-49. http://dx.doi.org/10.1016/j.mycres.2006.12.001

[52] Arocena, J.M. and Glowa, K.R. (2000) Mineral Weathering in Ectomycorrhizosphere of Subalpine Fir (*Abies lasiocarpa* (Hook.) Nutt.) as Revealed by Soil Solution Composition. *Forest Ecology and Management*, **133**, 61-70. http://dx.doi.org/10.1016/S0378-1127(99)00298-4

[53] Gobran, G.R., Clegg, S. and Courchesne, F. (1998) Rhizospheric Processes Influencing the Biogeochemistry of Forest Ecosystems. *Biogeochemistry*, **42**, 107-120. http://dx.doi.org/10.1023/A:1005967203053

[54] Wang, X.P. and Zabowski, D. (1998) Nutrient Composition of Douglas-Fir Rhizosphere and Bulk Soil Solutions. *Plant and Soil*, **200**, 13-20. http://dx.doi.org/10.1023/A:1004240315308

[55] Esperschütz, J., Pritsch, K., Gattinger, A., Welzl, G., Haesler, J., Buegger, F., Winkler, J.B., Munch, J.C. and Schloter, M. (2009) Influence of Chronic Ozone Stress on Carbon Translocation Pattern into Rhizosphere Microbial Communities of Beech Trees (*Fagus sylvatica* L.) during a Growing Season. *Plant and Soil*, **323**, 85-95. Http://dx.doi.org/10.1007/S11104-009-0090-2

[56] Turpault, M.P., Righi, D. and Utérano, C. (2008) Clay Minerals: Precise Markers of the Spatial and Temporal Variability of the Biogeochemical Soil Environment. *Geoderma*, **147**, 108-115. http://dx.doi.org/10.1016/j.geoderma.2008.07.012

[57] van der Heijden, G., Legout, A., Pollier, B., Mareschal, L., Turpault, M.P., Ranger, J. and Dambrine, E. (2013) Assessing Mg and Ca Depletion from Broadleaf Forest Soils and Potential Causes—A Case Study in the Morvan Mountains. *Forest Ecology and Management*, **293**, 65-78. http://dx.doi.org/10.1016/j.foreco.2012.12.045

[58] Legout, A., Walter, C. and Nys, C. (2008) Spatial Variability of Nutrient Stocks in the Humus and Soils of a Forest Massif (Fougeres, France). *Annals of Forest Science*, **65**, 108-118. http://dx.doi.org/10.1051/forest:2007080

[59] Nys, C., Ranger, D., Ranger, J., Bonnaud, P., Gelhaye, D., Lhomme, J., Masar, L. and Vairelles, D. (1983) Comparative Study of Two Forest Ecosystems in French Primary Ardennes 3. Bioelement Content and Biological Cycle. *Annals of Forest Science*, **40**, 41-66. http://dx.doi.org/10.1051/forest:19830102

[60] Nys, C. and Ranger, J. (1988) The Consequence of the Substitution of Tree Species on the Biogeochemical Mechanism of a Forest Ecosystem—The Sulfur Cycle. *Annals of Forest Science*, **45**, 169-188. http://dx.doi.org/10.1051/forest:19880301

[61] Hinsinger, P. (2000) Bioavailability of Trace Elements as Related to Root-Induced Changes in the Rhizosphere. In: Gobran, G.R., Wenzel, W.W. and Lombi, E., Eds., *Trace Elements in the Rhizosphere*, CRC Press LCC, Boca Raton, 25-41. http://dx.doi.org/10.1201/9781420039993.ch2

[62] Liu, C. and Huang, P.M. (2005) Kinetics of Cadmium Desorption from Iron Oxides Formed under the Influence of Citrate. In: Huang, P.M. and Gobran, G.R., Eds., *Biogeochemistry of Trace Elements in the Rhizosphere*, Elsevier, Amsterdam, 183-196. http://dx.doi.org/10.1016/B978-044451997-9/50008-8

[63] Sipos, P., Németh, T., Kovács Kis, V. and Mohai, I. (2008) Sorption of Copper, Zinc and Lead on Soil Mineral Phases. *Chemosphere*, **73**, 461-469. http://dx.doi.org/10.1016/j.chemosphere.2008.06.046

[64] Violante, A., Ricciardella, M., Pigna, M. and Capasso, R. (2005) Effects of Organic Ligands on the Adsorption of Trace Elements onto Metal Oxides and Organo-Mineral Complexes. In: Huang, P.M. and Gobran, G.R., Eds., *Biogeochemistry of Trace Elements in the Rhizosphere*, Elsevier, Amsterdam, 157-182. http://dx.doi.org/10.1016/B978-044451997-9/50007-6

[65] Séguin, V., Courchesne, F., Gagnon, C., Martin, R.R., Naftel, S.J. and Skinner, W. (2005) Mineral Weathering in the Rhizosphere of Forested Soil. In: Huang, P.M. and Gobran, G.R., Eds., *Biogeochemistry of Trace Elements in the Rhizosphere*, Elsevier, Amsterdam, 29-55. http://dx.doi.org/10.1016/B978-044451997-9/50004-0

[66] Craul, P.J. (1992) Urban Soil in Landscape Design. Wiley, New York.

[67] Leyval, C. (2005) Efect of Arbuscular Mycorhhizal (AM) Fungi on Heavy Metal and Radionuclide Transfer to Plants. In: Huang, P.M. and Gobran, G.R., Eds., *Biogeochemistry of Trace Elements in the Rhizosphere*, Elsevier, Amsterdam, 419-429. http://dx.doi.org/10.1016/B978-044451997-9/50016-7

[68] Rufyikiri, G., Thiry, Y. and Declerck, S. (2005) Uptake and Translocation of Uranium by Arbuscular Mycorrhizal Fungi under Monoxenic Culture Conditions. In: Huang, P.M. and Gobran, G.R., Eds., *Biogeochemistry of Trace Elements in the Rhizosphere*, Elsevier, Amsterdam, 431-455. http://dx.doi.org/10.1016/B978-044451997-9/50017-9

[69] Kabata-Pendias, A. (2000) Trace Elements in Soils and Plants. 3rd Edition, CRC Press, Boca Raton.
      http://dx.doi.org/10.1201/9781420039900

[70] Rodriguez-Navarro, A. (2000) Potassium Transport in Fungi and Plants. *Biochimica et Biophysica Acta*, **1469**, 1-30.
      http://dx.doi.org/10.1016/S0304-4157(99)00013-1

15

Petroleum Products in Soil Mediated Oxidative Stress in Cowpea (*Vigna unguiculata*) and Maize (*Zea mays*) Seedlings

Fidelis Ifeakachuku Achuba

Department of Biochemistry, Delta State University, Abraka, Nigeria
Email: achubabch@yahoo.com

Abstract

The effects of petroleum products (kerosene, diesel, engine oil and petrol) treatment of soil at various sublethal concentrations (0.0%, 0.1%, 0.25%, 0.5%, 1.0%, 1.5% and 2.0%) on oxidative stress markers (lipid peroxidation, superoxide dismutase activity, catalase activity and xanthine oxidase) were studied in cowpea and maize seedlings. The results indicated that the petroleum products caused a significant increase in lipid peroxidation and a significant decrease in the activities of the antioxidant enzymes: Superoxide dismutase, catalase and xanthine oxidase activities. Kerosene had a greater effect on these indicators of oxidative stress than did the other petroleum products. The effects on lipid peroxidation and antioxidant enzymes were more pronounced in cowpea seedlings than in maize seedlings.

Keywords

Cowpea Seedlings, Maize Seedlings, Catalase Activity, Superoxide Dismutase Activity, Xanthine Oxidase Activity

1. Introduction

In photosynthetic plants, two forms of activated oxygen are formed from superoxide anion. These are hydrogen peroxide and hydroxyl radicals. Activated oxygen is often formed as a component of metabolism to enable "complex" chemical reactions such as the oxidation of xenobiotics or the polymerisation of lignin [1]. It is also formed by the dysfunctioning of enzymes or electron transport systems as a result of perturbations in the metabolism caused by chemical or environmental stress [1]. Activated oxygen species are produced in various sites in plants. These include the chloroplast, peroxisomes and mitochondria [2]-[4]. Various Fe-S proteins and NADH dehydrogenase have also been implicated as possible sites of superoxide and hydrogen peroxide formation [1]. The activated oxygen species generated either during normal metabolic activity or during metabolism of xenobiotics, when produced more than the body can accommodate leads to lipid peroxidation.

This process of lipid peroxidation precedes oxidative damage in plants and animals. However, living organisms are endowed with antioxidant defence systems. The defense mechanisms against oxidative damage include enzymes such as superoxide dismutase, catalase as well as non enzymes such as ascorbic acid. Alterations in the level of these antioxidants represent a measure of oxidative stress. In addition, the activity of xanthine oxidase is also an example of defense mechanism as well as a measure of oxidative stress.

Crude petroleum has been reported to alter oxidative stress indices [5]-[8]. These oxidative stress markers include lipid peroxidation and changes in the activities of anti-oxidant enzymes such as superoxide dismutase, catalase as well as xanthine oxidase The aim of this study was to investigate the effects of petroleum products: Kerosene, diesel, engine oil and petrol contaminated soil on some oxidative stress markers in cowpea and maize seedlings.

2. Materials and Methods

2.1. Refined Petroleum Products and Planting Materials

The refined petroleum products of known specific gravities (kerosene = 0.81; diesel = 0.85; engine oil = 0.87; petrol = 0.75) were obtained from Warri Refining and Petrochemical Company, Warri, Nigeria. Improved varieties of maize (*Zea mays*) were obtained as single batch from Delta Agricultural Development Project (DTADP) Ibusa Delta State, Nigeria. Improved varieties of *Vigna unguiculata* (L.) Walp were obtained from International Institute of Tropical Agriculture IITA, Ibadan, Nigeria. The soil (sand 84%, silt 5.0%, clay 0.4% and organic matter 0.6%, pH 6.1) was obtained from a fallow land in Delta State University, Abraka. The nutrient content of the soil used is shown in **Table 1**. The experiment was carried out in a laboratory condition of temperature 28°C and 12 hr day/night regime.

2.2. Soil Treatment and Planting of Seeds

One thousand six hundred grams of soil was added to each small size planting bags (1178.3 cm^3, 15 cm deep) and divided into six groups of five replicates. Groups 1 to 5 contained 0.1%, 0.25%, 0.5%, 1.0% and 2.0% (v/w) respectively of each of the petroleum products while group six served as control (0.0%). To the first bag, 1.6 ml of kerosene, corresponding to 0.1%, was added. The petroleum product treated soil sample was mixed vigorously with hand to obtain homogeneity of the mixture. The procedure was repeated for 0.25%, 0.5%, 1.0%, 1.5% and 2.0%. This same procedure was applied to diesel, engine oil and petrol. Each treatment including control was replicated five times. The treatments were watered every day in order to keep the soil moist. The design of

Table 1. Physicochemical properties of test soil.

Parameters	Value
pH	6.09
Total organic carbon, %	2.90
Phosphorus, mg/kg	<0.01
Nitrogen, mg/kg	8.47
Nitrate, mg/kg	9.86
Cation exchange capacity, meq/100g	0.74
Sodium, mg/kg	9.06
Potassium, mg/kg	6.72
Calcium, mg/kg	2.98
Magnesium, mg/kg	0.31

the experiment was completely randomized design (CRD).

Damaged seeds were determined by floatation. All seeds that floated on water were discarded and others that remained at the bottom of water were deemed potentially plantable. Three seeds were planted in each test bag to an approximate depth of 2 cm immediately after pollution and kept under partial shade. During the experiment 80 cm^3 of water was supplied to the set up as at when needed to keep the soil moist. Germination [which is indicated by the appearance of epicotyls (for cowpea) and hypocotyls (for maize) above the soil level] records was taken at 4 days interval up to 12 days. Seeds, which failed to sprout after 12 days were regarded as not germinable. At the end of each experimental period, the seedlings were carefully removed from the bags by destroying the bags while the bulk soil containing the seedling was placed under slow running tap water to wash off the soil particles.

2.3. Preparation of Extracts for the Determination of Oxidative Stress Markers in the Leaves of Cowpea and Maize Seedlings

The leaves (0.5 g) of four day old cowpea seedlings was measured and homogenized in pre-hilled mortar with pestle in the presence of 0.05 M phosphate buffer pH 7.5 and few drops of butylated hydroxyl toluene (BHT), filtered with double layered cheese cloth and then centrifuge at 5000 g for 10 min. The supernatant obtained was finally used for the determination of oxidative stress markers. The same procedure was adopted in the preparation of extract for eight and twelve-day-old cowpea seedlings respectively. This same procedure was followed in the preparation of leave extract of four, eight and twelve day's old maize seedlings.

2.4. Determination of Lipid Peroxidation Markers in the Leaves of Cowpea and Maize Seedlings

This assay is based on the reaction of malondialdehyde (MDA) with thiobabituric acid (TBA); forming a MDA-TBA$_2$ adduct that absorbs strongly at 532 nm. Acetic acid (1.0 ml) was placed in a test tube and to the test tube 1.0 ml of 10% TBA was added followed by 0.1 ml of the supernatant. The test tube was covered and immersed in boiling water for 15 min. the mixture was cool, thereafter centrifuged at 5000 g for 10 min. The spectrophotometer was zeroed and absorbance of test sample was read at 532 nm against the reagent blank.

2.5. Determination of Superoxide Dismutase (SOD) Activity in the Leaves of Cowpea and Maize Seedlings

SOD inhibits the auto-oxidation of epinephrine to adrenochrome [9]. To 2 ml of the homogenate, 2.5 ml of 0.05 M phosphate buffer, pH 7.4 was added. The reaction was initiated by the addition of 0.5 ml of freshly prepared 0.3 nm epinephrine to the buffer-supernatant mixture. This was mixed by inversion. The reference cuvette contained 2.5 ml of the buffer 0.5 ml of epinephrine and 2 ml of deionized water. The increase in absorbance at 480 nm was monitored every second for 150 seconds. One unit of superoxide dismutase activity is defined as the amount of enzyme required for 50% inhibition of the oxidation of epinephrine to adrenochrome at 480 nm per minute [10]. The enzyme activity was assayed with an Sp 1800 UV/VIS Spectrophptometer.

2.6. Determination of the Activity of MnSOD and Cu/Zn SOD in the Leaves of Cowpea and Maize Seedlings

Manganese dependent superoxide dismutase (MnSOD) was analysed in the presence of 1 mM NaCN, to suppress Cu/Zn SOD activity. The cytosolic copper/zinc superoxide dismutase (Cu/Zn SOD) activity was determined, as the difference between total SOD and cyanide sensitive enzyme activity [11]. The enzyme activity was assayed with a spectrometer S22.

The assay mixture contained 1.0 ml of the homogenate, 10 ml of 0.05 M phosphate buffer pH 7.4 and 0.05 ml of 1 mM NaCN. The reaction was initiated by adding 0.5 ml of 0.3% freshly prepared epinephrine. This was mixed by inversion. The absorbance was monitored at 480 nm for every 30 seconds for 150 seconds.

2.7. Determination of Catalase Activity in the Leaves of Cowpea and Maize Seedlings

Catalase breaks down hydrogen peroxide to give oxygen that oxidizes potassium dichromate. The oxidation of chromate gives a chromophore that absorbs maximally at 610 nm. The enzyme extract (0.5 ml) was added to the reaction mixture containing 1 ml of 0.05 M phosphate buffer (pH 7.5), 0.5 ml of 0.2 M H_2O_2, 0.4 ml H_2O and incubated for different time period t_1, t_2 and t_3 for 1 minute, 2 minutes and 3 minutes respectively. The reaction was terminated after each time interval by the addition of 2 ml of acid reagent (dichromate/acetic acid mixture) which was prepared by mixing 5% potassium dichromate with glacial acetic acid (1:3 by volume). To the control, the enzyme was added after the addition of acid reagent. All the tubes were heated for 10 minutes in boiling water and the absorbance was read at 610 nm with an Sp 1800 UV/VIS Spectrophptometer. Catalase activity was expressed in terms of moles of H_2O_2 consumed/min [12].

2.8. Determination of Xanthine Oxidase Activities in the Leaves of Cowpea and Maize Seedlings

Xanthine oxidase catalyses the conversion of methylene blue to the reduced clourless forms. Enzyme activity is proportional to the reciprocal of time taken for methylene blue to change to colourless. Two test tubes labeled control and test were placed in a test tube rack, one milliter of 0.05% neutral formaldelyde was pipetted into each test tubes. The 0.02% methylene blue solution was added in the test tube labeled test and followed by the addition of 1 ml of the supernatant to the respective test tube. 1 ml of distilled water was added to the control test tube and 2 drops of liquid paraffin was also added in both test tubes to prevent atmospheric oxidation.

2.9. Statistical Analysis

The results were expressed as mean ± SEM. All results were compared with respect to the control. Comparisons between the test and control were made by using Analysis of Variance (ANOVA), Least Significant Difference (LSD) was used to conduct Post Hoc test for the significant difference. Differences at $p < 0.05$ were considered as significant.

3. Results

Lipid peroxidation in the leaves of cowpea and maize seedlings grown in kerosene, diesel, and engine oil and petrol treated soil after four days of germination are shown in **Figure 1**. Lipid peroxidation in the leaves of cowpea seedling grown in kerosene treated soil was found to increase significantly ($p < 0.05$) from 0.25% through 2%. In maize seedlings grown in kerosene treated soil , there was a significant ($p < 0.05$) increase in lipid peroxidation in the leaves at 1.5% and 2% concentrations relative to control. Comparing lipid peroxidation in the leaves of cowpea and maize seedlings grown in kerosene treated soil; lipid peroxidation in the leaves was significantly higher in cowpea seedlings than in maize seedlings at 1.5% and 2% concentrations. Lipid peroxidation in the leaves of cowpea seedlings grown in diesel treated soil was found to increase significantly ($p < 0.05$) from 0.25% through 2% concentration compared with control. In maize seedlings grown in diesel treated soil, there was a significant ($p < 0.05$) increase in lipid peroxidation in the leaves from 0.25% to 2% concentration when compared with control. Lipid peroxidation was significantly ($p < 0.05$) lower in the leaves of cowpea seedlings than in the leaves of maize seedlings at 0.1% and slightly higher in the leaves of cowpea seedling than in the leaves of maize seedling at all other levels of concentrations tested. Also, lipid peroxidation in the leaves of cowpea seedlings grown in engine oil treated soil were found to increase significantly ($p < 0.05$) from 0.25% to 2% concentration when compared with control. Lipid peroxidation in the leaves of maize seedling grown in engine oil treated soil was also found to increase significantly from 0.25% to 2%. In comparison, lipid peroxidation was lower in the leaves of cowpea seedlings than in the leaves of maize seedlings. In all the concentrations tested, petrol treated soil resulted in significant ($p < 0.05$) increase of lipid peroxidation from 0.25% to 2% in the leaves of cowpea seedling when compared with control. Lipid peroxidation in the leaves of maize seedlings grown in petrol treated soil also exhibited significant increase from concentration level of 0.5% to 2% when compared with control. When compared, lipid peroxidation in the leaves of cowpea and maize seedlings grown in petrol treated soil was not significantly different. Lipid peroxidation in the leaves of cowpea and maize seedlings grown in kerosene, diesel, and engine oil and petrol treated soil after eight days of germination are shown in **Figure 1**. Lipid peroxidation in the leaves of cowpea seedlings grown in kerosene treated soil after eight days was found to increase significantly from 0.25% through 2% when compared with control. In maize seedlings grown in kerosene treated soil, there were significant ($p < 0.05$) increases in lipid peroxidation in the leaves at 1.5% and 2% concentrations compare to control. Comparing lipid

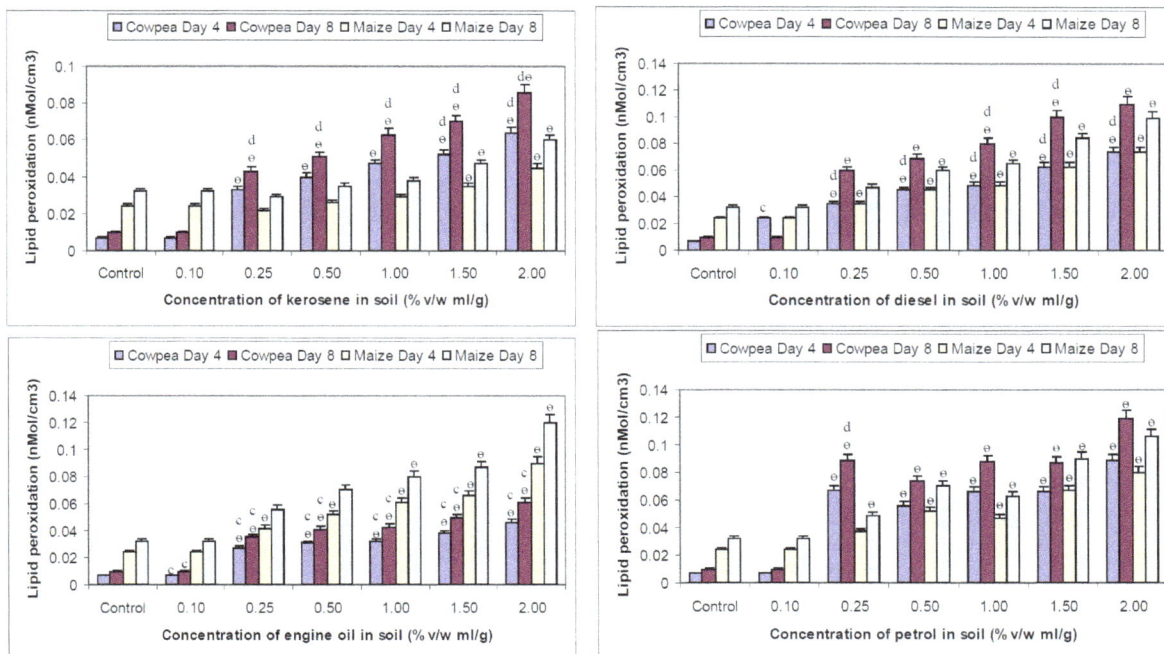

Figure 1. Effect of concentration of petroleum products on lipid peroxidation in the leaves of cowpea and maize after four and eight days of germination. *Significantly lower as compared to control; +Significantly lower as compared to engine oil; ++Significantly lower as compared to kerosene; eSignificantly lower relative to control; aSignificantly lower relative to other petroleum products; bSignificantly higher relative to other petroleum products; cSignificantly lower in cowpea relative to maize seedlings; dSignificantly higher in cowpea relative to maize seedlings.

peroxidation in the leaves of cowpea and maize seedlings grown in kerosene treated soil; lipid peroxidation was significantly (p < 0.05) higher in the leaves of cowpea seedling than in the leaves of maize seedlings from 0.25% through 2% concentrations. Lipid peroxidation in the leaves of cowpea seedlings grown in diesel treated soil were found to be significantly (p < 0.05) higher from 0.25% to 2% when compared with control. In maize seedlings grown in diesel treated, lipid peroxidation in the leaves showed a significant (p < 0.05) increase from 0.5% through 2% concentrations. Lipid peroxidation were significantly (p < 0.05) higher in the leaves of cowpea seedling at 1% and 1.5% concentration than in the leaves of maize seedling. Also, lipid peroxidation in the leaves of cowpea seedling grown in engine oil treated soil were found to increase significantly (p < 0.01) from 0.25% to 2% engine oil contaminated soil relative to control. Lipid peroxidation in the leaves of maize seedlings grown in engine oil treated soil showed a significant (p < 0.05) increase from 1% to 2% concentrations relative to control. In comparison, lipid peroxidation was significantly lower in the leaves of cowpea seedling than in the leaves of maize seedlings. Lipid peroxidation in the leaves of cowpea seedling grown in petrol treated soil was found to increase significantly (p < 0.05) from 0.25% to 2% concentrations compared with control. Lipid peroxidation in the leaves of maize seedlings grown in petrol treated soil showed a significant (p < 0.05) increase from 0.25% to 2% compared to control. When lipid peroxidation in the leaves of cowpea seedlings and maize seedlings grown in petrol treated soil are compared, lipid peroxidation level was found to be significantly (p < 0.05) higher at 0.25% in the leaves of cowpea than in the leaves of maize seedlings. Lipid peroxidation in the leaves of cowpea and maize seedlings grown in kerosene, diesel, and engine oil and petrol treated soil after twelve days of germination are shown in **Figure 2**. Lipid peroxidation in the leaves of cowpea seedlings grown in kerosene treated soil was found to increase significantly at 0.25% through 2%. In maize seedlings grown in kerosene treated soil, there was a significant increase in leaves lipid peroxidation at 1.0% to 2.0% concentrations respectively of kerosene in soil. Comparing lipid peroxidation in the leaves of cowpea and maize seedlings grown in kerosene treated soil; lipid peroxidation was significantly higher in the leaves of cowpea seedlings than in the leaves of maize seedlings at 0.25% and 2% concentrations. Similarly, lipid peroxidation in the leaves of cowpea seedlings grown in diesel treated soil was found to increase significantly at 0.25% to 2% concentrations compared to control. In maize seedlings

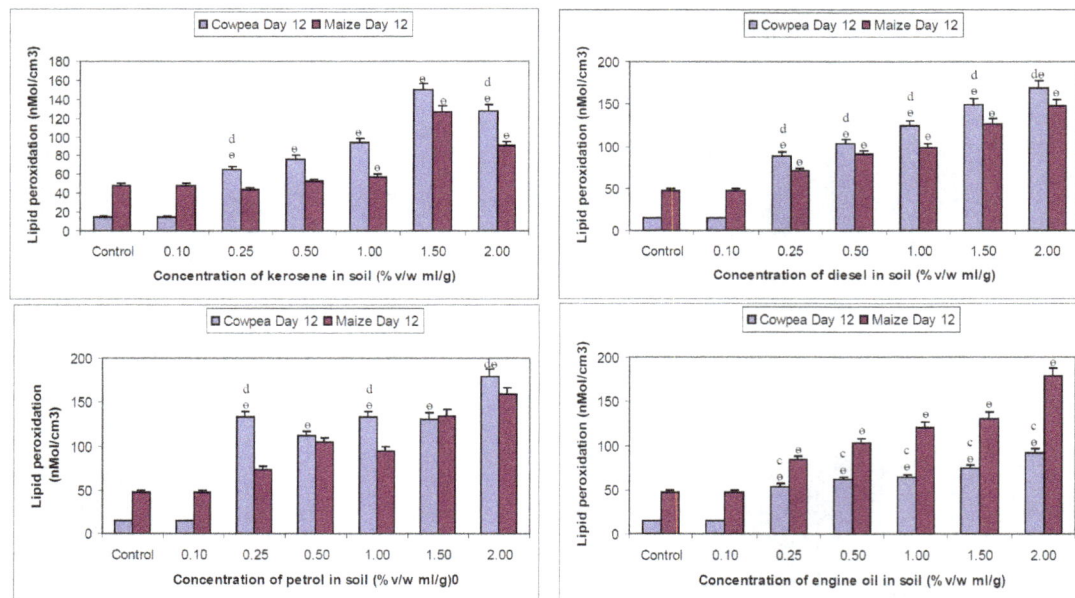

Figure 2. Effect of concentration of petroleum products on lipid peroxidation in the leaves of cowpea and maize after twelve days of germination. [*]Significantly lower as compared to control; [+]Significantly lower as compared to engine oil; [++]Significantly lower as compared to kerosene; [e]Significantly higher relative to control; [a]Significantly lower relative to other petroleum products; [b]Significantly higher relative to other petroleum products; [c]Significantly lower in cowpea relative to maize seedlings; [d]Significantly higher in cowpea relative to maize seedlings.

grown in diesel treated soil, there was a significant increase in lipid peroxidation in the leaves at 0.25% to 2% concentrations compared to control. Lipid peroxidation was significantly higher from 0.25% to 2% concentrations in the leaves of cowpea seedlings compared to that in the leaves of maize seedling. Also, lipid peroxidation in the leaves of cowpea and maize seedlings grown in engine oil treated soil were found to increase significantly from 0.25% to 2% relative to their respective control values. Maize seedlings grown in engine oil treated soil showed a significant increase in lipid peroxidation in the leaves from 0.25% to 2% concentrations relative to the control. In comparison, lipid peroxidation was significantly lower in the leaves of cowpea seedlings grown in engine oil contaminated soil than in the leaves of maize from 0.25% to 2% concentrations. Lipid peroxidation in the leaves of cowpea and maize seedlings grown in petrol treated soil were found to increase significantly from 0.25% to 2% concentrations. When lipid peroxidation in the leaves of cowpea and maize seedlings are compared in petrol treated soil, lipid peroxidation was found to be significantly higher at 0.25%, 1.0%, and 2.0% concentrations in cowpea seedlings over maize seedlings.

The activities of SOD in the leaves of cowpea and maize after four days of germination in kerosene, diesel, and engine oil and petrol treated soil are presented in **Figure 3**. Kerosene treated soil resulted in significant decrease ($p < 0.05$) in SOD activity in the leaves of cowpea seedlings compared with control. Maize seedling grown in kerosene treated soil resulted in significant ($p < 0.05$) decrease in SOD activity in the leaves in all the concentrations tested except at 1.5% where a slight decrease was observed. When the activities of SOD in the leaves of cowpea and maize seedlings grown in kerosene treated soil are compared, the enzyme was found to be significantly higher at 0.5% and 1% concentration in the leaves of cowpea seedlings than in the leaves maize seedlings. SOD activities in the leaves of cowpea seedlings grown in diesel treated soil showed a significant ($p < 0.05$) decrease across all the concentrations tested except at 1.5% compared with control. In maize seedlings grown in diesel treated soil, SOD activities in the leaves decreased significantly ($p < 0.05$) across all the concentrations tested relative to the control. When the activities of SOD in the leaves of cowpea and maize seedlings grown in diesel treated soil are compared, it was found to be slightly higher in cowpea seedling than maize seedlings. The activities of SOD in the leaves of cowpea seedling exposed to engine oil treated soil caused a significant ($p < 0.05$) increase in the activity of the enzyme above 1% levels of concentration compared with control. It only caused a slight decrease at 0.25% concentration. In maize seedlings

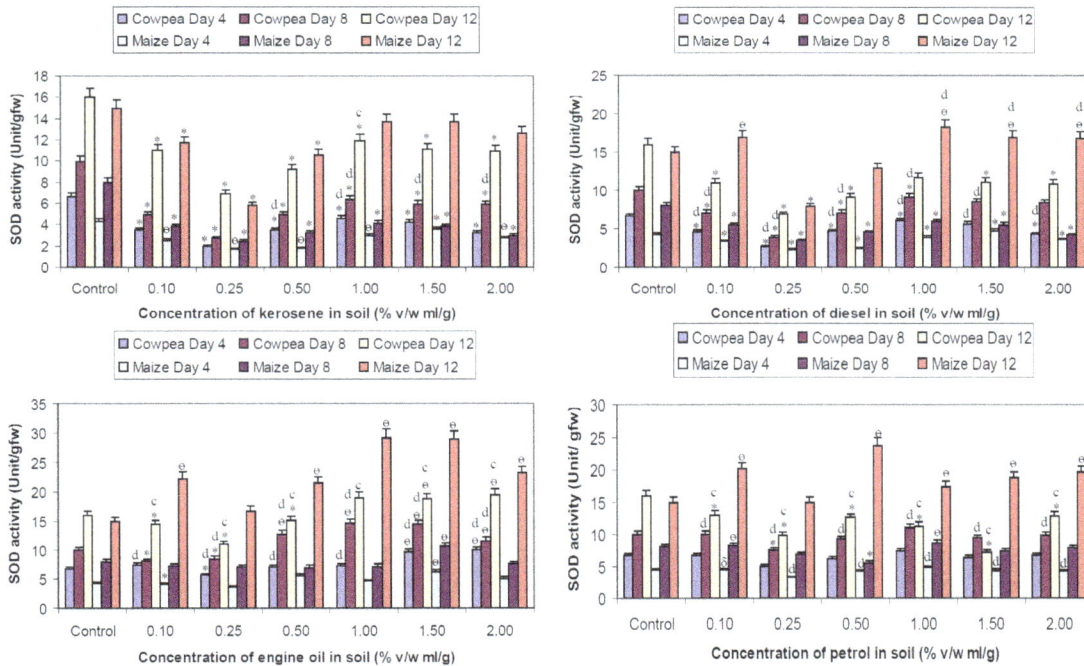

Figure 3. Effect of concentration of petroleum products on total SOD activities in leaves of cowpea and maize after four, eight and twelve days of germination. *Significantly lower as compared to control; +Significantly lower as compared to engine oil; ++Significantly lower as compared to kerosene; eSignificantly higher relative to control; aSignificantly lower relative to other petroleum products; bSignificantly higher relative to other petroleum products; cSignificantly lower in cowpea relative to maize seedlings; dSignificantly higher in cowpea relative to maize seedlings.

grown in diesel treated soil SOD activity significantly (p < 0.05) increased at 1.5% concentration and also showed a slight decrease at 0.1% and 0.25% concentration when compared with control. In all the concentrations, the activities of the SOD in the leaves were significantly higher in cowpea seedlings when compared to maize seedlings. Petrol treated soil had a slight increase in the activity of SOD in the leaves of cowpea seedlings at 1% concentration but showed slight reduction at 0.25% and 0.5% concentrations when compared with control. In maize seedlings however, only slight decrease in the activity of leaves SOD was observed at 0.25% concentration. No reasonable change was shown at all other levels of concentration. In all the concentrations tested, the activities of the enzyme were found to be significantly (p < 0.05) higher in the leaves of cowpea seedlings compared to maize seedlings. After eight days, kerosene treated soil resulted in significant decrease in SOD activity in the leaves of cowpea seedlings compared with control. Similarly, kerosene treated soil caused a significant decrease in SOD activity in the leaves of maize seedlings. When the activities of SOD in the leaves of cowpea and maize seedlings grown in kerosene treated soil are compared, the enzyme was found to be significantly higher in the leaves of cowpea seedlings at kerosene concentration of 1% to 2% than in the leaves of maize seedlings. The activities of SOD in the leaves of cowpea seedlings grown in diesel treated soil showed a significant (p < 0.05) decrease at 0.1% to 0.5% concentrations when compared with control. Slight decrease of leaves SOD activity was observed at other levels of concentration except 1%. Kerosene treated soil resulted in significant (p < 0.05) decrease of SOD activity in the leaves of maize seedlings in all the concentrations tested compared with control. When the activities of SOD in the leaves of cowpea and maize seedlings grown in diesel treated soil are compared; it was found to be significantly (p < 0.05) higher in the leaves of cowpea seedlings than in the leaves of maize seedlings. SOD activity in the leaves of cowpea seedling grown in engine oil treated soil showed a significant (p < 0.01) decrease up to 0.25%. However, further increase in engine oil concentration led to significant (p < 0.05) increase of leaves SOD activity. However, in maize seedlings grown in engine oil treated soil, there was a significant (p < 0.05) increase in the activity of leaves SOD at 1.5% concentration while a slight decrease of leaves SOD activity at 0.5% concentration was observed. In all the concentrations, the activities of leaves SOD were significantly higher in cowpea seedlings at concentration levels above 0.25% when compared to the leaves SOD of maize seedlings. Petrol treated soil had a slight increase in the activities of total SOD in the leaves of cowpea seedlings at 0.1% and 1% levels of soil contamination. However, a significant decrease of leaves SOD activity in cowpea seedling at 0.25% concentration was observed. In maize seedlings, significant ($p < 0.05$) increases of leaves total SOD activity was observed at 0.1% and 1% concentrations; while a significant ($p < 0.05$) decrease was recorded at 0.5%. When compared, leaves total SOD activity in cowpea seedlings was significantly ($p < 0.05$) higher than in maize seedlings. The activities of superoxide dismutase (SOD) in the leaves of cowpea and maize seedlings after twelve days of germination in kerosine, diesel, engine oil and petrol treated soil are shown in **Figure 3**. In all the concentrations tested, kerosene treated soil resulted in significant decrease ($p < 0.05$) of SOD activity in the leaves of cowpea compared with control. However, in maize seedling, kerosene treated soil only resulted in significant decrease ($p < 0.05$) up to 0.5% concentration; thereafter further increase in kerosene concentration did not result in significant decrease in SOD activity. When the activities of SOD in the leaves of cowpea and maize seedlings grown in kerosene treated soil are compared, the enzyme was found to be significantly lower in the leaves of cowpea seedlings at kerosene concentration of 1.0% than in the leaves of maize seedlings. Similarly, in all the concentrations tested, diesel treated soil resulted in significant decrease ($p < 0.05$) of SOD in the leaves of cowpea seedlings compared with control. However, in diesel treated soil, there was significant decrease in the activity of SOD in the leaves of maize seedlings at 0.25%. Moreover, there was a significant increase in the activity of SOD in the leaves of maize seedlings at 0.1, 1.0% 1.5% and 2% diesel concentrations. When the activities of total SOD in the leaves of maize seedlings are compared with those of cowpea seedlings grown in diesel treated soil, the enzyme was found to be significantly ($p < 0.05$) higher in the leaves of maize seedling at diesel concentrations above 0.5%. Unlike kerosene treated soil, the changes in the activities of the enzyme in diesel treated soil are not concentration dependent.

The activities of superoxide dismutase (SOD) in the leaves of cowpea seedling grown in engine oil treated soil significantly ($p < 0.05$) decreased up to 0.5%. At higher concentrations of engine oil above 1.0% there was a significant increase in the activity of the enzyme in the leaves of cowpea seedlings. However, in maize seedling grown in engine oil treated soil, there was a significant ($p < 0.05$) increase in the activity of leaves SOD in all the concentrations except at 0.25% engine oil in soil. In all the concentrations, the activities of the enzymes were significantly ($p < 0.05$) lower in the leaves of cowpea seedlings when compared to maize seedling. Petrol treated soil had a significant ($p < 0.05$) reduction in the activities of (SOD) in the leaves of cowpea seedlings grown in all the concentrations tested relative to the control. In maize seedlings grown in petrol treated soil, there was a significant ($p < 0.05$) increase in the activity of leaves SOD at 0.1%, 0.5%, 1.0%, 1.5% and 2.0% petrol in soil. However, there was no change in the activity of the enzyme in maize seedlings grown in 0.25% petrol contaminated soil. In all the concentrations tested the activity of leaves SOD was found to be significantly lower in the leaves of cowpea seedlings compared to that in the leaves of maize seedlings.

The activities of Cu/Zn SOD in the leaves of cowpea and maize seedlings after four days of germination in kerosene, diesel, and engine oil and petrol treated soil are shown in **Figure 4**. In all the concentrations tested, kerosene treated soil showed a significant decrease of Cu/Zn SOD activities in the leaves of cowpea and maize seedlings when compared with their respective control values. When the activities of Cu/Zn SOD in the leaves of cowpea seedlings grown in Kerosene treated soil were compared to that in maize seedlings, it was found to be significantly higher in the leaves of cowpea seedlings at 1% to 2.0% than in the leaves of maize seedlings. In cowpea seedlings grown in diesel treated soil, the activities of Cu/ZnSOD in the leaves showed a significant ($p < 0.05$) decrease in all the concentrations tested compared with control. However, in maize seedlings a significant ($p < 0.05$) decrease in Cu/ZnSOD activity in the leaves at 0.1%, 0.25%, 0.5% and 2% concentrations was observed. However, leaves Cu/ZnSOD exhibited a slight decrease at 1% and 1.5% concentrations compared with control. In contrast, the activities of Cu/ZnSOD in the leaves of cowpea seedlings grown in diesel treated soil was significantly ($p < 0.05$) lower than the activity in the leaves of maize seedlings in all the concentrations above 0.5%. At the concentrations tested the activities of Cu/Zn SOD in the leaves of cowpea seedlings grown in engine oil treated soil slightly decreases at 0.25% and 0.5% concentrations. Maize seedlings grown in engine oil treated soil exhibited a significant ($p < 0.01$) increase of Cu/Zn SOD activities in the leaves at 1.5% concentration. A slight decrease however, was observed at 0.25% and 1% concentrations. By way of contrast, the activities of Cu/Zn SOD in the leaves of cowpea seedlings grown in engine oil treated soil appeared to be significantly ($p < 0.05$) higher than in the leaves of maize seedlings. In all the concentrations tested, cowpea seedlings grown in petrol treated soil showed a slight decrease of Cu/Zn

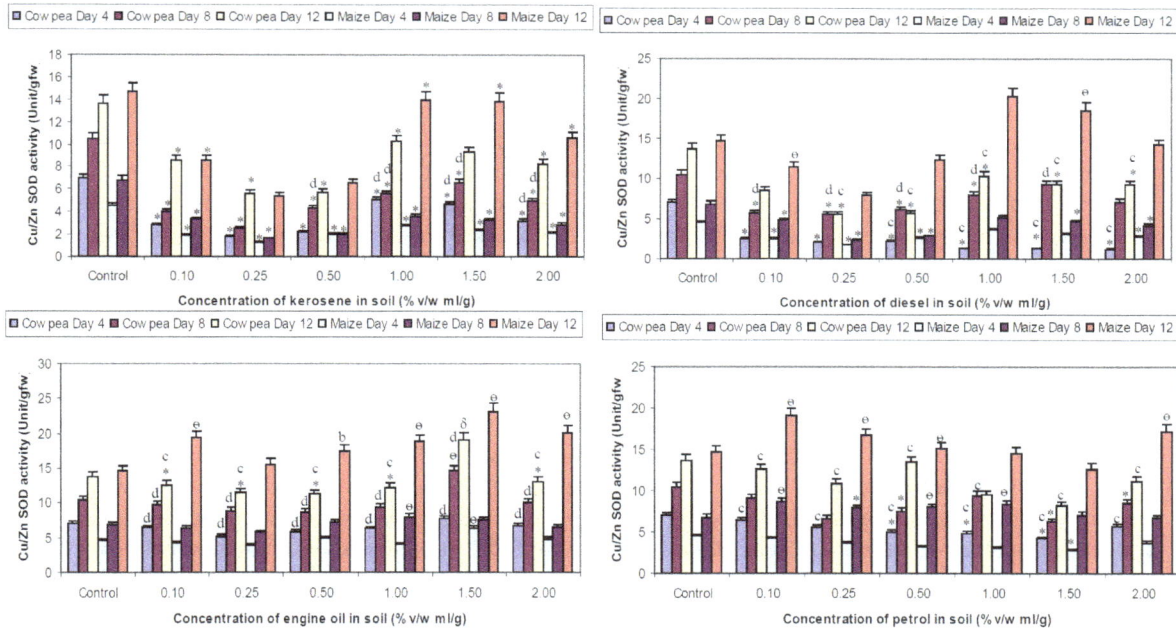

Figure 4. Effect of concentration of petroleum products on Cu/Zn SOD activities in leaves of cowpea and maize after four, eight and twelve days of germination *Significantly lower as compared to control; +Significantly lower as compared to engine oil; ++Significantly lower as compared to kerosene; eSignificantly higher relative to control; aSignificantly lower relative to other petroleum products; bSignificantly higher relative to other petroleum products; cSignificantly lower in cowpea relative to maize seedlings; dSignificantly higher in cowpea relative to maize seedlings.

SOD activities in the leaves at 0.1%, 0.25% and 2% concentrations. However, a significant ($p < 0.01$) decrease in the activity of leaves Cu/Zn SOD was observed at other levels of concentrations tested. In maize seedlings grown in petrol treated soil the activities of leaves Cu/Zn SOD decreased slightly at all concentrations tested except at petrol concentration of 1.5% where it showed a significant ($p < 0.05$) decrease in the activities of the enzyme. When the activities of Cu/Zn SOD in the leaves of cowpea seedlings are compared with the activities of the enzyme in maize seedlings, it was observed to be significantly ($p < 0.05$) lower in the leaves of cowpea seedling. The activities of Cu/Zn SOD activity in the leaves of cowpea and maize seedlings after eight days of germination in kerosene, diesel, engine oil and petrol treated soil are shown in **Figure 4**. In all the concentration tested, kerosene treated soil showed a significant ($p < 0.05$) decrease of Cu/Zn SOD activities in the leaves of both cowpea and maize seedlings compared with control. When the activities of Cu/Zn SOD in the leaves of cowpea and maize seedlings grown in kerosene treated soil are compared, it was found to be significantly ($p < 0.05$) higher in cowpea seedlings at levels of soil contamination above 0.25%. Diesel treated soil, resulted in significant ($p < 0.05$) decrease of Cu/Zn SOD activity in the leaves of cowpea seedlings in all the concentrations tested except 1.5% compared with control. Similarly, maize seedlings grown in diesel treated soil showed a significant ($p < 0.05$) decrease in the activities of Cu/Zn SOD in the leaves at all levels of concentration tested except 1% diesel concentration. When compared, diesel treated soil resulted in a significantly ($p < 0.05$) higher activity of Cu/Zn SOD in the leaves of cowpea seedlings than in maize seedlings. At the various concentrations; the activity of Cu/Zn SOD in the leaves of cowpea seedlings grown in engine oil treated soil significantly ($p < 0.05$) increase at 1.5% but slight decreases at concentration levels below 1.5% compared with control were observed. The activity of Cu/Zn SOD in the leaves of maize seedlings grown in engine oil treated soil significantly ($p < 0.05$) increase at 1% concentration. By way of contrast, the activities of Cu/Zn SOD were significantly higher in the leaves of cowpea seedlings than in maize seedling. In all the concentrations tested, cowpea seedlings grown in petrol treated soil had Cu/Zn SOD activities in the leaves slightly lower at 0.1%, and 1% concentrations, but significantly ($p < 0.05$) decreased at other levels of concentration when compared with control. In maize seedlings grown in petrol treated soil, a significant ($p < 0.05$) increase in the activities of Cu/Zn SOD in the leaves at 0.1%, 0.5% and 1% concentration was observed but minor increases were recorded at 0.25% and 1.5%. When compared, there was no significant difference in the activities of Cu/Zn SOD in the leaves of both cowpea and maize seedlings grown in petrol treated soil. The activity of copper/zinc superoxide dismutase (Cu/ZnSOD) in the leaves of cowpea and maize seedlings after twelve days of germination in kerosene, diesel, engine oil and petrol treated soil are shown in **Figure 4**. In cowpea seedlings grown in kerosene treated soil a significant decrease of Cu/ZnSOD activity in the leaves was observed at all concentrations tested. Similarly, in maize seedlings grown in kerosene contaminated soil a significant decrease of Cu/ZnSOD activity in the leaves at 0.1%, 1.0%, 1.5% and 2.0% was observed. When the activities of Cu/Zn SOD in the leaves of cowpea seedlings grown in kerosene treated soil were compared to those of maize seedlings, non significant lower values in cowpea across the various levels of concentrations was observed. In cowpea seedlings grown in diesel treated soil, the activities of Cu/Zn SOD in the leaves were found to be significantly lower at all concentrations tested compared to the control. At the various concentrations, the activities of Cu/Zn SOD in the leaves of maize seedling grown in diesel treated soil exhibited an increase at 0.1%, and 1.5%; a decrease in the activities of Cu/Zn SOD in the leaves of maize seedlings grown in diesel contaminated soil was shown at other concentrations except at 0.25% concentration. The activities of Cu/Zn SOD in the leaves of cowpea seedling grown in diesel treated soil exhibited decreases at various concentrations compared to the activity in the leaves of maize seedlings. The activities of Cu/ZnSOD in the leaves of cowpea seedling grown in engine oil treated soil significantly (p < 0.05) decreased at 0.1%, to 2%, but increased at 1.5%, compared to control. However, in maize seedling grown in diesel treated soil, very high increases of Cu/ZnSOD activities in the leaves at 0.1%, 0.5% to 2% concentrations were observed. The activities of Cu/Zn SOD in the leaves of cowpea seedlings grown in engine oil treated soil, were significantly (p < 0.05) lower across the various concentrations than in maize seedlings. Cowpea seedlings grown in petrol treated soil had Cu/Zn SOD activities in the leaves relatively lower at 0.1%, and 0.5% concentrations respectively. However, petrol treated soil resulted in a decrease in Cu/Zn SOD in the leaves activities at 0.25%, 1%, 1.5% and 2.0% concentrations. Maize seedlings grown in petrol treated soil only resulted in significant increase in the activities of Cu/Zn SOD in the leaves at all concentrations except at 1% and 1.5% relative to control. When compared, the activities of the enzyme in the leaves of cowpea seedlings significantly (p < 0.05) than in the leaves of maize seedlings.

The activities of MnSOD in the leaves of cowpea and maize seedlings after four days of germination in kerosene, diesel, and engine oil and petrol treated soil are shown in **Figure 5**. Kerosene treated soil resulted in significant (p < 0.05) decrease of MnSOD activities in the leaves of cowpea seedlings except at kerosene concentration of 1%. Similarly, kerosene treated soil gave rise to a significant decrease of MnSOD activities in the leaves of maize seedlings at all the levels of concentrations tested except 1.5% where a slight increase was observed. There was a higher increase in MnSOD activities in the leaves of cowpea seedlings than in the leaves of maize seedlings when compared. The activities of MnSOD in the leaves of cowpea seedlings grown in diesel treated soil significantly (p < 0.05) decreased at 0.1% 0.25% and 0.5% concentrations compared to control. However, a slight decrease was recorded at concentration levels above 1%. The activities of MnSOD in the leaves of maize seedlings exposed to diesel showed a slight decrease at all levels of concentrations tested except at 0.25% and 0.5% where significant (p < 0.05) reduction was observed. At 1.5% concentration MnSOD activity in the leaves of maize seedling showed a slight increase when compared with control. Nonetheless, there was a higher increase in MnSOD activities in the leaves of cowpea seedlings than in the leaves of maize seedlings. Engine oil treated soil resulted in significant (p < 0.05) increase in MnSOD activities in the leaves of cowpea seedlings at concentration levels above 1%. It was also found that MnSOD activities in the leaves activities exhibited a significant (p < 0.05) decrease at 0.1% and 0.25% concentrations. Also, MnSOD activities in the leaves of maize seedlings grown in engine oil treated soil significantly (p < 0.05) increased at 1.5% and 2% concentrations. However, a slight decrease was recorded at 0.25% concentration. In addition, the activities of MnSOD were significantly (p < 0.05) higher in the leaves of cowpea seedlings than in the leaves of maize seedlings when compared. Petrol treated soil resulted in slight decrease of MnSOD activities in the leaves of cowpea in all the concentrations except at 0.1% and 0.5% concentrations where significant (p < 0.05) decreases were observed. However, it showed a significant decrease at 0.25%, 1% and 1.5% concentrations. Also, there was an insignificant increase of MnSOD activities in the leaves at 2% level of petrol concentration. In maize however, petrol treated soil only gave rise to slight increase in MnSOD activities in the leaves in all the levels of concentrations tested except at 1.5% where a minor decrease was observed. When compared, MnSOD activities were slightly higher in the leaves of cowpea seedlings than in the leaves of maize seedlings. The activities of MnSOD in the leaves of cowpea and maize seedlings after eight days of germination in kerosene, diesel, engine

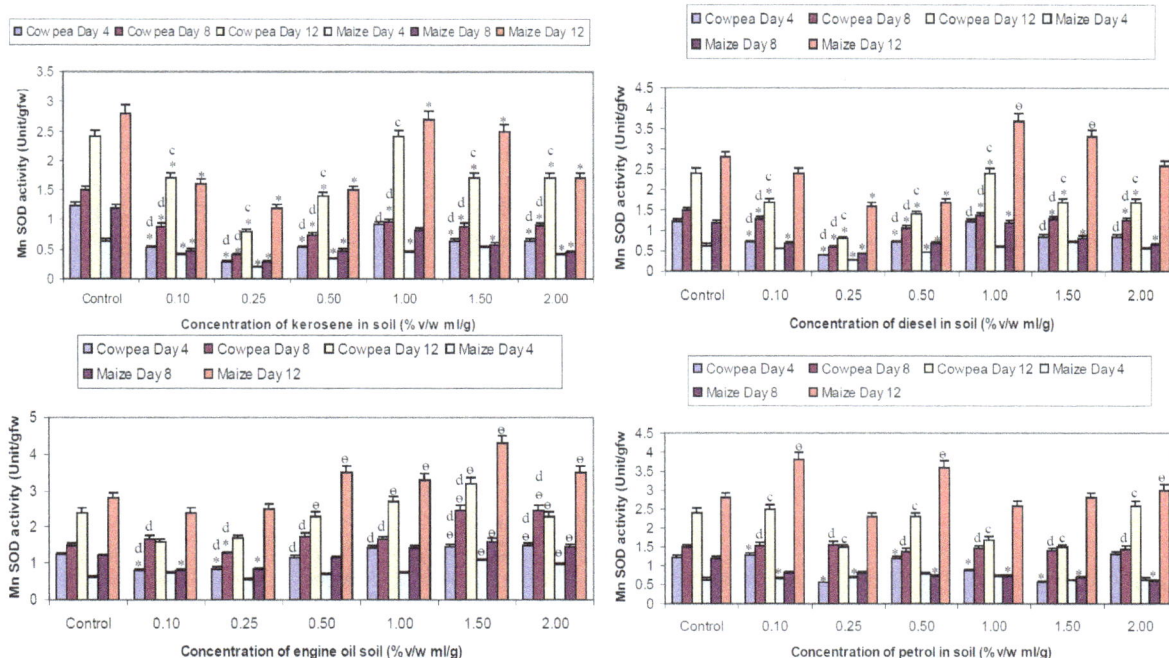

Figure 5. Effect of concentration of petroleum products on MnSOD activities in leaves of cowpea and maize after four, eight and twelve days of germination. *Significantly lower as compared to control; +Significantly lower as compared to engine oil; ++Significantly lower as compared to kerosene; eSignificantly higher relative to control; aSignificantly lower relative to other petroleum products; bSignificantly higher relative to other petroleum products; cSignificantly lower in cowpea relative to maize seedlings; dSignificantly higher in cowpea relative to maize seedlings.

oil and petrol treated soils are shown in **Figure 5**. Kerosene treated soil resulted in significant (p < 0.05) decrease of MnSOD activities in the leaves of cowpea seedlings grown in kerosene treated soil compared with control. In maize seedlings, however, MnSOD activities in the leaves decreased significantly (p < 0.01) at all the levels of concentrations tested except 1% where a slight reduction was observed. In comparison, the activities of MnSOD were significantly (p < 0.05) higher in the leaves of cowpea seedling than in the leaves of maize seedlings. At the concentrations tested, the activities of Mn SOD in the leaves of cowpea seedlings grown in diesel treated soil resulted in a significant (p < 0.05) decrease in the activity of the enzyme relative to control value. In maize seedlings exposed to diesel treated soil. MnSOD activity in the leaves decreased significantly (p < 0.05) in all the concentrations tested. However, at 1% concentration a slight decrease in MnSOD activity in the leaves was recorded. When the activities of MnSOD in the leaves of cowpea and maize seedlings grown in diesel treated soil were compared, it was found to be significantly (p < 0.05) higher across the concentrations tested in cowpea seedlings than in maize seedlings. The activities of MnSOD in the leaves of cowpea seedlings grown in engine oil treated soil significantly (p < 0.05) increased at 1.5% and 2% concentrations compared to control value. However, MnSOD activities in the leaves of cowpea seedlings grown at 0.1%, 0.5% and 1% concentration showed a slight increase compared to control. Nevertheless, a significant decrease was observed at 0.25%. In maize seedlings grown in engine oil treated soil, MnSOD activities in the leaves activities showed a significant (p < 0.05) increase at 1.5% and 2%. However, there was significant (p < 0.05) decrease at 0.1% and 0.25%. Moreover, a slight increase of MnSOD activities in the leaves was also observed at 0.5% and 1% concentrations when compared to the control. When MnSOD activities in the leaves of cowpea and maize seedlings grown in engine oil treated soil were compared, it was observed to be significantly (p < 0.05) higher at all the levels of concentrations tested in cowpea than in maize seedlings. Petrol treated soil gave rise to a slight decrease in MnSOD activities in the leaves at concentration levels above 0.25% in cowpea seedlings when compared with control. However, in maize seedling grown in petrol treated soil, a significant decrease of MnSOD activities in the leaves was exhibited in all the concentrations tested. When compared, MnSOD activities were significantly higher in the leaves of cowpea seedlings than in maize seedlings. The activities of MnSOD in the leaves of cowpea and maize seedlings after twelve days of germination in kerosene, diesel, engine oil and petrol treated soils are shown in **Figure 5**. Kerosene treated soil gave rise to a significant ($p < 0.05$) decrease in Mn SOD activities in the leaves of cowpea seedling grown in all the concentrations except 1% kerosene contaminated soil. The results also show that MnSOD activities in the leaves of cowpea seedlings grown in kerosine contaminated significantly ($p < 0.05$) decreased at 0.25% and 0.5% level of contamination. In maize seedlings grown in kerosene contaminated soil, the MnSOD activities significantly ($p < 0.05$) decreased at concentrations corresponding to those of cowpea seedlings and vice-versa. Nonetheless, MnSOD activities in the leaves of cowpea seedlings were lower than in the leaves of maize seedlings. Diesel treated soil resulted in a significant ($p < 0.05$) decrease in MnSOD activities in the leaves of cowpea seedlings at 0.1%, 1%, 1.5% and 2%. It was also found that MnSOD activities exhibited a significant decrease at 0.25% and 0.5%. Also, MnSOD activities in the leaves of maize seedlings grown in diesel treated soil significantly increased at 1% and 1.5% but decreased at 0.25% and 0.5%. When compared MnSOD activities were lower in the leaves of cowpea than in the leaves of maize seedlings, MnSOD activities in the leaves of cowpea seedling in engine oil treated soil showed a significant increase from 0.5% - 2% concentrations. On the other hand, the activities of Mn SOD in the leaves of maize seedling grown in engine oil treated soil showed a sharp increase at 0.5% - 2% concentrations. The activities of Mn SOD in the leaves of cowpea seedling grown in petrol treated soil resulted in a slight increase at 0.1%, 0.5% and 2% concentrations but decreases were recorded at 0.25%, 1% and 1.5%. The activities MnSOD in the leaves of maize seedling showed a sharp increase at concentrations levels of 0.1%, 0.5% and 2%. Conversely, the result shows a decrease in Mn SOD activities in the leaves at 0.25%, 1% and 1.5% concentrations. However, MnSOD activities were lower in the leaves of cowpea seedlings than in the leaves of maize seedling.

The activities of catalase in leaves of cowpea and maize seedling in kerosene, diesel, and engine oil and petrol treated soil after four days of germination are shown in **Figure 6**. At the various concentrations tested, kerosene treated soil brought about a successive decrease of catalase activity in the leaves of cowpea seedlings as well as in the leaves of maize seedlings compared with control. In comparison, however, there was no significant difference in the decrease across the various concentrations tested between the activities of catalase in

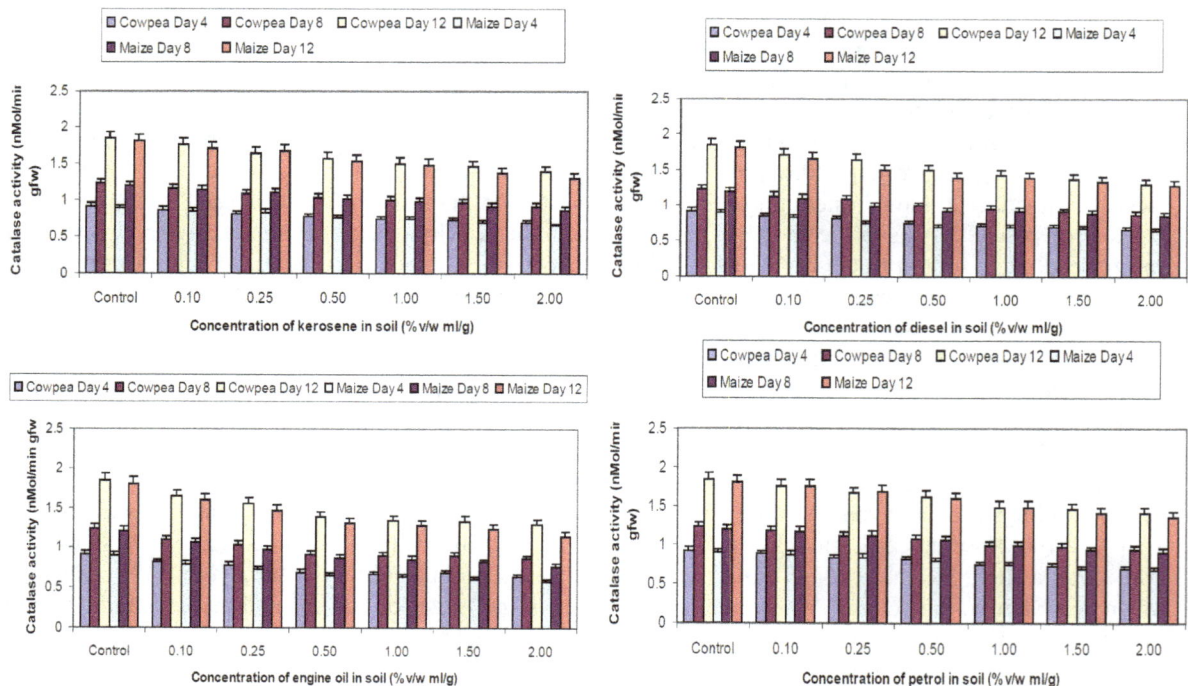

Figure 6. Effect of concentration of petroleum products on catalase activities in leaves of cowpea and maize after four, eight and twelve days of germination. *Significantly lower as compared to control; +Significantly lower as compared to engine oil; ++Significantly lower as compared to kerosene; eSignificantly higher relative to control; aSignificantly lower relative to other petroleum products; bSignificantly higher relative to other petroleum products; cSignificantly lower in cowpea relative to maize seedlings; dSignificantly higher in cowpea relative to maize seedlings.

the leaves of cowpea and maize seedlings. Catalase activities tested in the leaves of cowpea seedlings grown in diesel treated showed a decrease in activities at all levels of soil contamination compared to control. Similarly, catalase activity in the leaves of maize seedlings grown in diesel treated indicated a progressive decrease across all levels of concentrations tested. However, catalase activity appeared to be slightly higher in the leaves of cowpea seedlings than in the leaves of maize. In all the concentrations tested, engine oil treated soil resulted in successive decrease of catalase activity in the leaves of cowpea seedlings as well as in the leaves of maize seedlings relative to control. Nonetheless, the decrease of catalase activity was slightly higher in the leaves of cowpea seedlings compared to that of maize seedlings. Petrol treated soil resulted in slight decrease of catalase activity in the leaves of cowpea seedlings at all levels of concentration compared with control. Similarly, in maize seedlings, catalase activity in the leaves showed a minor decrease in all the concentration tested compared to control. There was no significant difference in the activities of catalase in the leaves of cowpea and maize seedlings.

The activities of catalase in the leaves of cowpea and maize seedling after eight days of germination in kerosene, diesel, engine oil and petrol treated soil are shown in **Figure 6**. At the various concentrations tested, kerosene treated soil brought about a successive decrease of catalase activity in the leaves of cowpea seedlings as well as in maize seedlings compared with control. In comparison, however, the decreases across the various concentrations tested were more pronounced in the leaves of cowpea seedlings than in the leaves of maize seedlings. Similarly, catalase activity in the leaves of both cowpea seedlings and maize seedlings grown in diesel treated soil decreased throughout the various levels of concentrations tested relative to control. No significant differences existed in the activities of catalase in the leaves of cowpea seedlings and maize seedlings grown in diesel treated soil. In all the concentrations tested, engine oil treated soil resulted in a slight decrease of catalase activity in the leaves of cowpea seedlings compared with control. In maize seedlings, catalase activity in the leaves insignificantly reduced across all the concentrations tested relative to control. By contrast, the decrease of catalase activity was slightly more in the leaves of maize seedling than cowpea seedlings. In all the concentrations tested, petrol treated soil gave rise to a decrease of catalase activity in the leaves of cowpea seedling compared with control. Similarly, petrol treated soil brought about decrease of catalase activity in the leaves of maize seedlings. When compared, the decrease in the activities of catalase was more in the leaves of maize seedlings than in the leaves of cowpea seedlings

The activities of catalase in the leaves of cowpea and maize seedlings after twelve days of germination in kerosene, diesel, engine oil and petrol treated soil are shown in **Figure 6**. At the various concentrations tested, kerosene treated soil brought about a successive decrease of catalase activity in the leaves of cowpea seedlings as well as in the leaves of maize seedlings compared with control. There were no distinct differences between the activities of the enzyme across the various concentrations in the leaves of cowpea seedlings as well as in the leaves of maize seedlings. Catalase activities in the leaves of cowpea seedlings grown in diesel treated soil showed a decrease in activities in all the concentrations tested. Similarly, catalase activities in all the concentrations tested in the leaves of maize seedling indicated a progressive decrease. However, catalase activities decreased less in the leaves of cowpea seedlings relative to that in the leaves of maize seedlings. Also in all the concentrations tested, engine oil treated soil resulted in significant decrease of catalase activity in the leaves of cowpea seedlings compared with control. Similarly, catalase activities in the leaves of maize seedling decreased successively as concentration of engine oil increase. However, the decrease in catalase activity in the leaves of cowpea seedlings was slightly lower compared to that of maize seedlings. In all the concentrations tested, petrol treated soil gave rise to a significant ($p < 0.05$) decrease in catalase activity in the leaves of cowpea seedling. Similarly, catalase activities decreased in the leaves of maize seedling in all the concentrations tested relative to control. However, the reduction in the catalase activity at 0.1%, 1% and 2% concentrations was less in the leaves of cowpea seedlings than in the leaves of maize seedlings. On the contrary, the reduction of catalase activities in the leaves of cowpea seedlings grown in petrol treated soil at 0.25% was more than that of maize seedlings.

The activities of xanthine oxidase in the leaves of cowpea and maize seedlings after four days of germination in kerosene, diesel, engine oil and petrol treated soil are shown in **Figure 7**. In kerosene treated soil, of the various concentrations tested, it was found that xanthine oxidase activities in the leaves of cowpea and maize seedlings decreased significantly ($p < 0.05$) compared with control. Comparing xanthine oxidase activities in the leaves of cowpea and maize seedlings, there was no significant difference between the activities of the enzyme at each level of concentration. In diesel treated soil, xanthine oxidase activity decreased significantly ($p < 0.05$) across all the concentrations tested in the leaves of both cowpea and maize seedlings. By comparison, the

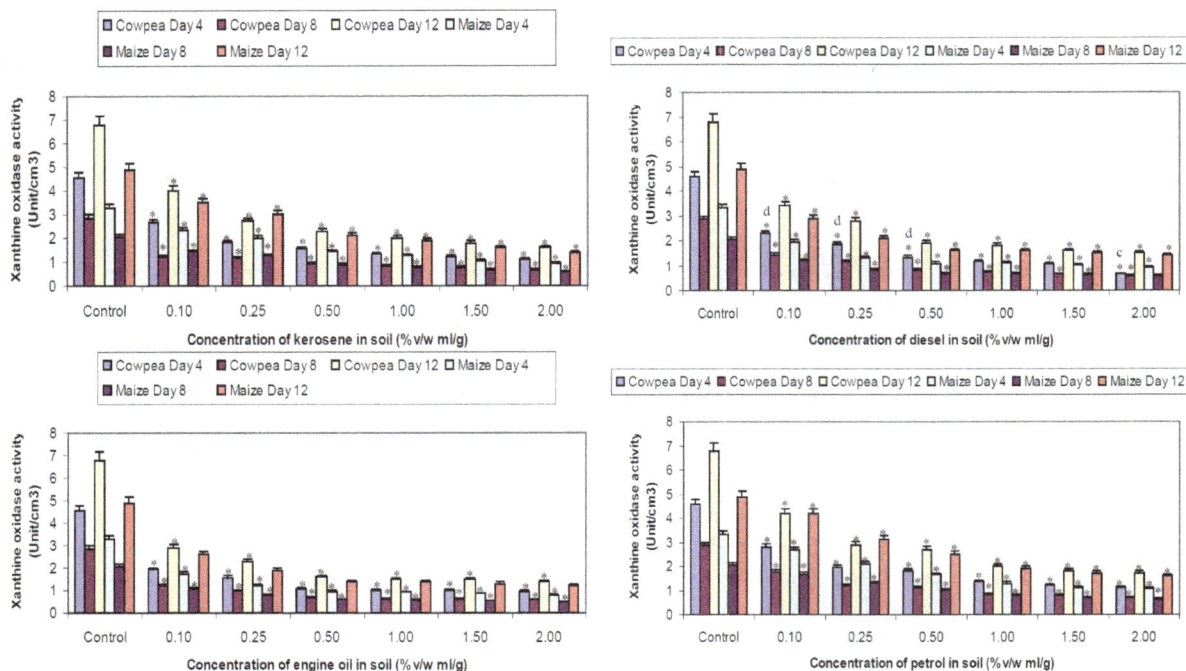

Figure 7. Effect of concentration of petroleum products on xanthine oxidase activities in the leaves of cowpea and maize after four, eight and twelve days of germination. *Significantly lower as compared to control; +Significantly lower as compared to engine oil; ++Significantly lower as compared to kerosene; eSignificantly higher relative to control; aSignificantly lower relative to other petroleum products; bSignificantly higher relative to other petroleum products; cSignificantly lower in cowpea relative to maize seedlings; dSignificantly higher in cowpea relative to maize seedlings.

activity of xanthine oxidase in the leaves of cowpea seedlings appeared to be higher at 0.1% - 0.5% and lower at 2% concentrations in relation to the xanthine oxidase activity in the leaves of maize seedling. In all the concentrations tested, engine oil treated soil resulted in significant decrease in xanthine oxidase activity in the leaves of cowpea seedlings. In same vain, xanthine oxidase activity in the leaves of maize seedlings decreased significantly across all concentrations tested compared with control. In comparison, xanthine oxidase activity was slightly higher in the leaves of cowpea seedlings than in the leaves of maize seedlings, but a significant difference in the enzyme activity was not observed. Petrol treated soil caused a significant ($p < 0.05$) decrease in the activities of xanthine oxidase activity in the leaves of both cowpea and maize seedlings compared with control. In comparison, xanthine oxidase activity was slightly higher in the leaves of cowpea leaves but not significantly different from the activity in the leaves of maize seedlings.

The activities of xanthine oxidase in the leaves of cowpea and maize seedlings after eight days of germination in kerosene, diesel, engine oil and petrol treated soil are shown in **Figure 7**. In kerosene treated soil, of the various concentrations tested, it was found that xanthine oxidase activities in the leaves of maize seedlings decreased significantly ($p < 0.05$) compared with control. Similarly, xanthine oxidase activity in the leaves of cowpea seedlings grown in kerosene treated soil significantly ($p < 0.05$) decrease in all the concentrations tested. Comparing xanthine oxidase activities in the leaves of cowpea and maize seedlings, it was found to be slightly lower in cowpea seedlings at all concentration except at 0.1% and 0.25%, but no significant difference existed in xanthine oxidase activity between cowpea and maize seedlings. In diesel treated soil, xanthine oxidase activities decreased significantly ($p < 0.05$) across all the concentration tested in the leaves of both cowpea and maize seedlings. By comparison the activities of xanthine oxidase increased slightly in the leaves of cowpea seedlings than in the leaves of maize seedlings across all the concentrations tested. In all the concentrations tested, engine oil treated soil resulted in significant ($p < 0.05$) decrease in xanthine oxidase activity in the leaves of cowpea seedlings. Similarly, xanthine oxidase activity decreased significantly in the leaves of maize seedlings grown in engine oil treated soil compared with control. In comparison, the activities of xanthine oxidase in the leaves of cowpea seedlings are higher than in maize seedlings. However, there is no significant difference between the xanthine oxidase activity in the leaves of cowpea and maize seedlings. Petrol treated soil gave rise to significant (p < 0.05) decrease in xanthine oxidase activity in the leaves of cowpea seedlings compared with control. Similarly, the leaves of maize seedlings grown in petrol treated soil witnessed a significant (p < 0.05) decrease in xanthine oxidase activities compared with control. When compared, the activities of xanthine oxidase activities in the leaves of cowpea seedlings were slightly higher than in the leaves of maize seedlings at 0.1%, 0.5% and1.5%, but slightly lower at 0.25%. The activities of xanthine oxidase in the leaves of cowpea and maize seedlings after twelve days of germination in kerosene, diesel, engine oil and petrol treated soil are shown in **Figure 7**. In kerosene treated soil, of the various concentrations tested, it was found that xanthine oxidase activities in the leaves of cowpea and maize seedlings decrease significantly (p < 0.05) compared with control. Comparing xanthine oxidase activities in the leaves of cowpea and maize seedlings, there was no significant difference between the activities of the enzyme at each level of concentration. In diesel treated soil, xanthine oxidase activity decreased significantly (p < 0.05) across all the concentration tested in the leaves of both cowpea and maize seedlings. By comparison, the activity of xanthine oxidase in the leaves of cowpea seedling grown in diesel treated soil was higher than in the leaves of maize seedling. But at each concentration there was no significant difference between the activity of the enzyme in the leaves of both cowpea and maize seedlings. In all the concentrations tested, engine oil treated soil resulted in a significantly (p < 0.05) decrease in xanthine oxidase in the leaves of cowpea seedling. Similarly, it was found that the activity of xanthine oxidase in the leaves of maize seedling was significantly (p < 0.05) inhibited across all concentrations relative to control. In comparison, xanthine oxidase activity was slightly higher in the leaves of cowpea seedling but not statistically different from the activity in the leaves of maize seedling. Petrol treated soil caused a significant (p < 0.05) decrease of xanthine oxidase activity in the leaves of both cowpea and maize seedlings. In comparison, xanthine oxidase activity was slightly higher in the leaves of cowpea seedling except at 0.25% but not statistically different from the activity in the leaves of maize seedlings.

4. Discussion

The formation of lipid peroxidation products in plants exposed to adverse environmental conditions is an indication of free radical formation in tissue and it may be used as index of lipid peroxidation in biological system [13] [14]. Previous report indicated that metal exposure results in the generation of reactive oxygen species in plant [15]. The present results show that an increasing level of refined petroleum products in soil increased the level of lipid peroxidation in the leaves of both cowpea and maize seedlings exposed to petroleum products treated soil for four and eight days after germination (**Figure 1**) and twelve days (**Figure 2**). An increase in the level of lipid peroxidation has been reported for plants exposed to metal ion [14] [16]-[18]. Elevated levels of reactive oxygen species initiate lipid peroxidation [19] [20] that culminate in oxidative stress [21]. An increase in the level of lipid peroxidation is the evidence most frequently cited in support of the involvement of oxidative stress in tissues [22]-[24].

This study shows that refined petroleum products (Kerosene, diesel, and petrol and engine oil) could induce lipid peroxidation in exposed plants (**Figure 1** and **Figure 2**). This is because refined petroleum toxicity could result in the production of reactive oxygen species (ROS); which in turn can cause membrane damage. The relationship between lipid peroxidation and ROS has been reported [25] [26]. The process of lipid peroxidation was more pronounced in cowpea seedlings compared to maize seedlings (**Figure 1** and **Figure 2**), which also lends further credence to the report that monocotyledonus seeds are less affected by toxicants than dicotyledonous seeds.

The production of reactive oxygen species during stress has been reported to results from pathways such as photorespiration, from the photosynthetic apparatus and mitochondria respiration. Environmental stresses have been shown to trigger the active production of reactive oxygen species by NADPH oxidase [27]-[30]. The enhanced production of reactive oxygen species during stress can pose threat to cells but plant have inbuilt mechanism to counteract the reactive oxygen species. The scavenging mechanisms of plant include superoxide dismutase (SOD) and catalase [31]-[33]. The present investigation indicated that the exposure of cowpea and maize seedlings to refined petroleum products, kerosene, diesel engine oil and petrol affected superoxide dismutase activity after four, eight days and twelve days of germination (**Figure 3**). After twelve days of exposure to kerosene, diesel and petrol treated soil the activity of total SOD decreased relative to control but there was increase in the activity of the enzyme in the leaves of cowpea and maize seedlings grown in engine oil treated soil compared to the other petroleum products (**Figure 3**). Decrease in the activity of the enzyme portends reduction in the capacity of the exposed plant to handle reactive oxygen species. This is because increase in the activity of the antioxidant enzyme may have a role in imparting tolerance against any type of environmental stress [33]. Moreover, the petroleum products caused a reduction in Cu/Zn-dependent superoxide dismutase after four, eight days and twelve days of germination (**Figure 4**) as well as Mn-containing superoxide dismutase (MnSOD) after four, eight days and twelve days of growth in contaminated soil (**Figure 4**). These enzymes are involved in the general defense system against natural or chemically induced production of reactive oxygen species [34]. It has been suggested that a decrease in the activity in these detoxification mechanisms can generate severe cell damage due to imbalance in the production of toxic oxygen radicals. Conversely, increased activity could contribute to cell protection from chemical toxicants [35]-[38]. Smirnoff [39] established that an increase in the capacity of antioxidant defenses in response to an increased level of reactive oxygen represents an indirect measure of oxidative stress.

The results indicated that kerosene affected the seedlings more than the other refined petroleum products. Similarly, the activity of the antioxidant enzymes in the leaves of maize seedlings appears to be less affected by the petroleum products compared to cowpea seedlings. This is exhibited by the effects of petroleum products on CuZnSOD (**Figure 4**) and MnSOD (**Figure 5**) in cowpea and maize seedlings. This is in agreement with previous reports in which maize seedling has been shown to be resistant to crude oil contamination [40] while beans possess the ability to uptake crude oil [41]. However, this observation is in contrast to the report of Baek *et al.* [42] in which *Zea mays* was found to be more sensitive to oil exposure than red beans. Previous reports of interspecies differences in sensitivity to petroleum hydrocarbon, which may be related to differences in systemic uptake of hydrocarbon and cell wall structure, had been documented [43] [44].

Catalase is another important enzyme that protect living system against oxidative stress, being able to scavenge hydrogen peroxide, which is the major product produced by superoxide dismutase [45]. The inhibition of catalase activity as concentration of petroleum product increases after four, eight days and twelve days of germination (**Figure 6**) corroborates the effect of petroleum hydrocarbon on superoxide dismutase activity in the leaves of both cowpea and maize seedlings. The similarity of catalase to superoxide dismutase is not suprising because the two enzymes have been reported to act in tandem [5] [46]. However, there is no marked difference in the activity of catalase in the leaves of both cowpea and maize seedlings at each of the concentration tested (**Figure 6**).

A wealth of information is available confirming that xanthine oxidase is involved in the metabolism of heterocyclic and polycyclic aromatic hydrocarbons [47] [48] as well as the involvement of the enzyme in the conversion of hypoxanthine via xanthine into uric acid. The role of uric acid as an antioxidant has been documented [49] [50]. The significant ($p < 0.05$) decreases in the activity of the enzyme as the concentration of petroleum products increased after four, eight and twelve days of germination (**Figure 7**) portends that petroleum mediated decrease in xanthine oxidase activity in cowpea and maize seedlings could contribute to oxidative stress by its decreased ability to metabolize aromatic hydrocarbon as well as its reduced ability to generate uric acid, a potent antioxidant. This effect is more pronounced in the leaves of cowpea seedlings compared to maize seedlings (**Figure 7**). The decrease in the activities of superoxide dismutase, catalase and xanthine oxidase as a result of increased formation of reactive oxygen species was earlier reported [51]-[54]. Therefore, it is pertinent to note that refined petroleum products could impose oxidative stress in exposed plant.

5. Conclusion

This study has indicated that exposure of plant to refined petroleum products in soil could impose oxidative stress on plant. The effect was more on cowpea relative to maize seedlings. Moreover, kerosene affected the exposed plant more than the other petroleum products

References

[1]   Mittler, R. (2002) Oxidative Stress, Antioxidants and Stress Tolerance. *Trends in Plant Science*, 7, 405-410. http://dx.doi.org/10.1016/S1360-1385(02)02312-9

[2]   Elstner, E.F. (1991) Mechanisms of Oxygen Activation in Different Compartments of Plant Cells. In: Pell, E.J. and Steffen, K.L., Eds., *Active Oxygen/Oxidative Stress and Plant Metabolism*, American Society of Plant Physiologists, Rockville, 13-25.

[3]   Voet, D. and Voet, J.G. (1995) Biochemistry. 2nd Edition, John Wiley and Sons, New York, 816-820.

[4]   Rich, P.R. and Bonner Jr., W.D. (1978) The Site of Superoxide Anion Generation in Higher Plant Mitochondria. *Arc-*

hive of Biochemistry and Biophysics, **188**, 206-213. http://dx.doi.org/10.1016/0003-9861(78)90373-9

[5]   Achuba, F.I. and Osakwe, S.A. (2003) Petroleum Induced Free Radical Toxicity in African Catfish (*Clarias garieponus*). *Fish Physiology and Biochemistry*, **29**, 97-103. http://dx.doi.org/10.1023/B:FISH.0000035905.14420.eb

[6]   Anozie, O.I. and Onwurah, I.N. (2001) Toxic Effect of Bonny Light Crude Oil in Rat after Ingestion of Contaminated Diet. *Nigerian Journal of Biochemistry and Molecular Biology* (*Proceeding Supplement*), **16**, 1035-1085.

[7]   Downs, C.A., Shigenska, G., Fauth, J.E., Robinson, C.E. and Huang, A. (2002) Cellular Physiological Assessment of Bivalves after Chronic Exposure to Spolled Exon Valdes Crude Oil Using a Novel Molecular Diagnostic Biotechnology. *Environmental Science and Technology*, **36**, 2987-2993. http://dx.doi.org/10.1021/es011433k

[8]   Achuba, F.I., Peretiomo-Clarke, B.O. and Ebokaiwe, P. (2005) Pollution Induced Oxidative Stress in African Catfish (*Clarias heterobranchus*). *European Journal of Scientific Research*, **8**, 52-73.

[9]   Aksnes, A. and Njaa, R.L. (1981) Catalase, Glutathione Peroxidase and Superoxide Dismutase in Different Fish Species. *Comparative Biochemistry and Physiology*, **69B**, 893-896.

[10] Misra, H.P. and Fridovich, I. (1972) The Role of Superoxide in the Auto-Oxidation of Epinephrine and a Simple Assay for Superoxide Dismutase. *Biochemical Journal*, **247**, 3170-3175.

[11] Crapo, J.D., McCord, J.M. and Fridovich, I. (1978) Preparation and Assay of Superoxide Dismutases. *Methods in Enzymology*, **53**, 382-393. http://dx.doi.org/10.1016/S0076-6879(78)53044-9

[12] Rani, P., Meena Unni, K. and Karthikeyan, J. (2004) Evaluation of Antioxidant Properties of Berries. *Indian Journal of Clinical Biochemistry*, **19**, 103-110. http://dx.doi.org/10.1007/BF02894266

[13] Heath, R.L. and Packer, L. (1968) Photoperoxidation in Isolated Chloroplasts. I. Kinetics and Stoichiometry of Farry Acid Peroxidation. *Archives Biochemistry and Biophysics*, **125**, 189-198. http://dx.doi.org/10.1016/0003-9861(68)90654-1

[14] Srivastava, M., Lena, Q.M., Singh, N. and Singh, S. (2005) Antioxidant Responses of Hyper-Accumulator and Sensitive Fern Species to Arsenic. *Experimental Botany*, **56**, 1335-1342. http://dx.doi.org/10.1093/jxb/eri134

[15] Hartley-Whitaker, J., Ainsworth, G. and Mehary, A.A. (2001) Copper and Arsenate-Induced Oxidative Stress in *Hocus lanatus* L. Clones with Differential Sensitivity. *Plant Cell and Environment*, **24**, 713-722. http://dx.doi.org/10.1046/j.0016-8025.2001.00721.x

[16] Somashekaraiah, B.V., Padmaja, K. and Prasad, A.R.K. (1992) Phytotoxicity of Cadmium Ions on Germinating Seedlings of Mung Bean (*Phaseolus vulgaris*): Involvement of Lipid Peroxides in Chlorophyll Degradation. *Physiologia Plantarum*, **85**, 85-89. http://dx.doi.org/10.1111/j.1399-3054.1992.tb05267.x

[17] Gallego, S.M., Benavides, M.P. and Tomaro, M.L. (1996) Effect of Heavy Metal Ion Excess on Sunflower Leaves: Evidence for Involvement of Oxidative Stress. *Plant Science*, **121**, 151-159. http://dx.doi.org/10.1016/S0168-9452(96)04528-1

[18] Lozano-Rodriguez, E. Hernandez, C.E., Bonay, P. and Carpena-Ruiz, R.O. (1997) Distribution of Cadmium in Shoots and Root Tissues of Maize and Pea Plants: Physiological Disturbances. *Journal of Experimental Botany*, **48**, 123-128. http://dx.doi.org/10.1093/jxb/48.1.123

[19] Frei, B. (1994) Reactive Oxygen Species and Antioxidant Vitamins: Mechanism of Action. *American Journal of Medicine*, **97**, S5-S13. http://dx.doi.org/10.1016/0002-9343(94)90292-5

[20] Val, A.L. and Almeida-Val, V.F. (1999) Effects of Crude Oil on Respiratory Aspect of Some Fish Species of the Amazon. In: Val., A.L. and Almerda-Val, V.M.F., Eds., *Biology of Tropical Fish*, INPA, Manaus, 227-291.

[21] Yu, Q. and Rengel, Z. (1999) Micronutrient Deficiency Influences Plant Growth and Activities of Superoxide Dismutase in Narrow Leafed Lupins. *Annals of Botany*, **8**, 175-182. http://dx.doi.org/10.1006/anbo.1998.0811

[22] Halliwell, B. (1989) Oxidants and the Central Nervous System: Some Fundamental Questions. *Acta Neurologica Scandinavica*, **126**, 23-33. http://dx.doi.org/10.1111/j.1600-0404.1989.tb01779.x

[23] Halliwell, B. and Gutteridge, J.M.C. (1990) The Antioxidant of Human Extra Cellular Fluids. *Archives of Biochemistry and Biophysics*, **280**, 1-8. http://dx.doi.org/10.1016/0003-9861(90)90510-6

[24] Liu, J. and Mori, A. (1994) Involvement of Reactive Oxygen Species in Emotional Stress: A Hypothesis Based on the Immobilization Stress Induced Oxidative Damage and Antioxidant Defense Changes in Rat Brain and the Effect of Antioxidant Treatment with Reduced Glutathione. *International Journal of Stress Management*, **1**, 249-263.

[25] Achuba, F.I. (2010) Spent Engine Oil Mediated Oxidative Stress in Cowpea (*Vigna unguiculata*) Seedlings. *Electronic Journal of Environment, Food and Agricultural Chemistry*, **9**, 910-917.

[26] Gutteridge, M.C. (1995) Lipid Peroxidation and Antioxidants as Biomarker of Tissue Damage. *Clinical Chemistry*, **41**, 1819-1828.

[27] Hammond-Kosack, K.E. and Jones, J.D.G. (1996) Resistance Gene-Dependent Plant Defense Responses. *The Plant Cell Online*, **8**, 1773-1791. http://dx.doi.org/10.1105/tpc.8.10.1773

[28] Orozo-Cardenase, M. and Ryan, C.A. (1999) Hydrogen Peroxide Is Generated Systematically in Plant Leaves by Wounding and Systemin via the Octadecanoid Pathway. *Proceedings of the National Academy of Sciences of the United States of America*, **96**, 6553-6557.

[29] Cazale, A.C., Droillard, M.J., Wilson, C., Heberle-Bors, E., Barbier-Brygoo, H. and Lauriere, C. (1999) MAP Kinase Activation by Hypoosmotic Stress of Tobacco Cell Suspensions: Towards the Oxidative Burst Response? *Plant Journal*, **19**, 297-307. http://dx.doi.org/10.1046/j.1365-313X.1999.00528.x

[30] Pei, Z.M., Murata, Y., Benning, G., Thomine, S., Klüsener, B., Allen, G.J., Grill, E. and Schroeder, J.I. (2000) Calcium Channels Activated by Hydrogen Peroxide Mediate Abscisic Acid Signalling in Guard Cells. *Nature*, **406**, 731-734.

[31] Asada, K. and Takahashi, M. (1987) Production and Scavenging of Active Oxygen in Chloroplasts. In: Kyle, D.J., Osmond, C.B. and Arntzen, C.J., Eds., *Photoinhibition*, Elsevier, Amsterdam, 227-287.

[32] Bowler, C., Van Montague, M. and Inze, D. (1994) Superoxide Dismutase in Plants. *Critical Review of Plant Science*, **13**, 199-218. http://dx.doi.org/10.1080/07352689409701914

[33] Jayakumar, K., Jaleel, A.C. and Viayarengan, P. (2007) Changes in Growth, Biochemical Constituents and Antioxidant Potentials in Radish (*Raphasus sativus* L.) under Cobalt Stress. *Turkish Journal of Biology*, **31**, 127-136.

[34] Fridovich, I. (1986) Biological Effects of Superoxide Radical. *Archive of Biochemistry and Biophysics*, **247**, 1-11. http://dx.doi.org/10.1016/0003-9861(86)90526-6

[35] Sevenian, A. and Hochistein, P. (1985) Mechanism and Consequences of Lipid Peroxidation in Biological Systems. *Annual Review of Nutrition*, **5**, 365-390. http://dx.doi.org/10.1146/annurev.nu.05.070185.002053

[36] Stern, A. (1985) Red Cell Oxidative Damage. In: Sies, H., Ed., *Oxidative Stress*, Academic Press, New York, 331-349.

[37] Parke, V.O. (1987) Role of Enzymes in Protection against Lipid Peroxidation. *Regulatory Toxicology and Pharmacology*, **7**, 222-235. http://dx.doi.org/10.1016/0273-2300(87)90035-3

[38] Saltman, P. (1989) Oxidative Stress: A Radical View. *Seminar in Hematology*, **26**, 249-256.

[39] Smirnoff, N. (1993) The Role of Active Oxygen in the Response of Plants to Water Deficit and Desiccation. *New Phytologist*, **125**, 27-58. http://dx.doi.org/10.1111/j.1469-8137.1993.tb03863.x

[40] Ayotarmuno, J.M. and Kogbara, R.B. (2007) Determining the Tolerance Level of *Zea mays* (Maize) to Crude Oil Polluted Agricultural Soil. *African Journal of Biotechnology*, **6**, 1332-1337.

[41] Aboaba, O.A., Aboaba, O.O., Nwachuku, N.C., Chukwu, E.E. and Nwachukwu, S.C.U. (2007) Evaluation of Bioremediation of Agricultural Soils Polluted with Crude Oil by Planting Beans Seeds, *Phaseolus vulgaris*. *Nature and Science*, **5**, 53-60.

[42] Baek, K., Kom, H., Oh, H., Young, B., Kim, J. and Lea, I. (2004) Effect of Crude Oil Components and Bioremediation on Plant Growth. *Journal of Environmental Science and Health*, **A39**, 2465-2477. http://dx.doi.org/10.1081/ESE-200026309

[43] Albert, P.H. (1995) Petroleum and Individual Polycyclic Aromatic Hydrocarbons. In: Hoffman, D.J., Rattner, B.A., Burton Jr., G.A. and Cairons Jr., J., Eds., *Handbook of Ecototoxicology*, CRC Press, Boca Raton, 330-355.

[44] Ogbo, E.M. (2009) Effects of Diesel Fuel Contamination on Seed Germination of Four Crop Plants. *Arachis hypogala, Vigna unguiculata, Sorghum bicolor* and *Zea mays*. *African Journal of Biotechnology*, **8**, 250-253.

[45] Asada, K. (1992) Ascorbate Peroxidase: A Hydrogen Peroxide Scavenging Enzyme in Plants. *Physiologia Plantarum*, **85**, 235-241. http://dx.doi.org/10.1111/j.1399-3054.1992.tb04728.x

[46] Iyawe, H.O.T. and Onigbinde, A.O. (2004) Effect of an Anti-Malarial and Micronutrient Supplementation on Respiration Induced Oxidative Stress. *Pakistan Journal of Nutrition*, **3**, 318-321. http://dx.doi.org/10.3923/pjn.2004.318.321

[47] Panoutsopoulos, G.I. and Beedham, C. (2004) Enzymatic Oxidation of Phthalazine with Guinea Pig Liver Aldehyde Oxidase and Liver Slices: Inhibition by Isovanillin. *Acta Biochemistry Polonica*, **51**, 953-951.

[48] Panoutsopolous, G.I., Kauretas, D. and Beedham, C. (2004) Contribution of Aldehyde Oxidase, Xanthine Oxidase and Aldehyde Dehydrogenase on the Oxidation of Aromatic Aldehyde. *Chemical Research in Toxicology*, **17**, 1368-1376. http://dx.doi.org/10.1021/tx030059u

[49] Peden, D.B., Hohman, R., Brown, M.E., Masont, R.T., Berkebilef, C., Falest, H.M. and Kaliner M.A. (1990) Uric Acid Is a Major Antioxidant in Human Nasal Airway Secretions. *Proceedings of National Academy of Science*, **87**, 7638-7642. http://dx.doi.org/10.1073/pnas.87.19.7638

[50] Nieto, F.J., Iribarren, C., Gross, M.D., Comstock, G.W. and Cutler, R.G. (2000) Uric Acid and Serum Antioxidant Capacity: A Reaction to Atherosclerosis? *Atherosclerosis*, **148**, 131-139. http://dx.doi.org/10.1016/S0021-9150(99)00214-2

[51] Trush, M.A. and Kensler, T.W. (1991) An Overview of the Relationship between Oxidative Stress and Chemical Car-

cinogenesis. *Free Radical Biology and Medicine*, **10**, 201-209. http://dx.doi.org/10.1016/0891-5849(91)90077-G

[52] Sies, H. (1991) Oxidative Stress: Introduction. In: Sies, H., Ed., *Oxidative Stress: Oxidant and Antioxidants*, Academic Press, San Diego, 15-22.

[53] Pigeolet, E., Corbisler, P., Houbion, A., Lambert, D., Michiels, C., Raes, M., *et al.* (1990) Glutathione Peroxidases, Superoxidase and Oxygen Derived Free Radicals. *Mechanisms of Ageing and Development*, **51**, 283-297. http://dx.doi.org/10.1016/0047-6374(90)90078-T

[54] Achuba, F.I. and Otuya, E.O. (2006) Protective Influence of Vitamins against Petroleum-Induced Free Radical Toxicity in Rabbit. *The Environmentalist*, **26**, 295-300. http://dx.doi.org/10.1007/s10669-006-0158-y

16

Analysis of Post-Burial Soil Developments of Pre-AD 79 Roman Paleosols near Pompeii (Italy)

Sebastian Vogel[1]*, Michael Märker[2,3]

[1]University of Tübingen c/o German Archaeological Institute, Berlin, Germany
[2]Heidelberg Academy of Sciences and Humanities c/o University of Tübingen, Tübingen, Germany
[3]Department of Earth Sciences (DST), University of Florence, Florence, Italy
Email: seb_vogel@gmx.de

Abstract

The AD 79 eruption of Somma-Vesuvius completely buried the ancient landscape around Pompeii (Italy) to some extent conserving the pre-AD 79 Roman paleosols of the Sarno River plain. To estimate potential post-burial soil developments of these paleosols detailed soil liquid and solid phase analysis were carried out. Firstly, an *in-situ* soil hydrological monitoring was conducted within a pre-AD 79 paleosol in natural undisturbed stratification. The results show that soil water flow and nutrient transport from the overlying volcanic deposits into the pre-AD 79 paleosol take place. Secondly, to estimate their influence on the paleosol's mineral soil properties, the solid phase of four pre-AD 79 paleosols and associated modern unburied soils were analysed and compared. By combining the data from the soil liquid and solid phase analysis, potential post-burial changes in the paleosols were estimated. Finally, a rise of the mean groundwater table was determined since AD 79. This distinguishes the Sarno River plain into two different zones of post-burial soil developments: 1) lower altitudes where formerly terrestrial paleosols are now influenced by groundwater dynamics and 2) higher altitudes where the paleosols are still part of the vadose zone and rather influenced by infiltration water or interflow. Thus, the mechanism of potential post-burial soil development being active in the pre-AD 79 paleosols is not uniform for the entire Sarno River plain but strongly depends on the paleotopographic situation.

Keywords

Soil Liquid Phase, Soil Solid Phase, Post-Burial Soil Developments, Pre-AD 79 Roman Paleosol, Pompeii, Somma-Vesuvius

---

*Corresponding author.

1. Introduction

Paleosols are geoarchives of a past soil formation since diverse mechanisms of burial can contribute to a preservation of original soil properties. Consequently, paleosols can yield important information on paleopedological and paleoenvironmental conditions before burial. However, a critical use of paleosol characteristics in interpreting paleoenvironments must be applied because pronounced diagenetic processes, *i.e.* post-burial soil developments, can take place, such as (Crowther *et al.*, 1996 [1]; Retallack, 1998 [2], 2001 [3]; French, 2003 [4]).

1) Decomposition of soil organic matter (SOM)

Within well drained terrestrial soils SOM is decomposed after burial by aerobic microorganisms. That can be detected by a change in the Munsell soil colour towards brighter values (higher chroma). However, the general depth function of the SOM concentration of the paleosol stays mostly unaltered.

2) Leaching of nutrients from the overlying strata

In highly permeable substrate seepage water can lead to a dislocation of bases from the upper layers into the paleosol and hence to an alteration of the paleosol's nutrient status.

3) Soil compaction after burial

The superimposed load of the overlying deposits can cause soil compaction and a reduction of pore space.

Due to the probability of post-burial soil developments detailed field descriptions and laboratory analyses are necessary to understand both past and present pedogenetic processes that lead to the present-day appearance and properties of a paleosol (Scudder *et al.*, 1996 [5]). Combined soil solid and liquid phase studies are a powerful tool to study post-burial soil developments. If the analysis of the soil solid phase provides good information about the factors and processes that influence a soil since the beginning of pedogenesis, soil liquid phase chemistry helps to understand active pedogenic processes. Gravitational water particularly represents the mobile fraction of the soil liquid phase that percolates through the soil profile and interacts and communicates with the entire stratigraphy. Consequently, measuring the chemical composition of the gravitational water most notably reflects the active redistribution and mobility of substances within the soil profile by vertical or lateral transport processes (Ugolini and Dahlgren, 1987 [6]; Wolt, 1994 [7]; Snakin *et al.*, 2001 [8]). The analysis of the soil liquid phase within a buried paleosol enables the identification of those processes that are recently active and are likely to influence the paleosols solid phase properties after burial. Finally, by combining soil liquid and solid phase studies an estimate about post-burial soil developments can be made.

The explosive eruption of Somma-Vesuvius AD 79 almost completely buried the ancient landscape around Pompeii to some extent conserving the pre-AD 79 Roman paleosols. Foss, 1988 [9] and Foss *et al.*, 2002 [10], investigated pre-AD 79 paleosols in the Pompeii area and stated that they show little disturbance or destruction. The paleosols were very similar to the modern soils suggesting a similar environment for soil development before and after AD 79. The paleosols developed in pumice and ash analog to the present-day soils. However, the paleosols had a higher pH value and a higher nutrient content (Foss, 1988 [9]). Thus, Foss, 1988 [9], and Foss *et al.*, 2002 [10], presumed this to be the result of bases (Ca^{2+}, Mg^{2+}) leaching from the overlying volcanic sediments and recharging the surface horizons of the paleosols.

Inoue *et al.*, 2009 [11], investigated the pre-AD 472 paleosols at Somma Vesuviana around 15 km north of Pompeii. It developed from the AD 79 pyroclastic deposits and was buried during the AD 472 eruption of Somma-Vesuvius. These paleosols also showed higher pH values compared to the modern soils which they also attributed to leaching of bases from the overlying tephra deposits and accumulation in the paleosols.

As from today's state of research little is known about the development of volcanic paleosols after their burial (Zehetner *et al.*, 2003 [12]; Agnelli *et al.*, 2007 [13]) the objective of this study is to analyse both the soil liquid and solid phase of pre-AD 79 paleosols near Pompeii to get deeper insight into potential post-burial soil developments. Special attention is given to the evaluation of Foss' und Inoue's hypothesis of nutrient transport by soil water flow from the overlying volcanic deposits into the paleosols. Furthermore, present-day groundwater data and the pre-AD 79 topography (Vogel and Märker, 2010 [14]; Vogel *et al.*, 2011 [15]) will be used to estimate the post-AD 79 groundwater trend with respect to the pre-AD 79 paleosol. This aims at confining the area of the Sarno River plain where the pre-AD 79 paleosol is no longer part of the vadose zone but recently influenced by groundwater dynamics.

2. Research Area

The soil liquid and solid phase was studied in the vicinity of Pompeii. It is located in the Sarno River plain in the southern part of the Campanian plain and southwest of Somma-Vesuvius volcanic complex (**Figure 1**). The stratigraphic sequence of the study area is particularly composed of volcaniclastic deposits from the periodic activity of Somma-Vesuvius as well as soil formations during phases of volcanic quiescence. Soils develop on ash, pumice or scoria lapilli fallout, pyroclastic surges and lava flows whereas pedogenesis especially depends on deposition modalities of the volcanic material, pedoclimatic conditions, topography, anthropogenic activity and time (Lulli, 2007 [16]).

The climate of the study area is Mediterranean with almost 70% of the annual precipitation falling between October and March and a pronounced dry summer season. On a long term average the mean annual precipitation is 865 mm whereas November is the wettest month with 129 mm and July is dryest with 16 mm. The mean annual temperature is 17.4°C. The hottest month is August with a mean temperature of 25.7°C whereas January is coolest with 9.6°C (Osservatorio Meteorologico, Università di Napoli Federico II, from 1870 to today).

The soil liquid phase was sampled within the archaeological excavation of Villa Regina (Boscoreale) around 1.3 km northwest of ancient Pompeii. Topographically, this Roman farm (villa rustica) is situated in a longitudinal depression between the western footslopes of the Pompeiian hill and the southern footslopes of Somma-

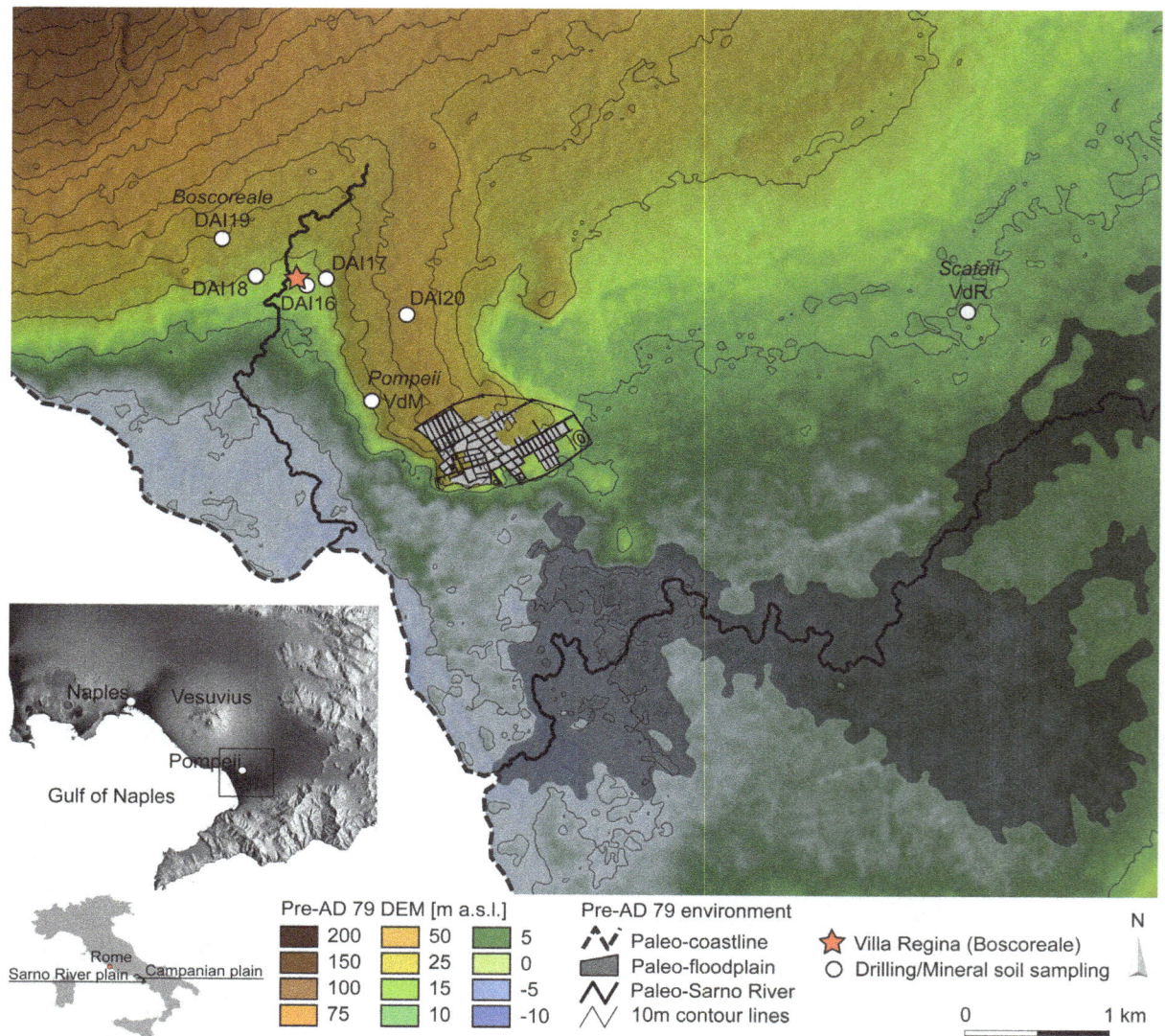

Figure 1. Paleotopographic location of the soil water monitoring (Villa Regina) northwest of ancient Pompeii and surrounding soil sampling sites. The stratigraphic section of DAI16 shows that at Villa Regina the pre-AD 79 paleosol is situated in a depth of around 8 m and overlain by recurring pyroclastic material, fluvial deposits and soil substrate.

Vesuvius (**Figure 1**). There are two main reasons why Villa Regina was chosen for the soil liquid phase analysis:

1) The post-AD 79 deposits at the footslopes of Somma-Vesuvius have an above-average thickness of 8 m compared to other parts of the plain. Consequently, in case soil water flow into the paleosol can be detected at Boscoreale it is likely to be occurring in most areas of the Sarno River plain where the post-AD 79 deposits are usually thinner.

2) At the archaeological excavation of Villa Regina the pre-AD 79 paleosol was easily accessible to install the soil hydrological monitoring system in natural and undisturbed stratigraphy. Furthermore, it was protected from vandalism over the study period of 20 months.

The soil solid phase was sampled at three different sites incorporating different topographic situations as well as slightly different stratigraphic conditions (**Figure 1**):

1) Scafati (Via della Resistenza (VdR)): This site is located in the river plain approximately 3 km east of ancient Pompeii. The pre-AD 79 paleosol is situated in a depth of about 3.3 m. It is covered by 2 m of pumice fallout and 0.9 m of pyroclastic density current (PDC) deposits from the explosive AD 79 eruption of Somma-Vesuvius followed by a medieval paleosol and the modern soil.

2) Pompeii (Villa dei Misteri (VdM)): It is situated at the southwestern slope of the Pompeiian hill near the archaeological site of Villa dei Misteri and about 350 m northwest of the city walls of ancient Pompeii. The pre-AD 79 paleosurface appears in a depth of 5.7 m. The thicknesses of overlying AD 79 pumice fallout and PDC deposits are about 2.3 m and 1.7 m, respectively. This is followed by about 1.7 m of a medieval paleosol and the modern soil (Rispoli *et al.*, 2008 [17]).

3) Boscoreale (Villa Regina (VR)): Five mechanical core drillings were carried out near the archaeological excavation of Villa Regina (VR) along a NW-SE transect crossing the longitudinal depression between the Pompeiian hill and Somma-Vesuvius. The drillings DAI18 and DAI19 are situated in the northwest at the footslope of Somma-Vesuvius whereas DAI17 is located at the footslope of the Pompeiian hill. DAI16 is situated at the bottom of the depression at the exact location of the soil liquid phase analysis whereas DAI20 is located on top of the Pompeiian hill in the southeast (Vogel and Märker, 2012 [18]). At DAI16 the pre-AD 79 paleosol is covered by approximately 2.5 m of AD 79 pumice fallout and PDC deposits. This is overlain by volcanoclastics from later eruptions and intercalated fluvial deposits and pedogenised soil material representing phases of volcanic quiescence. This results in a total thickness of post-AD 79 deposits of around 8 m lying on top of the pre-AD 79 paleosurface.

According to groundwater data of the "Autorità di Bacino del Sarno" the pre-AD 79 Roman layer of the three study sites is not influenced by groundwater. Thus, the pre-AD 79 paleosols are part of the vadose zone and soil water within the profile derives vertically from infiltration water or laterally from interflow.

3. Material and Methods

To assess the post-burial soil developments of the pre-AD 79 Roman paleosols near Pompeii, we conducted soil liquid and solid phase analysis combined with an estimation of the post-AD 79 groundwater trend. Consequently, the following sections will be subdivided accordingly.

3.1. Groundwater Table

Between the eruption of Somma-Vesuvius in AD 79 and today a relative rise of the groundwater table can be determined for the Sarno River plain. Consequently, in many parts of the inner plain, although terrestrial before AD 79, the pre-AD 79 paleosol has come under the influence of groundwater. To calculate this approximate rise of the groundwater table since AD 79 and to confine the zone where the primal terrestrial pre-AD 79 paleosol is recently influenced by groundwater dynamics, present-day groundwater data were combined with the pre-AD 79 topography (Vogel and Märker, 2010 [14]; Vogel *et al.*, 2011 [15]).

At first, the depth of the present-day mean annual groundwater table was modelled from more than 5600 groundwater observation points (by courtesy of the Autorità di Bacino del Sarno). Thereafter, this groundwater table was subtracted from the pre-AD 79 digital elevation model (DEM) of the Sarno River plain to determine the area where the groundwater table recently lies above the pre-AD 79 soil surface. Vogel *et al.*, 2011 [15], modelled the pre-AD 79 floodplain or wetlands related to the paleo-Sarno River which are characterised by fluvial/palustrine deposits before AD 79. In general, wetlands show a water table that stands at or near the terrain surface for a sufficient period of the year to allow the development of palustrine deposits (Brinson, 1993 [19]). Consequently, the reconstructed paleofloodplain can be considered as the area where, before AD 79, the mean groundwater table was at or near the pre-AD 79 paleosurface (**Figure 1**).

From these two groundwater levels (at the pre-AD 79 surface, today and before AD 79) the mean "altitude above channel network" index (AACN; SAGA terrain analysis module; 0.06% channel network density) was deduced from the pre-AD 79 DEM. The AACN calculates the vertical distance to the pre-AD 79 Sarno River and thus directly reflects the groundwater table. Finally, by subtracting the arithmetic means of the two indices the approximate rise of the groundwater table since AD 79 was determined.

3.2. Soil Liquid Phase

At the archeological excavation of Villa Regina (Boscoreale) a soil hydrological monitoring system was installed within the pre-AD 79 paleosol in natural stratification. It aims at determining if the pre-AD 79 paleosol is subject to soil water flow from the overlying volcanic deposits either vertically by percolation or laterally by interflow. The soil moisture and the soil liquid phase were determined *in-situ* by means of frequency domain reflectometer (FDR) and tension lysimeters (suction cups), respectively. Tension lysimeters were used since this methodology predominantly samples the mobile fraction of soil water that moves through inter-aggregate pores or preferential flow channels and thus reflects the chemical transport within the stratigraphy (Wolt, 1994 [7]). Due to the depth of the pre-AD 79 paleosol of about 8 m a vertical installation from the present-day surface was not feasible. Consequently, the installation was implemented laterally inside the natural undisturbed stratification of the side walls of the archaeological excavation of Villa Regina (**Figure 2**).

Altogether four sampling systems were installed, two within the lower section of the AD 79 pumice fallout layer and two within the pre-AD 79 paleosol underneath. The composition of the soil liquid phase may not directly refer to specific chemical processes active in the pre-AD 79 paleosol. Nevertheless, the net effect of recent

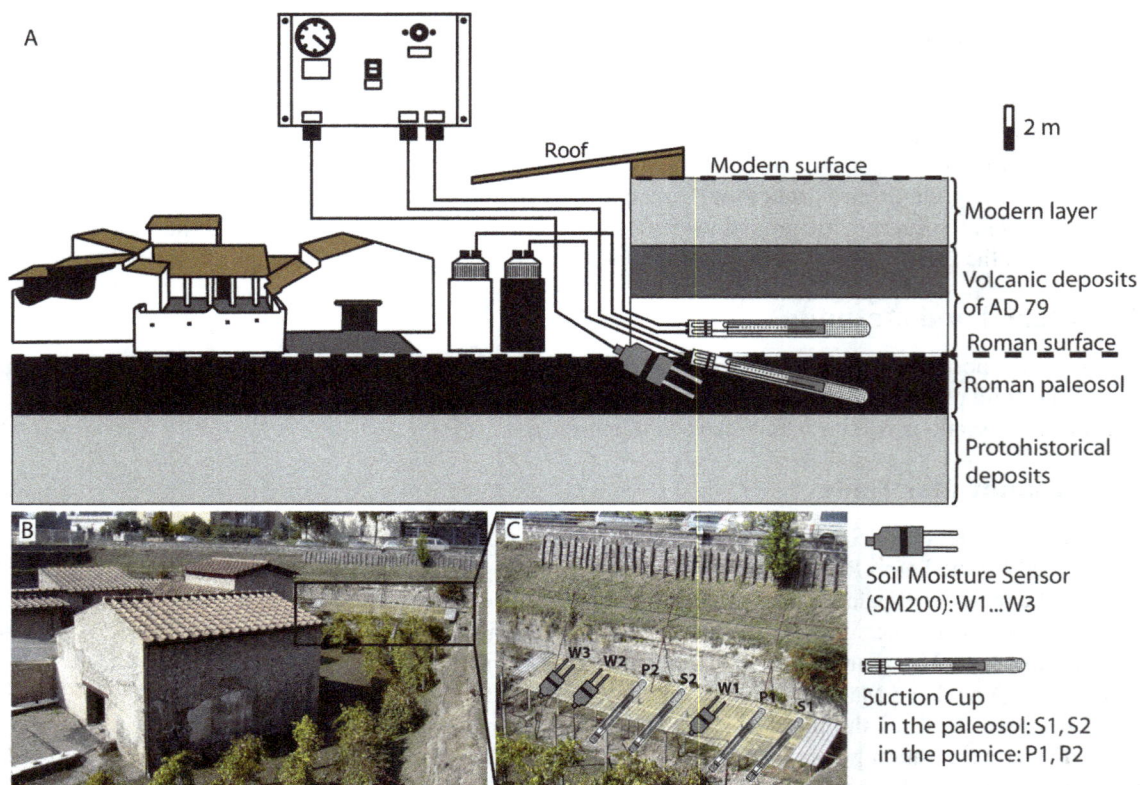

Figure 2. Schematic view of the experimental setup using suction cups and soil moisture sensors (SM200) (A). The study site is situated at the northern edge of the archaeological excavation of Villa Regina (Boscoreale) (B). A roof was build to protect the installation from precipitation water entering the excavation (C).

transport of nutrients into the paleosol is manifested in the concentration of substances in the soil water extract (Wolt, 1994 [7]). Consequently, the main objective of the given sampling design is to determine if a soil water flow to the pre-AD 79 paleosol occurs. Moreover, we provide data of the general trends in the soil liquid phase chemistry, i.e. the occurrence of certain ions in the lysimeter solution at the point of sampling (Grossmann and Moss, 1994 [20]; de Vries and Leeters, 1994 [21]; Wolt, 1994 [7]; Manderscheid and Matzner, 1995 [22]).

Due to the remarkable depth of burial and the absence of interactions with present-day plant roots the composition of the soil liquid phase within the pre-AD 79 paleosol will be less heterogeneous compared to the rooting zone of the modern soil. This may also be favoured by the relative chemical homogeneity of the AD 79 pumice fallout layer overlying the pre-AD 79 paleosol. Consequently, four sampling systems are expected to yield qualitative data on the soil liquid phase chemistry of the sampling area.

In the paleosol the suction cups were installed in slurry of the paleosol material to ensure an optimal contact between the instrument and the surrounding soil and a minimum of disturbance. Because of the coarseness of the AD 79 pumice lapilli layer above, the suction cups could not be installed in its own material. Hence, silica slurry of fine sand was used having a low adsorption capacity to minimise the influence of the slurry material on the soil liquid phase.

To extract the soil liquid phase a transient tension (vacuum) of 550 mbar (pF 2.7) was established biweekly to the suction cups by evacuation with an external pump. Due to a slight decrease of tension during the two weeks between evacuations an applied tension of 550 mbar enables the sampling of the soil liquid phase near field capacity. Even though the gradient generated by the suction cups will not exclusively act on macropores but to some extent also on smaller pores (Grossmann and Udluft, 1991 [23]) the applied tension predominantly samples the mobile soil water fraction. The approximate pore diameter that is drained at a certain soil water potential can be calculated by means of the law of capillary rise or the Young-Laplace equation:

$$r = \frac{2\gamma \cos\alpha}{h p_w g} \tag{1}$$

where r is the radius of the capillary tube, γ the surface tension of water, α the contact angle, h the water potential, p_w the density of water and g the gravitational acceleration (Scheffer and Schachtschabel, 2010 [24]). Hence, applying a tension of 550 mbar and assuming complete wettability ($\alpha = 0$) of the substrate the suction cups extract water from soil pores of a minimum diameter of 5.4 μm which corresponds to intermediate sized pores.

Through the introduction of oxygen into the soil during installation of the suction cups the biological activity of the paleosol may be stimulated causing a mobilization of nutrients. This can influence the composition of the soil liquid phase directly after installation (Grossmann, 1988 [25]). For this reason, the obtained lysimeter solution was rejected for the first months after installation. From the installation to the first soil water sampling the system had about three months to re-equilibrate.

The suction cups were installed 1.5 m inside the stratigraphy to prevent them from being influenced by water entering the soil profile from the open excavation side of Villa Regina. Furthermore, the installation was carried out on the upslope part of the excavation assuring that rain water directly entering the excavation is diverging. Additionally, the entire installation was covered by a roof over its whole length (**Figure 2**).

Suction cups were used that collect the solution in their interior. That has the advantage that the extracted water stays cool and protected from sunlight until the sampling takes place, minimising chemical transformations within the lysimeter solution. Due to its specific adsorption characteristics the filter material of the suction cup may have an influence on the composition of the sample. Consequently, the investigation of a certain analyte requires a particular filter material. As this study focusses on dissolved inorganic substances polyamide filters were used (Litaor, 1988 [26]; Grossmann, 1988 [25]; DVWK, 1990 [27]; Haberhauer, 1997 [28]; Spangenberg et al., 1997 [29]; Tischner et al., 1998 [30]).

The composition of the soil liquid phase was monitored over a period of 20 months to cover the variability throughout the year. The autumn and winter period was studied twice. This period is believed to be most relevant because the precipitation maximum and relatively low evapotranspiration between October and March is most likely to cause soil water flow down to the pre-AD 79 paleosol. The soil liquid phase in the suction cups was controlled biweekly, sampled and analysed for: 1) pH value, 2) major cations (Ca^{2+}, Mg^{2+}, K^+, Na^+) and anions (SO_4^{2-}, HPO_4^{2-}, NO_3^-, Cl^-) because of their importance for soil fertility and nutrient conditions and 3) Al^{3+} because of its relevance in terms of the acidity status and plant physiology (Al-toxicity) (Hendershot and

Courchesne, 1991 [31]; Wolt, 1994 [7]; Fillion *et al.*, 1999 [32]; Derome, 2002 [33]; Roman *et al.*, 2002 [34]; Strahm *et al.*, 2005 [35]).

Since the composition and amount of the soil liquid phase sampled by the tension lysimeters varies with soil water content, additionally, three soil moisture sensors (SM200) where installed in the paleosol measuring the volumetric water content (θ) every six hours. The SM200 soil moisture sensor is a frequency domain reflecto-meter (FDR) measuring θ indirectly by determining the permittivity of the soil (Delta-T Devices Ltd., 2006 [36]).

The soil liquid phase of the vadose zone strongly depends on water deriving from precipitation. To get a rough estimate of how the precipitation regime interacts with the soil moisture of the pre-AD 79 paleosol and its soil water chemistry throughout the year, meteorological data were collected (Campania Region). It may not be possible to relate a single precipitation event to the soil water sampled in the paleosol at a depth of 8 m. Never-theless, especially the time between the first precipitations in autumn and the first response in the paleosol's soil moisture as well as the first soil water sampling will be of particular interest. Moreover, the first soil water flow after the dry summer season is expected to have a strong influence on the composition of the soil liquid phase in terms of higher element concentrations that are leaching into the pre-AD 79 paleosol (Arthur and Fahey, 1993 [37]).

3.3. Soil Solid Phase

At Scafati, Pompeii and Boscoreale modern soils and associated pre-AD 79 paleosols were identified in four stratigraphies (DAI16, DAI18, VdM, VdR). They were described in the field and the soil solid phase was sam-pled for the subsequent soil physical and chemical laboratory analyses.

Following generally Retallack, 1998 [2], 2001 [3] and French, 2003 [4], as well as in particular Foss, 1988 [9], Foss *et al.*, 2002 [10] and Inoue *et al.*, 2009 [11], we hypothesise that post-burial soil changes in the pre-AD 79 paleosols may above all be associated with nutrient transport from the overlying strata and decomposition of soil organic matter (SOM). Hence, to determine post-burial soil developments, the soil solid phase of both the pre-AD 79 paleosols and the associated modern soils were analysed and compared. Thereby, the modern soils were taken as a reference for an unburied volcanic soil.

To assess the effects of soil water flow and associated nutrient transport on the mineral soil properties the fol-lowing analyses were carried out: 1) soil pH measured in 0.01 M $CaCl_2$ (DIN ISO 10390 [38]); 2) effective cation exchange capacity (CEC_{eff}) and the amount of exchangeable cations (K, Na, Ca, Mg, Mn, Al and Fe) by sequential extraction with unbuffered ammonium chloride; and 3) total inorganic carbon (TIC) by elementary analysis using the dry combustion reference method (DIN EN 13137 [39]).

To study the post-burial decomposition of soil organic matter in the paleosols total organic carbon (TOC) was determined by elementary analysis using the dry combustion reference method (DIN EN 13137 [39]) as well as total nitrogen using the modified method of Kjeldahl (DIN ISO 11261 [40]).

As mentioned earlier Foss, 1988 [9], stated that the environment for soil development before and after AD 79 was similar resulting in a relative similarity between the paleosols and the modern soils around Pompeii. Hence, for this comparison we presuppose similar climatic conditions before AD 79 and today. This is indicated by Sa-dori and Narcisi, 2001 [41], from lacustrine sediments in Sicily as well as from Buccheri *et al.*, 2002 [42], from sea core data of the Salerno Gulf (southern Tyrrhenian Sea). However, similar vegetational conditions at the in-vestigated sites have to be hypothesized as they are more difficult to determine. Isotope analysis of $\delta 13C$ and $\delta 15N$ of the pre-AD 79 paleosols and the modern soils revealed a predominance of C_3 vegetation before AD 79 and today (Oelmann and Ruppenthal, 2012, written communication). Moreover, the comparison between the pre-AD 79 paleosols and the modern soils was only carried out at the same topographic location of each sampling point. This eliminates the influence of the relief as a soil forming factor in explaining soil variations.

The particular chemical composition of the modern soils and the pre-AD 79 paleosols may differ due to a slightly different geology of the parent volcanic material as well as due to anthropogenic input in terms of fertil-izers or waste materials. This may falsify the comparison of the two soils. To characterise the chemical compo-sition of the soils, total amounts of Na, Ca, K, Mg, Mn, Al and Fe were determined by element analysis using aqua regia (nitrochloric acid) and AAS /ICP-AES. To eliminate the effect of a different geology of the soil sub-strate and anthropogenic input ratios were introduced correcting a particular measured element concentration by its total concentration (see Equations (2) and (3)).

The CEC ratio allows for a comparison of the nutrient status of the paleosols and the modern soils even if both soils exhibit different total element concentrations. A consistently higher CEC ratio in the paleosols may for instance indicate nutrient input via leaching. As CEC is the sum of exchangeable cations the CEC ratio was calculated by dividing the cation exchange capacity by the sum of the total concentrations of K, Na, Ca and Mg (Equation (2)).

$$CEC\ ratio = \frac{CEC}{\left(K_{tot} + Na_{tot} + Ca_{tot} + Mg_{tot}\right)} \qquad (2)$$

The nutrient ratio is used for the comparison of a particular nutrient cation, hence dividing the amount of an exchangeable cation (K_{exch}, Na_{exch}, Ca_{exch}, Mg_{exch}) by the total amount of the respective element (K_{tot}, Na_{tot}, Ca_{tot}, Mg_{tot}) (Equation (3)). Thus, the nutrient ratio helps to identify which nutrients are predominantly leaching from the volcanic deposits into the paleosols.

$$Nutrient\ ratio = \frac{Ion_{exch}}{Element_{tot}} \qquad (3)$$

By means of division by the total element concentrations, the CEC ratio and the nutrient ratios also eliminate methodical defects of the CEC method with respect to $CaCO_3$ contents > 10 $g \cdot kg^{-1}$ which may cause dissolution of $CaCO_3$ and increased values of CEC and exchangeable Ca^{2+} (Dohrmann and Kaufhold, 2009 [43]).

For the comparison between the modern soils and the pre-AD 79 Roman paleosols only topsoils were considered as in general they are characterised by high SOM and nutrient turnover rates as well as by major soil chemical reactions due to plant roots and microorganisms. Finally, considering the results of both the soil liquid and soil solid phase analysis major differences between the unburied modern soils and the buried paleosols are utilised to gain valuable insights into potential post-burial soil developments of the pre-AD 79 paleosols.

4. Results

4.1. Groundwater Table

The combination of the present-day groundwater table with the pre-AD 79 topography reveals that a large area of the inner Sarno River plain is affected by a groundwater table lying at or above the pre-AD 79 paleosurface (**Figure 3**). Altogether this comprises the area near the river network which amounts to around 75 km^2 or 37% of the entire Sarno River plain. However, this includes an area of about 25 km^2 or 32 % where the pre-AD 79 layer was originally (before burial) characterised as a terrestrial paleosol. Consequently, there must have been a considerable rise of the groundwater table from AD 79 to the present day. This post-AD 79 groundwater rise was determined to approximately 1.8 m.

4.2. Soil Liquid Phase

4.2.1. Soil Water Flow

Figure 4 shows the soil moisture (θ) of the pre-AD 79 Roman paleosol at Villa Regina and the precipitation regime for the period 2008/09. At the beginning of the study period in October 2008 the pre-AD 79 paleosol was still dry from the preceding summer season. This is supported by a constant and low initial θ of around 22%. First single rainfall events started in September 2008 whereas the main rainy season with more regular and intense precipitation began in early November. However, it took about one month, until the beginning of December, for a first response of θ in the pre-AD 79 paleosol. Thus, the soil water needed about one month to percolate from the present-day surface to the pre-AD 79 paleosol in a depth of 8 m.

The phase of rewetting of the pre-AD 79 paleosol after the dry summer season lasted from early December 2008 until February 2009 when θ continuously increased to its maximum of 25.8%. The main rainy season terminated in mid-February with more sporadic rainfall events until the end of April. From mid March and again with a delayed response of about one month θ started to decrease, at first rather slightly and more explicitly from the end of April. Then θ decreased to its minimum of 13.6% in the end of September 2009. Finally, the soil moisture data reveal that the pre-AD 79 paleosol at Villa Regina was at no time of the year water saturated.

The first soil liquid phase was extracted by the suction cups at the end of December, another month after the first response of θ. By the time of the first soil water sampling θ was 24.3%. Henceforward, soil liquid phase could

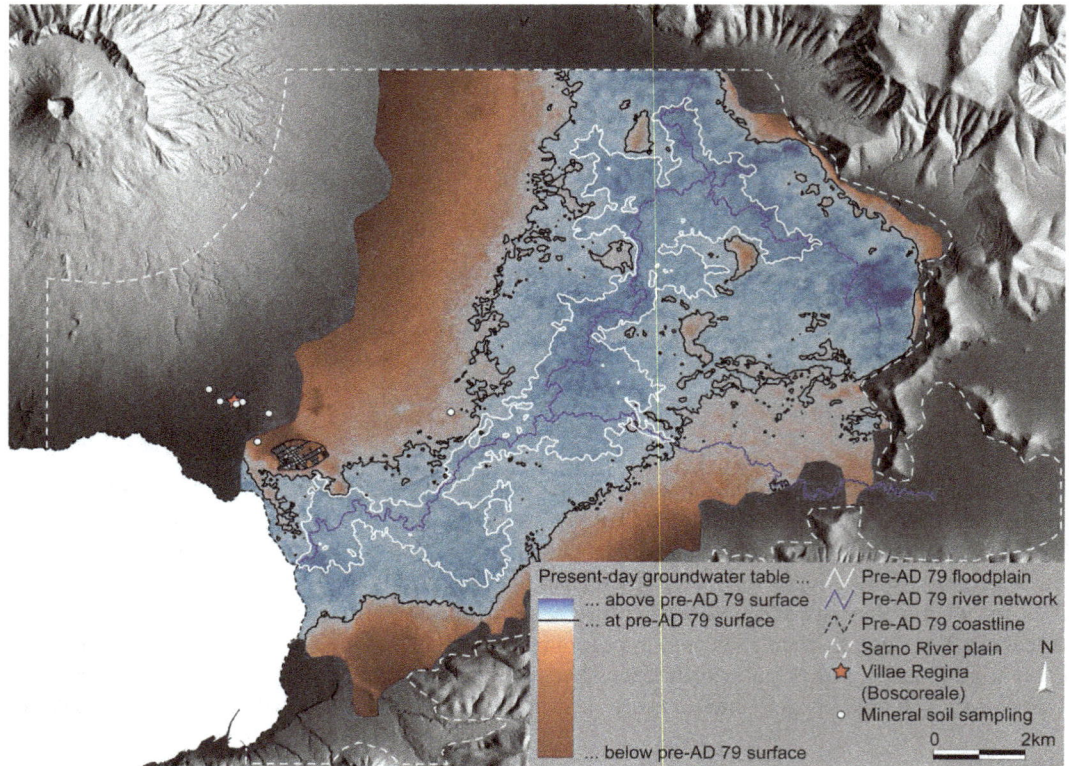

Figure 3. Model of the present-day mean groundwater table with respect to the pre-AD 79 paleosurface. The black line confines the area where the groundwater table is at or above the pre-AD 79 paleosurface (blue colours). Accordingly, the white line of the pre-AD 79 floodplain approximately describes the area where before AD 79 the mean groundwater table was at or above the pre-AD 79 paleosurface.

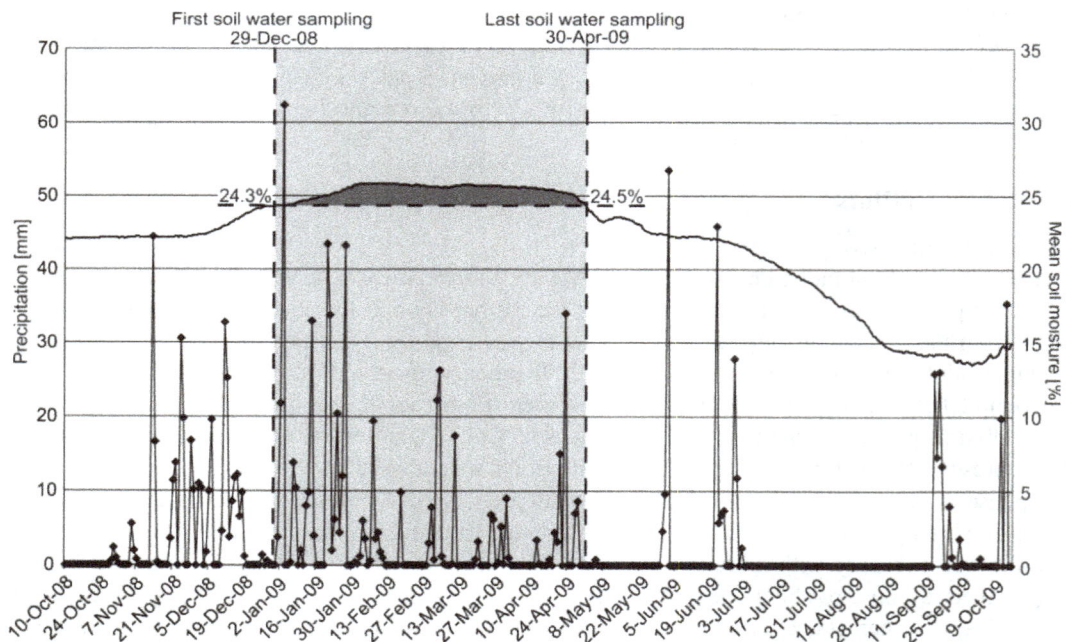

Figure 4. Soil moisture (θ) of the pre-AD 79 paleosol at Villa Regina (Boscoreale) and precipitation chart of Napoli-Osservatorio Meteo D.G.V. for the study period 2008/09. Highlighted is the period when mobile soil water could be extracted from the paleosol (light gray) when θ exceeded about 24.5% (dark gray).

be sampled biweekly for the next four months until the end of April 2009. By the end of the sample period θ had decreased to 24.5%.

In contrast to the pre-AD 79 paleosol no soil water could be extracted from the AD 79 pumice lapilli fallout directly above.

During the winter season of 2009/10 the first lysimeter sample was taken at the beginning of February 2010 about one month later than 2009. This coincides with the paleosol having a much lower initial soil moisture (θ) in the summer of 2009 (13.6%) in comparison to 2008 (22%) (**Figure 4**). Furthermore, in the first half of the rainy season, from October to January 2008/09, there were a total of 735 mm of precipitation in comparison to only 492 mm in 2009/10. Consequently, in 2009/10 the process of rewetting of the soil substrate took much longer until percolation took place and θ of the pre-AD 79 paleosol was sufficiently high for soil water to be sampled by the suction cups. Consequently, in 2009/10 the period of soil water sampling was shorter, lasting only 100 days in comparison to 120 days in 2008/09.

4.2.2. Soil Water Chemistry

The soil water chemistry of the 2008/09 and 2009/10 seasons is summarised in **Tables 1** and **2**. The pH value of the soil liquid phase represents the active acidity of the pre-AD 79 paleosol (Mengel and Kirkby, 2001 [44]). With a mean pH value of 7.5 the soil water extract is slightly alkaline, varying only marginally throughout the year. The mean cation concentrations are: 42.9 mg/l for Ca^{2+}, 35.6 mg/l for K^+, 23.3 mg/l for Na^+ and 5.7 mg/l for Mg^{2+}. The mean anion concentrations are: 60.5 mg/l for Cl^-, 23.4 mg/l for NO_3^- and 22.7 mg/l for SO_4^{2-}. The mean calculated electrical conductivity (EC_{calc}) at 25°C is 441 µS/cm. Only the first two soil water extracts at the beginning of the sample period contained very small amounts of aluminium whereas later on no Al^{3+} could

Table 1. Chemical properties of the soil water extract of the pre-AD 79 paleosols and the soil moisture (θ) at the date of sampling (sample period 2008/09). Soil water chemistry corrected for soil moisture is put in parentheses (ND, not detectable).

Date of sampling	29-Dec 2008	14-Jan 2009	26-Jan 2009	9-Feb 2009	24-Feb 2009	9-Mar 2009	23-Mar 2009	6-Apr 2009	15-Apr 2009	30-Apr 2009	Arithmetic mean
θ [%]	24.3	24.8	25.4	25.8	25.4	25.6	25.7	25.6	25.4	24.5	**25.3**
pH	7.7	7.6	7.5	7.8	7.6	7.6	7.2	7.3	7.7	7.4	**7.5**
EC_{calc} [µS cm^{-1}] at 25°C (EC_{calc}/θ)	526.2 (21.7)	433.0 (17.5)	388.1 (15.3)	405.6 (15.7)	398.2 (15.7)	401.9 (15.7)	463.7 (18.0)	464.4 (18.1)	465.0 (18.3)	459.7 (18.8)	**440.6** **(17.5)**
Cations [mg l^{-1}]											
Ca^{2+} (Ca^{2+}/θ)	41.1 (1.69)	40.4 (1.63)	38.2 (1.50)	38.3 (1.48)	38.1 (1.50)	39.3 (1.54)	48.0 (1.87)	48.0 (1.88)	48.6 (1.91)	48.7 (1.99)	**42.9** **(1.70)**
K^+ (K^+/θ)	45.6 (1.88)	39.3 (1.58)	37.1 (1.46)	33.8 (1.31)	32.9 (1.30)	32.8 (1.28)	33.7 (1.31)	33.7 (1.32)	33.6 (1.32)	33.8 (1.38)	**35.6** **(1.41)**
Na^+ (Na^+/θ)	24.5 (1.01)	25.1 (1.01)	23.1 (0.91)	19.8 (0.77)	20.5 (0.81)	21.6 (0.84)	24.1 (0.94)	24.2 (0.95)	24.8 (0.98)	25.0 (1.02)	**23.3** **(0.92)**
Mg^{2+} (Mg^{2+}/θ)	5.1 (0.21)	4.8 (0.19)	4.4 (0.17)	5.3 (0.20)	5.1 (0.20)	5.2 (0.20)	6.6 (0.26)	6.6 (0.26)	6.7 (0.26)	6.8 (0.28)	**5.7** **(0.22)**
Al^{3+} (Al^{3+}/θ)	0.3 (0.01)	0.1 (0)	ND	ND	ND	ND	ND	ND	ND	ND	-
Anions [mg·l^{-1}]											
Cl^- (Cl^-/θ)	75.3 (3.10)	53.5 (2.16)	43.8 (1.72)	58.8 (2.28)	56.2 (2.21)	55.2 (2.16)	67.0 (2.61)	67.1 (2.62)	65.2 (2.57)	63.4 (2.59)	**60.5** **(2.4)**
NO_3^- (NO_3^-/θ)	44.3 (1.82)	27.1 (1.09)	20.9 (0.82)	19.5 (0.76)	19.3 (0.76)	19.1 (0.75)	20.8 (0.81)	20.8 (0.81)	21.7 (0.85)	20.2 (0.82)	**23.4** **(0.93)**
SO_4^{2-} (SO_4^{2-}/θ)	33.2 (1.36)	24.2 (0.98)	23.2 (0.91)	21.0 (0.81)	20.8 (0.82)	21.2 (0.83)	20.9 (0.81)	21.0 (0.82)	21.4 (0.84)	20.6 (0.84)	**22.7** **(0.90)**
HPO_4^{2-} (HPO_4^{2-}/θ)	ND	ND	ND	ND	ND	ND	ND	ND	ND	ND	-

Table 2. Chemical properties of the soil water extract of the Roman paleosols at the date of sampling (sample period 2009/10) (ND, not detectable).

Date of sampling	2-Feb 2010	18-Feb 2010	2-Mar 2010	18-Mar 2010	1-Apr 2010	15-Apr 2010	29-Apr 2010	13-May 2010	Arithmetic mean
pH	7.4	7.8	7.7	7.7	7.8	7.8	7.8	7.8	**7.7**
EC_{calc} [$\mu S \cdot cm^{-1}$] at 25°C	528.2	495.3	465.0	470.1	478.2	498.9	489.3	502.1	**490.9**
Cations [$mg \cdot l^{-1}$]									
Ca^{2+}	50.3	46.3	40.2	40.8	41.3	45.7	46.0	46.8	**44.7**
K^+	46.7	42.3	39.7	40	41	40.7	40.5	40.8	**41.5**
Na^+	18.3	20.4	21.4	21.9	22	23.8	22.5	24.6	**21.9**
Mg^{2+}	6.9	7	6.7	6.72	6.75	6.8	6.7	6.8	**6.8**
Al^{3+}	0.2	0.1	ND	ND	ND	ND	ND	ND	-
Anions [$mg \cdot l^{-1}$]									
Cl^-	70.3	60.2	58.3	58.7	60.2	62.3	61.5	61.9	**61.7**
NO_3^-	38.7	39.6	37.5	37.9	38	38.4	39.0	38.7	**38.5**
SO_4^{2-}	32.7	33.5	30.8	31.1	32	32.5	31.6	31.8	**32.0**
HPO_4^{2-}	1.1	0.7	0.5	0.5	0.6	0.7	0.6	0.7	**0.68**

be detected. Moreover, no HPO_4^{2-} ions could be determined in the soil water extract.

The amount of water available in the soil has a distinct influence on the liquid phase chemistry in terms of concentration or dilution effects during dry or wet conditions, respectively (Mengel and Kirkby, 2001 [44]). Consequently, to gain a corrected course of the soil water composition, EC_{calc} and the cation and anion concentrations were devided by the soil moisture (θ) of the pre-AD 79 paleosol at the date of sampling. However, due to the relative small variation of θ during the sample period of only 1.5% the course of soil water composition does not change significantly. EC_{calc}/θ has its maximum at the first soil solution sampling end of December 2008 and decreases for the first four weeks until the end of January 2009. In March it increases, stagnates and again slightly increases at the very end of the sample period. The increased EC_{calc}/θ ratio during the first weeks of soil water sampling coincides with increased ion concentrations. In the middle of the rainy winter season EC_{calc}/θ reaches its minimum. Thereafter, EC_{calc}/θ increases at the end of March and slightly at the end of April on account of an increase of Ca^{2+}, Cl^- and partly of Na^+.

During the winter season of 2009/10 the above described characteristics of the soil water chemistry of the pre-AD 79 paleosol recurred. The ion concentration is increased at the beginning, then decreases to its minimum in the middle of the sample period and re-increases at its end (**Table 2**). Even though the mean ion concentrations have slightly increased in 2009/10 compared to the previous sampling season, the relative proportions have not changed. Hence, the fact that in 2009/10 very small amounts of phosphate anions were detected in the soil water extract most likely results from that general trend. The reason for that slight increase is not known, however, it is not relevant for the present study.

Unfortunately, in autumn 2009 all three soil moisture sensors failed as they probably lost their attachment to the substrate possibly due to soil shrinkage during the dry summer season. Consequently, the ion concentrations could not be corrected for θ which would have even amplified the course of soil water chemistry to a small extent as it did in 2008/09.

4.3. Soil Solid Phase

Tables 3 and **4** show the results of the soil solid phase analysis for the modern soils and the associated pre-AD 79 paleosols. Only when a soil parameter shows the same trend in all four pairs of soils, we assumed this to be the result of post-burial effects.

Table 3. Soil solid phase properties of the pre-AD 79 Roman paleosols and modern soils near Pompeii I.

Site	Soil	Soil texture	Munsell soil colour	pH (CaCl$_2$)	CaCO$_3$ [g·kg^{-1}]	TOC [g·kg^{-1}]	N [g·kg^{-1}]	CEC [mmol$_c$·kg^{-1}]	Exchangeable cations [mmol$_c$·kg^{-1}]			
									K	Na	Ca	Mg
DAI16	modern soil	Loamy sand	10YR3/1	6.0	7.1	11.6	1.1	139.0	20.7	48.7	52.2	17.1
	pre-AD 79 paleosol	Sandy loam	10YR4/2	7.1	7.9	8.9	0.8	140.2	20.3	13.2	94.7	11.7
DAI18	modern soil	Sandy loam	10YR3/1	7.4	30.8	8.0	0.5	66.0	19.0	10.4	32.7	3.4
	pre-AD 79 paleosol	Sandy loam	10YR4/2	7.5	16.2	3.4	0.1	172.2	16.3	6.6	138.5	10.7
VdM	modern soil	Loamy sand	10YR3/2	6.8	2.7	14.3	1.4	216.9	33.6	5.6	164.0	14.3
	pre-AD 79 paleosol	Sandy loam	10YR4/3	7.7	4.0	3.1	0.3	222.9	22.1	8.3	177.1	15.1
VdR	modern soil	Loamy sand	10YR3/1	5.8	0.5	16.3	2.6	107.5	6.3	7.7	83.8	9.7
	pre-AD 79 paleosol	Sandy loam	10YR3/1	8.7	0.2	6.0	1.7	183.5	18.8	26.7	125.2	12.7

Table 4. Soil solid phase properties of the pre-AD 79 Roman paleosols and modern soils near Pompeii II.

Site	Soil	K$_{tot}$ [g·kg^{-1}]	Na$_{tot}$ [g·kg^{-1}]	Ca$_{tot}$ [g·kg^{-1}]	Mg$_{tot}$ [g·kg^{-1}]	CEC$_{pot}$ ratio	Nutrient ratio			
							$\frac{K_{exch}}{K_{tot}}$	$\frac{Na_{exch}}{Na_{tot}}$	$\frac{Ca_{exch}}{Ca_{tot}}$	$\frac{Mg_{exch}}{Mg_{tot}}$
DAI16	modern soil	37.4	12.8	29.2	11.4	0.16	0.55	3.80	1.79	1.50
	pre-AD 79 paleosol	25.8	8.4	31.7	9.2	0.16	0.79	1.57	2.99	1.27
DAI18	modern soil	38.6	13.5	38.2	13.1	0.07	0.49	0.77	0.86	0.26
	pre-AD 79 paleosol	33.2	16.7	30.3	8.3	0.15	0.49	0.40	4.57	1.29
VdM	modern soil	37.3	11.9	31.6	9.7	0.24	0.90	0.47	5.19	1.47
	pre-AD 79 paleosol	15.0	9.8	23.0	6.4	0.29	1.47	0.85	7.70	2.36
VdR	modern soil	35.4	10.7	32.5	10.5	1.21	0.18	0.72	2.58	0.92
	pre-AD 79 paleosol	21.1	6.6	23.2	7.8	3.13	0.89	4.05	5.40	1.63

The modern soils are well penetrated by roots and soil fauna and generally appear loosely structured. In contrast, the paleosols have a massive structure with little recognizable macro porosity. The mean bulk density of the paleosols is 1.4 g/cm³ and that of the modern soils is 1.1 g/cm³. The Munsell soil colour (on moist samples) of the modern soils is very dark gray and that of the paleosols is dark grayish brown. The pH values of the modern soils are between 5.8 and 7.4 and of the paleosols between 7.1 and 8.7. Furthermore, in contrast to the modern soils, in the paleosols the pH value is not consistently increasing with increasing depth. The CEC ratios (Equation (2)) of the paleosols are higher compared to the modern soils. This coincides with higher nutrient ratios (Equation (3)) for Ca and K. The modern soils have low to moderate soil organic matter (SOM) contents whereas in the pre-AD 79 paleosols it is very low to low. This corresponds to amounts of total organic carbon (TOC) of 8 to 16.3 g·kg^{-1} for the modern soils and 3.1 to 8.9 g·kg^{-1} for the paleosols. Accordingly, nitrogen ranges between 0.5 and 2.6 in the modern soils and between 0.1 and 1.7 in the paleosols.

5. Discussion

5.1. Groundwater Table

As mentioned above, after AD 79 the groundwater table increased by about 1.8 m. To some extent, this can be an indirect result of the AD 79 eruption. The burial of the Sarno River plain is believed to have left a bare landscape free of vegetation and covered with unconsolidated volcanic material. Consequently, due to these unstable conditions especially the surrounding steep mountain slopes were vulnerable to severe erosion processes during intense rainfall events causing debris flows and landslides (Cinque et al., 2000 [45]; Cinque and Robustelli, 2009 [46]). Even though the AD 79 pumice lapilli layer show a very high hydraulic conductivity, the above PDC deposits consist of fine tephra that, together with intense rainfall, can generate infiltration excess overland flow and thus erosion. Erosion and the destruction of the vegetation cover cause a reduction of the water retention capacity which, in turn, results in increased runoff and an increasing groundwater table in the adjacent river plain (Vogel and Märker, 2012 [18]). The latter may also be caused by the accelerated deforestation and the clearance of the natural vegetation since the Middle Ages (around 700 BP, see Schneider, 1985 [47]; Buccheri et al., 2002 [41]; Russo Ermolli and di Pasquale, 2002 [48]).

Considering this groundwater rise, the Sarno River plain must be divided into two different zones of potential post-burial soil developments:

1) the lower areas of the inner Sarno River plain where after AD 79 the originally terrestrial pre-AD 79 paleosol has come under the influence of a rising groundwater table; and

2) the higher areas of the Sarno River plain where the paleosol is still part of the vadose zone and more likely influenced by vertical or lateral soil water flow.

This demonstrates that the mechanism of post-burial soil developments being active in the pre-AD 79 paleosols is by no means uniform for the entire Sarno River plain but strongly depends on the paleotopography and the paleolandscape position.

Vogel and Märker, 2012 [18], studied the effect of the post-burial groundwater rise on the originally terrestrial pre-AD 79 paleosols. Under the present-day influence of groundwater dynamics the paleosols are characterised by: 1) higher amounts of organic carbon and nitrogen; 2) a darker Munsell colour of the buried topsoil; 3) lower amounts of calcium carbonate; 4) increased sulphate concentrations together with a decreased pH value; and 5) redoximorphic features in the overlying AD 79 pumice fallout layer.

5.2. Soil Liquid Phase

5.2.1. Soil Water Flow

Figure 4 illustrates that, at Villa Regina, the soil water needed about one month to percolate from the present-day surface to the pre-AD 79 paleosol in a depth of 8 m. That delay results from the fact that the overlying volcanic ash and pumice layers as well as the volcanic soils have a large water retention capacity (Shoji and Takahashi, 2002 [49]; Sahin et al., 2005 [50]). Moreover, as earlier mentioned, the stratigraphy overlying the pre-AD 79 paleosol at Villa Regina consists of a sequence of pyroclastic material, fluvial deposits and soil substrates showing different soil textural characteristics. In unsaturated multi-layered soil profiles water movement is not necessarily strictly vertical but considerably influenced by the stratification effect. Especially less permeable fine grained layers, like ash layers or soils interbedded in highly permeable pumice layers or fluvial deposits tend to induce unsteady flow patterns. Water is held in the finer material and does not percolate further down into coarser material until the former layer is saturated. Finally, not until the water potential of the two strata reach the same value the water movement is again controlled by the gravitational potential (Fiorillo and Wilson, 2004 [51]; Javaux and Vanclooster, 2004 [52]).

Soil water could be extracted from the paleosol in a four months period between the end of December 2008 and the end of April 2009. The paleosol required a approximate soil moisture of more than about 24.5% for mobile gravitational water to be available and to be extracted by the suction cups with the applied tension of 550 mbar. In contrast, when θ is below 24.5% the matric potential attracting the soil water to the mineral soil particles is increased and exceeds the tension of the suction cups.

These above mentioned results verify that a soil water flow from the overlying volcanic deposits into the pre-AD 79 paleosol does take place as it was assumed by Foss, 1988 [9], Foss et al., 2002 [10] and Inoue et al., 2009 [11]. However, it only occurs during the rainy winter period and, under the stratigraphic conditions at Villa Regina, with a delay of about one month.

In contrast to the pre-AD 79 paleosol no soil water could be extracted from the AD 79 pumice lapilli fallout directly above. This could be due to the following two reasons. Firstly, the porosity of the pumice layer has a bimodal structure. On one hand the pumice lapilli has a very coarse texture with a lot of external macropores between the clasts. On the other hand each pumiceous particle consists of a very fine internal porous system. Thus, the water may not move homogeneously through the pumice layer. In the macropores between the clasts it randomly follows preferential flow paths until it is absorbed by the pumiceous particles showing a high water retention capacity (Sahin et al., 2005 [50]). Hence, the first infiltration water after a dry summer season is expected to be absorbed by the pumice and almost no gravitational water flow may be extracted by the suction cups from the external macropores until the particles reach saturation point. Secondly, the preferential flow through the pumice layer is often restricted to a small fraction of the total macropore volume. Thus, the suction cups were simply not able to capture such a preferential flow path. However, the fact that soil water could be extracted from the pre-AD 79 paleosol underneath proves that water flow does take place, either vertically by infiltration or laterally by interflow. Infact, interflow on top of the pre-AD 79 paleosurface can also result in a "moist" paleosol but a "dry" pumice layer directly above when upslope water percolates through the pumice and laterally enters the paleosol. Interflow upon the pre-AD 79 soil surface is caused by the strong decrease of permeability at the transition from the AD 79 pumice layer to the pre-AD 79 paleosol. Hood infiltrometer measurements of saturated hydraulic conductivity (K_{sat}) yielded K_{sat} values of around 0.005 cm·s^{-1} for the paleosol. In contrast, K_{sat} of the white pumice layer was beyond instrumentation range of the hood infiltrometer ($>0.01 \text{ cm·s}^{-1}$). Hence, it was minimum one order of magnitude higher compared to the paleosol (Vogel and Märker, 2011 [53]).

5.2.2. Soil Water Chemistry

The analysis of the soil water chemistry at Villa Regina showed that the dominant cation percolating into the pre-AD 79 paleosol is calcium. Despite the high concentration of Cl^-, the relatively low EC_{calc} of the soil water extract reveals that the pre-AD 79 paleosol can be considered non-saline (Schoeneberger et al., 2002 [54]). However, due to the relative vicinity of the study area to the Tyrrhenian Sea high amounts of Cl^- and Na^+ in the lysimeter extract may result from dry deposition of sea water aerosols during onshore winds (Scheffer and Schachtschabel, 2010 [24]). Nitrate may rather derive from anthropogenic sources. This corresponds with finding of Adamo et al., 2007 [55], who found increased amounts of NO_3^- in aquifers of the Sarno River plain due to intensive agricultural use. During the first two weeks of the sampling period very small amounts of aluminium were detected at slightly alkaline pH values of the soil water extracts. This seems to be an artefact of the sampling or analysis procedure. In fact, finely dispersed colloidal Al may have entered the soil solution sample by passing the membrane of the suction cups and at some stage between pH and Al measurement had the opportunity to be dissolved to Al^{3+} (Wolt, 2014, written communication; see Kennedy et al., 1974 [56]; Laxen and Chandler, 1982 [57]). However, the subsequent absence of free Al indicates that no soil acidification is active within the pre-AD 79 paleosol which agrees with the slightly alkaline pH of the soil water. Finally, the fact that no HPO_4^{2-} ions could be detected in the soil water extract is due to its weak solubility and usually strong retention by the soil solid phase and soil organic matter (Wolt, 1994 [7]). Furthermore, the retention of phosphate in a soil is pH-dependent leaving it more soluble at a slightly acid to moderately acid pH value. At alkaline pH on the other hand its solubility is reduced by the formation of Ca-phosphates (Welp et al., 1983 [58]).

The course of the soil water chemistry revealed increased ion concentrations at the beginning and at the end of the sample period. This can be explained by the "first flush effect" that can occur at the beginning of the rainy season when the first infiltration water reaches the pre-AD 79 paleosol after a dry summer season. An accumulation of ions during summer then results in higher element concentrations in the soil water extract. Arthur and Fahey, 1993 [37], observed the same effect but associated with snowmelt instead of precipitation water. At the initial stages of snowmelt the solute concentrations in the soil solution were also high and declined rapidly in the first four to six weeks.

It is striking that the "first flush effect" is much stronger for the anions than for the cations. This may be due to the fact that volcanic soils have a high fraction of variable charge (Madeira et al., 2007 [59]). Consequently, at a slightly alkaline pH value the net surface charge is negative resulting in a high cation exchange capacity and a negligible anion exchange capacity. This is particularly important for the retention and leaching of cations and anions (Nanzyo et al., 1993 [60]; Madeira et al., 2007 [59]). Hence, at slightly alkalkine pH, cations tend to be adsorbed to the surface of the soil colloids whereas anions are rather leached out of the profile by the mobile soil water. In the middle of the rainy winter season the ion concentrations reach their minimum because of continuous leaching from the soil (De Pascale and Barbieri, 1997 [61]). At the end of the sample period the ion concentrations start to reincrease. This is caused by the inversion of the "first flush effect", *i.e.* the decline of ion leaching from the soil and its accumulation due to a smaller amount of mobile soil water at the end of the rainy season (De Pascale and Barbieri, 1997 [61]). However, this effect is much less pronounced compared to the beginning of the sample period, probably due to retention effects of the soil matrix as well as a much slower decrease of precipitation at the end of the rainy season in contrast to a steep and fast increase at the beginning (Wagner, 1967 [62].

Beyond the described "first flush effect" no significant seasonal variation could be detected in the soil water chemistry. This may be due to the following reasons:

1) In the pre-AD 79 paleosol at about 8 m depth plants are absent. Thus, there is no influence on the soil water composition by nutrient uptake and ion secretion of plant roots;

2) The mobile soil water fraction predominantly sampled by tension lysimeters has much shorter contact times with the soil solid phase compared to capillary water. This results in lower ion concentrations and less seasonality (Wolt, 1994 [7]);

3) Since soil water could only be extracted from the paleosol in a short period of four months, there was no seasonal variation in the soil water composition.

5.3. Soil Solid Phase

From the relative stratigraphic position of the pre-AD 79 paleosols and the modern soils and by comparison with the eruption history of Somma-Vesuvius an estimation of their approximate soil age was made. With respect to the paleosols, the total soil age has to be subdivided into two main soil development periods, *i.e.* the duration of pre-burial and post-burial soil development. The parent volcanic material at Boscoreale (VR) and Pompeii (VdM) can most likely be ascribed to the Avellino eruption or the following AP1/AP2 eruption (3450 - 3000 BP; Andronico and Cioni, 2002 [63] and references therein). At Scafati (VdR) the pre-AD 79 paleosol developed from the later AP eruption products of Somma-Vesuvius (AP3 - AP6) which were dated between 2.710 ± 60 BP (Rolandi *et al.*, 1998 [64]) and 217 BC (Stothers and Rampino, 1983 [65]; Rolandi *et al.*, 1998 [64]). Consequently, the pre-burial age of the pre-AD 79 paleosols, *i.e.* the time they developed upon the ancient land surface, ranges between 1050 and 1500 years for Boscoreale and Pompeii and between 300 and 800 years for Scafati. In contrast, the post-burial soil age of all pre-AD 79 paleosols is approximately 1900 years. Thus, they existed for a much longer period of time under buried than under unburied conditions. This fact demonstrates the necessity to critically evaluate the possibility of post-burial soil developments in the paleosols before drawing conclusions on their ancient soil characteristics. Considering both, pre-burial and post-burial duration, the total soil age of the pre-AD 79 paleosols can be estimated between 2200 and 3400 years. According to its tephrostratigraphic position the age of the modern soils is between 1300 and 1900 years as they are underlain by the pyroclastic fallout of a medieval eruption of Somma-Vesuvius (DAI16, VdM, VdR) and reworked ash layers of the final stage of the AD 79 eruption (DAI18).

The results of the soil solid phase analysis show distinct characteristics that distinguish the topsoils of the buried pre-AD 79 paleosols from the unburied modern soils at the same topographic location. The paleosols have a pH value that is more than one unit higher compared to the modern soils. This corresponds with the observations of Foss, 1988 [9], Foss *et al.*, 2002 [10] and Inoue *et al.*, 2009 [11]. The lower pH values of the modern soils most likely result from:

1) Active accumulation and decomposition of soil organic matter (SOM) producing organic acids;

2) Nutrient uptake and proton release of plant roots;

3) Acid deposition from the atmosphere (acid rain); and

4) Leaching of basic cations from the upper horizons by percolating soil water.

In contrast, the results of the soil liquid phase study verify that the pre-AD 79 paleosols are rather subject to post-burial nutrient input by leaching from the overlying deposits which increased their pH value. This confirms the earlier mentioned assumption of Foss, 1988 [9], Foss *et al.*, 2002 [10] and Inoue *et al.*, 2009 [11]. However, as described above this process only takes place during the rainy winter season when mobile soil water is present. Furthermore, in the paleosols the pH decreasing influence of plants and active accumulation of SOM is missing.

It was determined that the pH of the paleosols often does not consistently increase with depth as can be seen for most of the modern unburied soils. After burial, the former A-horizons of the paleosols are subject to ion accumulations from the overlying strata. Hence, the boundaries between zones of leaching and accumulation can gradually blur (Holliday, 2004 [66]).

The post-burial nutrient accumulation in the paleosols also results in higher CEC ratios compared to the modern soils. With regard to the nutrient ratios only Ca and K are clearly enriched in the paleosols. Hence, Ca and K seem to be the dominant base cations to be transported through the profile which is congruent with the results from the soil water analysis. This only partly corresponds to Foss, 1988 [9] and Foss et al., 2002 [10], who presumed that especially Ca^{2+} and Mg^{2+} leach from the overlying volcanic sediments and recharge the surface horizons of the paleosols. Due to the neutral to slightly alkaline pH value of the pre-AD 79 paleosols the surface charge of the soil colloids is expected to be predominantly negative. Thus, adsorption of anions to the exchange sites of the paleosols is negligible. Hence, they are rather leached out of the soil profile.

The soil organic matter (SOM) contents are low to moderate for the modern soils and very low to low for the paleosols. Thus, the amount of organic carbon (TOC) is approximately 7 $g \cdot kg^{-1}$ lower in the paleosols. Macroscopically, this agrees with a Munsell soil colour (on moist samples) having higher value and chroma, i.e. very dark gray compared to dark grayish brown. According to TOC, the paleosols also contain lower amounts of nitrogen than the modern soils. The decreased concentrations of TOC and N in the paleosols may be due to post-burial decomposition of SOM. With burial in AD 79 the accumulation of SOM within the paleosols has stopped. As the AD 79 pumice fallout layer covering the pre-AD 79 paleosols is well aerated no air exclusion took place. Hence, microbial activity to gradually decompose SOM is not stopped even though it may have slowed down. This is supported by Crowther et al., 1996 [1], who studied post-burial change in a humic rendzina soil in England. They state that decomposition reduced the amounts of TOC by about 29 % in only 32 years after burial. Only in extremely anaerobic conditions SOM may be preserved from decomposition after burial (see Dimbleby, 1984 [67]). Hence, Crowther et al., 1996 [1], conclude that the present-day concentrations of TOC in the buried soils are not reflecting their original SOM status.

In the Sarno River plain, a second reason for reduced amounts of TOC in the paleosols may be due to interaction with hot volcaniclastic fallout during the AD 79 eruption. To some extent, heating of the pre-AD 79 paleosols during contact with the AD 79 pumice fallout may have caused sublimation of SOM. Thomas and Sparks, 1992 [68], modelled the process of tephra cooling during fallout from eruption columns. They found that the heat loss of clasts decreases with increasing grain size and decreasing fall height. Whereas smaller clasts are deposited cold, they conclude that larger clasts of Plinian fallout can retain enough heat to pose hazards to life and property by igniting fires.

Finally, soil structure and bulk density of the modern soils and the pre-AD 79 paleosols are used as an indicator of soil compaction. Already from the macroscopic comparison of the two soils it is evident that the pre-AD 79 paleosols show a massive soil structure with little recognizable secondary porosity, i.e. structure-dependent porosity. This may result from: 1) the superimposed load of 3.3 (VdR) to 8 m (VR) of overlying post-AD 79 deposits; 2) the absence of root penetration; and 3) the reduction of active soil biota such as earthworms. However, as shown by the soil moisture and hydraulic conductivity measurements, no waterlogging occurred in the paleosols as the texture-dependent primary porosity of the sandy loam still permits soil water movement. In contrast to the paleosols, the unburied modern soils are well penetrated by roots and soil fauna and generally appear more loosely structured. This is in accordance with a slightly increased mean bulk density of 1.4 g/cm^3 for the pre-AD 79 paleosols compared to 1.1 g/cm^3 for the modern soils.

6. Conclusions

A soil liquid and solid phase analysis was carried out in the pre-AD 79 Roman paleosols around Pompeii to estimate potential post-burial soil development. The combination of the present-day mean groundwater table of the Sarno River plain (Autorità di Bacino del Sarno) with the pre-AD 79 topography (Vogel et al., 2011 [15]) revealed that, in terms of post-burial soil developments, one has to subdivide the Sarno River plain into two different zones: 1) at around 37% of the Sarno River plain the Roman paleosols are influenced by a rise of the mean groundwater table of approximately 1.8 m since AD 79; and 2) at the remaining 63% the paleosols are still part of the vadose zone and more likely influenced vertically by infiltration water or laterally by interflow. Consequently, the mechanism of post-burial soil development being active in the pre-AD 79 paleosols is not uniform for the entire Sarno River plain but strongly depends on the paleotopographic situation.

The soil moisture and soil liquid phase study at the archaeological excavation of Villa Regina demonstrated that recently soil water flow down to the pre-AD 79 paleosol in a depth of 8 m takes place. Thus, since AD 79, this soil water can dissolve ions from the overlying volcanic deposits, incorporate them into the paleosol and influence its solid phase properties. However, mobile soil water was only extracted from the paleosol in a four months period when the soil moisture exceeded 24.5%. This restricts the potential influence of post-burial nutrient input to the rainy winter season. The soil liquid phase of the pre-AD 79 paleosol shows a distinct chemical composition. At the beginning and the end of the sample period increased ion concentrations were determined that can be explained by the "first flush effect" and its inversion, respectively.

Through the combination of the soil liquid phase study with the soil solid phase study an estimate of potential post-burial soil changes was made:

1) Leaching of nutrients from overlying deposits and accumulation in the paleosols led to a) higher CEC ratios, b) higher nutrient ratios for Ca^{2+} and K^+, and c) higher pH values in the paleosols. However, the results from the soil water and soil moisture study restrict the soil water flow to the rainy winter season.

2) Since the buried A-horizon of the paleosols is recently subject to ion accumulations from the overlying strata, former topsoil eluvial horizons gradually turn into subsoil illuvial horizons. This results in a reversion of the depth function of the pH value.

3) The paleosols show a massive soil structure with little recognizable macro porosity and have a slightly increased bulk density due to burial by several meters of post-AD 79 deposits.

4) Considering the particular conditions during and after burial in AD 79, decreased organic carbon and nitrogen contents and increased C/N ratios in the paleosols may derive from decomposition of soil organic matter or sublimation during contact with hot AD 79 pumice fallout.

It can be summarized that deeper insights were gained into the occurrence of mobile soil water within the pre-AD 79 paleosols and into the nature of potential post-burial soil developments. In the future, they should be taken into account when pre-AD 79 paleosols are characterised and interpreted with respect to paleoenvironmental conditions. Additionally, the presented results may provide hints regarding the susceptibility and hazard of the groundwater body to contamination by agricultural and industrial pollutants.

Acknowledgements

This study is part of the interdisciplinary SALVE-research project undertaken by the German Archaeological Institute (DAI) in cooperation with the Heidelberg Academy of Sciences and Humanities (HAW) and the University of Tübingen (www.salve-research.org). Project directors are Florian Seiler (DAI) and Michael Märker (HAW) and it was funded by the Deutsche Forschungsgemeinschaft (German Research Foundation). We would like to thank our local project partners and all their collaborators for their cooperation, particularly the Autorità di Bacino del Sarno and the Soprintendenza Speciale per i Beni Archaeologici di Napoli e Pompei. Particularly we thank Grete Stefani and her team at the archaeological excavation of Villa Regina for their support to install the soil water monitoring system. Special gratitude goes to Maria Teresa Pappalardo for sampling and maintenance as well as to Giovanni Di Maio for conducting the stratigraphic drillings and for his valuable local geological knowledge. Furthermore we would like to thank Peter Kühn, Philipp Hoelzmann, Stefan Wessel-Bothe, Silke Schweighoefer, Jörn Breuer, Yvonne Oelmann and Marc Ruppenthal, ECOLAB G.M. '65 S.R.L. and UABG GmbH for their various technical supports.

References

[1]   Crowther, J., Macphail, R.I. and Cruise, G.M. (1996) Short-Term, Post-Burial Change in a Humic Rendzina Soil, Overton down Experimental Earthwork, Wiltshire, England. *Geoarchaeology*, **11**, 95-117.
       http://dx.doi.org/10.1002/(SICI)1520-6548(199603)11:2<95::AID-GEA1>3.0.CO;2-4

[2]   Retallack, G.J. (1998) Core Concepts of Paleopedology. *Quaternary International*, **51-52**, 203-212.
       http://dx.doi.org/10.1016/S1040-6182(97)00046-3

[3]   Retallack, G.J. (2001) Soils of the Past, an Introduction to Paleopedology. Blackwell Science, Oxford.

[4]   French, C. (2003) Geoarchaeology in Action. Studies in Soil Micromorphology and Landscape Evolution. Routledge, London.

[5]   Scudder, S.J., Foss, J.E. and Collins, M.E. (1996) Soil Science and Archaeology. *Advances in Agonomy*, **57**, 1-76.

http://dx.doi.org/10.1016/S0065-2113(08)60922-0

[6] Ugolini, F.C. and Dahlgren, R.A. (1987) The Mechanism of Podzolization as Revealed by Soil Solution Studies. In: Righi, D. and Chauvel, A., Eds., *Podzols and Podzolisation* (*in French*), AFES et INRA, Plaisir et Paris, 195-203.

[7] Wolt, J.D. (1994) Soil Solution Chemistry: Applications to Environmental Science and Agriculture. Wiley, New York, 345 p.

[8] Snakin, V.V., Prisyazhnaya, A.A. and Kovács-Láng, E. (2001) Soil Liquid Phase Composition. Elsevier, Amsterdam.

[9] Foss, J.E. (1988) Paleosols of Pompeii and Oplontis. In: Curtis, R.L., Ed., *Studia Pompeiana and Classica*, Aristide D. Caratzas, Publisher, New Rochelle.

[10] Foss, J.E., Timpson, M.E., Ammons, J.T. and Lee, S.Y. (2002) Paleosols of the Pompeii Area. In: Jashemski, W.F., Ed., *The Natural History of Pompeii*, Cambridge University Press, Cambridge, 65-79.

[11] Inoue, Y., Baasansuren, J., Watanabe, M., Kamei, H. and Lowe, D.J. (2009) Interpretation of Pre-AD 472 Roman Soils from Physicochemical and Mineralogical Properties of Buried Tephric Paleosols at Somma Vesuviana Ruin, Southwest Italy. *Geoderma*, **152**, 243-251. http://dx.doi.org/10.1016/j.geoderma.2009.06.010

[12] Zehetner, F., Miller, W.P. and West, L.T. (2003) Pedogenesis of Volcanic Ash Soils in Andean Ecuador. *Soil Science Society of America Journal*, **67**, 1797-1809. http://dx.doi.org/10.2136/sssaj2003.1797

[13] Agnelli, A.E., Corti, G., Agnelli, A., Del Carlo, P., Coltelli, M. and Ugolini, F.C. (2007) Features of Some Paleosols on the Flanks of Etna Volcano (Italy) and Their Origin. *Geoderma*, **142**, 112-126. http://dx.doi.org/10.1016/j.geoderma.2007.08.003

[14] Vogel, S. and Märker, M. (2010) Reconstruction the Roman Topography and Environmental Features of the Sarno River Plain (Italy) before the AD 79 Eruption of Somma-Vesuvius. *Geomorphology*, **115**, 67-77. http://dx.doi.org/10.1016/j.geomorph.2009.09.031

[15] Vogel, S., Märker, M. and Seiler, F. (2011) Revised Modeling the Post-AD 79 Volcanic Deposits of Somma-Vesuvius to Reconstruct the Pre-AD 79 Topography of the Sarno River Plain (Italy). *Geologica Carpathica*, **62**, 5-16. http://dx.doi.org/10.2478/v10096-011-0001-3

[16] Lulli, L. (2007) Italian Volcanic Soils. In: Arnalf, Ó., Bartoli, F., Buurman, P., Óskarson, H., Stoops, G. and García-Rodeja, E., Eds., *Soils of Volcanic Regions in Europe*, Springer-Verlag, Berlin, Heidelberg, 51-67. http://dx.doi.org/10.1007/978-3-540-48711-1_7

[17] Rispoli, P., Di Maio, G. and Esposito, D. (2008) Explorative Excavations at Villa dei Misteri in Pompeii (in Italian). In: Guzzo, P.G. and Guidobaldi, M.P., Eds., *New Archaeological Research in the Vesuvian Area* (*Excavations* 2003-2006) (*in Italian*), International Congress, Rome, 542.

[18] Vogel, S. and Märker, M. (2012) Comparison of Pre-AD 79 Roman Paleosols in Two Contrasting Paleo-Topographical Situations around Pompeii (Italy). *Geografia Fisica e Dinamica Quaternaria*, **35**, 199-209.

[19] Brinson, M.M. (1993) A Hydrogeomorphic Classification for Wetlands. Wetlands Research Program, Technical Report WRP-DE-4, 79.

[20] Grossmann, J. and Moss, R. (1994) Variability of Water Quality in a Spruce Stand. *Journal of Plant Nutrition and Soil Science*, **157**, 47-51.

[21] de Vries, W. and Leeters, E.E.J.M. (1994) Effects of Acid Deposition on 150 Forest Stands in the Netherlands. Chemical Composition of the Humus Layer, Mineral Soil and Soil Solution. DLO Winand Staring Centre for Integrated Land, Soil and Water Research. Rep. 69.1, Wageningen, 84 p.

[22] Manderscheid, B. and Matzner, E. (1995) Spatial and Temporal Variation of Soil Solution Chemistry and Ion Fluxes through the Soil in a Mature Norway Spruce (*Picea abies* (L.) Karst.) Stand. *Biogeochemistry*, **30**, 99-114. http://dx.doi.org/10.1007/BF00002726

[23] Grossmann, J. and Udluft, P. (1991) The Extraction of Soil Water by the Suction-Cup Method: A Review. *Journal of Soil Science*, **42**, 83-93. http://dx.doi.org/10.1111/j.1365-2389.1991.tb00093.x

[24] Scheffer, F. and Schachtschabel, P. (2010) Textbook of Soil Science (in German). Spektrum Akademischer Verlag GmbH, Heidelberg, Berlin.

[25] Grossmann, J. (1988) Physical and Chemical Processes during Sampling of Seapage Water Using Suction Cups (in German). Ph.D. Dissertation, Institute of Hydrochemistry, Technische Universität München, Munich, 147 p.

[26] Litaor, M.I. (1988) Review of Soil Solution Samplers. *Water Resource Research*, **24**, 727-733. http://dx.doi.org/10.1029/WR024i005p00727

[27] DVWK (1990) Soil Water Sampling Using the Suction Cup Method (in German). Bd. Merkblatt 217, Berlin.

[28] Haberhauer, G. (1997) Adsorption Behaviours of Suction Cup Materials (in German). *Proceedings of the 7th Lysimeter Conference*, 7-9 April 1997, BAL Gumpenstein, 27-31.

[29] Spangenberg, A., Cecchini, G. and Lamersdorf, N. (1997) Analysing the Performance of a Micro Soil Solution Sampling Device in a Laboratory Examination and a Field Experiment. *Plant and Soil*, **196**, 59-70. http://dx.doi.org/10.1023/A:1004213006295

[30] Tischner, T., Nutzmann, G. and Pothig, R. (1998) Determination of Soil Water Phosphorus with a New Nylon Suction Cup. *Bulletin of Freshwater Ecology and Inland Fisheries*, **61**, 325-332.

[31] Hendershot, W.H. and Courchesne, F. (1991) Comparison of Soil Solution Chemistry in Zero Tension and Ceramiccup Tension Lysimeters. *Journal of Soil Science*, **42**, 577-583. http://dx.doi.org/10.1111/j.1365-2389.1991.tb00104.x

[32] Fillion, N., Probst, A. and Probst, J.L. (1999) Dissolved Organic Matter Contribution to Rainwater, through fall and Soil Solution Chemistry. *Analusis*, **27**, 409-413. http://dx.doi.org/10.1051/analusis:1999270409

[33] Derome, J. (2002) Submanual on Soil Solution Collection and Analysis. ICP Forest Manual Part III, Expert Panel on Soil, 111-161.

[34] Roman, R., Caballero, R. and Bustos, A. (2002) Variability of Soil Solution Ions in Fallowland Fields in Central Spain. *Edafologia*, **9**, 161-172.

[35] Strahm, B.D., Harrison, R.B., Terry, T.A., Flaming, B.L., Licata, C.W. and Petersen, K.S. (2005) Soil Solution Nitrogen Concentrations and Leaching Rates as Influenced by Organic Matter Retention on a Highly Productive Douglas-Fir Site. *Forest Ecology and Management*, **218**, 74-88. http://dx.doi.org/10.1016/j.foreco.2005.07.013

[36] Delta-T Devices Ltd. (2006) User Manual for the SM200 Soil Moisture Sensor. SM200-UM-1.1, 36.

[37] Arthur, M.A. and Fahey, T.J. (1993) Controls on Soil Solution Chemistry in a Subalpine Forest in North-Central Colorado. *Soil Science Society of America Journal*, **57**, 1122-1130. http://dx.doi.org/10.2136/sssaj1993.03615995005700040040x

[38] DIN ISO 10390 (2005) Soil Quality-Determination of pH. Beuth-Verlag, Berlin.

[39] DIN EN 13137 (2001) Characterization of Waste-Determination of Total Organic Carbon (TOC) in Waste, Sludges and Sediments. Beuth-Verlag, Berlin.

[40] DIN ISO 11261 (1997) Soil Quality-Determination of Total Nitrogen—Modified Kjeldahl Method. Beuth-Verlag, Berlin.

[41] Sadori, L. and Narcisi, B. (2001) The Postglacial Record of Environmental History from Lago di Pergusa, Sicily. *The Holocene*, **11**, 655-670. http://dx.doi.org/10.1191/09596830195681

[42] Buccheri, G., Capretto, G., Di Donato, V., Esposito, P., Ferruzza, G., Pecatore, T., Ermolli, E.R., Senatore, M.R., Sprovieri, M., Bertoldo, M., Carella, D. and Madonia, G. (2002) A High Resolution Record of the Last Deglaciation in the Southern Tyrrhenian Sea: Environmental and Climatic Evolution. *Marine Geology*, **186**, 447-470. http://dx.doi.org/10.1016/S0025-3227(02)00270-0

[43] Dohrmann, R. and Kaufhold, S. (2009) Three New, Quick CEC Methods for Determining the Amounts of Exchangeable Calcium Cations in Calcareous Clays. *Clay Minerals Society*, **57**, 338-352. http://dx.doi.org/10.1346/CCMN.2009.0570306

[44] Mengel, K. and Kirkby, E.A. (2001) Principles of Plant Nutrition. Kluwer Academic Publishers, Dordrecht, 849 p. http://dx.doi.org/10.1007/978-94-010-1009-2

[45] Cinque, A., Robustelli, G. and Russo, M. (2000) The Consequences of Pyroclastic Fallout on the Dynamics of Mountain Catchments: Geomorphic Events in the Rivo d'Arco Basin (Sorrento Peninsula, Italy) after the Plinian Eruption of Vesuvius in 79 AD. *Geografia Fisica e Dinamica Quaternaria*, **23**, 117-129.

[46] Cinque, A. and Robustelli, G. (2009) Alluvial and Coastal Hazards Caused by Long-Range Effects of Plinian Eruptions: the Case of the Lattari Mts. after the AD 79 Eruption of Vesuvius. *Geological Society, London, Special Publications*, **322**, 155-171. http://dx.doi.org/10.1144/SP322.7

[47] Schneider, R. (1985) Palynological Analysis in the Aspromonte of Calabria (Southern Italy) (in French). *Ligurian Notebooks of Prehistory and Protohistory*, **2**, 279-288.

[48] Ermolli, E.R. and di Pasquale, G. (2002) Vegetation Dynamics of South-Western Italy in the Last 28 kyr Inferred from Pollen Analysis of the Tyrrhenian Sea Core. *Vegetation History and Archaeobotany*, **11**, 211-219. http://dx.doi.org/10.1007/s003340200024

[49] Shoji, S. and Takahashi, T. (2002) Environmental and Agricultural Significance of Volcanic Ash Soils. *Global Journal of Environmental Research*, **6**, 113-135.

[50] Sahin, U., Ors, S., Ercisli, S., Anapali, O. and Esitken, A. (2005) Effect of Pumice Amendment on Physical Soil Properties and Strawberry Field Growth. *Journal of Central European Agriculture*, **6**, 361-366.

[51] Fiorillo, F. and Wilson, R.C. (2004) Rainfall Induced Debris Flows in Pyroclastic Deposits, Campania (Southern Italy). *Engineering Geology*, **75**, 263-289. http://dx.doi.org/10.1016/j.enggeo.2004.06.014

[52] Javaux, M. and Vanclooster, M. (2004) *In Situ* Long-Term Chloride Transport through a Layered, Nonsaturated Sub-

soil. 2. Effect of Layering on Solute Transport Processes. *Vadose Zone Journal*, **3**, 1331-1339. http://dx.doi.org/10.2136/vzj2004.1331

[53] Vogel, S. and Märker, M. (2011) Characterization of the Pre-AD 79 Roman Paleosol South of Pompeii (Italy): Correlation between Soil Parameter Values and Paleo-Topography. *Geoderma*, **160**, 548-558. http://dx.doi.org/10.1016/j.geoderma.2010.11.003

[54] Schoeneberger, P.J., Wysocki, D.A., Benham, E.C. and Broderson, W.D. (2002) Field Book for Describing and Sampling Soils. Version 2.0. Natural Resources Conservation Service, National Soil Survey Center, Lincoln, 228.

[55] Adamo, N., Imperatrice, M.L., Mainolfi, P., Onorati, G. and Scala, F. (2007) Water—The Monitoring in Campania 2002-2006 (in Italian). ARPAC, Napoli, 257.

[56] Kennedy, V.C., Zellweger, G.W. and Jones, B.P. (1974) Filter Pore-Size Effects on the Analysis of Al, Fe, Mn and Ti in Water. *Water Resource Research*, **10**, 785-790. http://dx.doi.org/10.1029/WR010i004p00785

[57] Laxen, D.P.H. and Chandler, I.M. (1982) Comparison of Filtration Techniques for Size Distribution in Freshwater. *Analytical Chemistry*, **54**, 1350-1355. http://dx.doi.org/10.1021/ac00245a023

[58] Welp, G., Herms, U. and Brümmer, G. (1983) Influence of Soil Reaction, Redox Conditions and Organic Matter on Phosphate Concentrations of the Soil Solution (in German). *Journal of Plant Nutrition and Soil Science*, **146**, 38-52.

[59] Madeira, M., Auxtero, E., Monteiro, F., García-Rodeja, E. and Nóvoa-Munoz, J.C. (2007) Exchange Complex Properties of Soils from a Range of European Volcanic Areas. In: Arnalf, Ó., Bartoli, F., Buurman, P., Óskarson, H., Stoops, G. and García-Rodeja, E., Eds., *Soils of Volcanic Regions in Europe*, Springer-Verlag, Berlin, Heidelberg, 369-385. http://dx.doi.org/10.1007/978-3-540-48711-1_27

[60] Nanzyo, M., Dahlgren, R. and Shoji, S. (1993) Chemical Characteristics of Volcanic Ash Soils. In: Shoji, S., Nanzyo, M. and Dahlgren, R., Eds., *Volcanic Ash Soils—Genesis, Properties and Utilization, Developments in Soil Science*, Elsevier, Amsterdam, 145-188. http://dx.doi.org/10.1016/S0166-2481(08)70267-8

[61] De Pascale, S. and Barbieri, G. (1997) Effect of Soil Salinity and Top Removeal on Growth and Yield of Broad Bean as Green Vegetable. *Scientia Horticulturae*, **71**, 147-165. http://dx.doi.org/10.1016/S0304-4238(97)00104-0

[62] Wagner, H.-G. (1967) The Cultural Landscape of Vesuvius. A Structural Analysis of Agriculture and Geography Considering Recent Transformations (in German). Annals of the Geographic Society of Hannover, 243.

[63] Andronico, D. and Cioni, R. (2002) Contrasting Styles of Mount Vesuvius Activity in the Period between the Avellino and Pompeii Plinian Eruptions and Some Implications for Assessment of Future Hazards. *Bulletin of Volcanology*, **64**, 372-391. http://dx.doi.org/10.1007/s00445-002-0215-4

[64] Rolandi, G., Petrosino, P. and McGeehin, J. (1998) The Interplinian Activity at Somma-Vesuvius in the Last 3500 Years. *Jouranl of Volcanology and Geothermal Research*, **82**, 19-52. http://dx.doi.org/10.1016/S0377-0273(97)00056-5

[65] Stothers, R.B. and Rampino, M.R. (1983) Volcanic Eruptions in the Mediterranean before AD 630 from Written and Archaeological Sources. *Journal of Geophysical Research*, **88**, 6357-6371. http://dx.doi.org/10.1029/JB088iB08p06357

[66] Holliday, V.T. (2004) Soils in Archaeological Research. Oxford University Press, Oxford, 448.

[67] Dimbleby, G.W. (1984) Anthropogenic Changes from Neolithic through Medieval Times. *New Phytologist*, **98**, 57-72. http://dx.doi.org/10.1111/j.1469-8137.1984.tb06098.x

[68] Thomas, R.M.E. and Sparks, R.S.J. (1992) Cooling of Tephra during Fallout from Eruption Columns. *Bulletin of Volcanology*, **54**, 542-553. http://dx.doi.org/10.1007/BF00569939

17

An Opportunity to Switch Energy Sources in Institutions in the Kilimanjaro Region, Tanzania and Benefit from Carbon Finance under the Sustainable Land Management Project

Stephen Mutimba[1]*, Francis X. Mkanda[2], Richard Kibulo[1]

[1]Camco Advisory Services (K) Ltd., Nairobi, Kenya
[2]Sustainable Land Management Project, Kilimanjaro, Tanzania
Email: *Stephen.Mutimba@camcocleanenergy.com, francis.mkanda@undp.org, rkibulo@yahoo.com

Abstract

The sustainable land management (SLM) project is seeking to engage with public institutions to explore the possibility of using the energy-switch principle presented by the carbon market to reduce emissions from inefficient use of biomass energy and discharge of human waste into the environment. Such a switch will be a triple-win situation that improves the natural environment, reduces deforestation, and provides avenues for revenue generation. As such, it commissioned a study of the pattern of energy consumption in the institutions and the type of cooking stoves they employ. Results show that firewood (51%) is the most widely used fuel because of availability, affordability, and reliability. The study also found that 56% of the institutions use energy-saving stoves, which is an opportunity that the project can seize to encourage use of alternatives sources of energy as opposed to biomass. In addition, 88% of the institutions expressed willingness to switch to biogas for cooking. This is yet another opportunity for scaling up the dissemination of renewable sources of energy in the region. Better adoption and wider use of renewable energy sources will take place when innovative financing mechanisms are devised to cover the high upfront cost of installing renewable energy systems. This has been one of the main barriers to scaling up the use of renewable in the region.

Keywords

Wood Fuel, Deforestation, Land Degradation, Carbon Finance, Renewable Energy, Cook Stoves

---

*Corresponding author.

1. Introduction

This study seeks to provide an understanding of the pattern of energy consumption and cooking technologies employed by institutions in the Kilimanjaro Region because through a sustainable land management project entitled reducing land degradation on the highlands of the Kilimanjaro Region, the government of the United Republic of Tanzania (URT) is seeking to engage with public institutions to explore the possibility of using the energy-switch principle presented by the carbon market to reduce emissions from inefficient use of biomass energy and discharge of human waste into the environment. Such a switch will be a triple-win situation that improves the natural environment, reduces deforestation, and provides avenues for revenue generation. The energy-switch principle is in line with UNDP's Millennium Development Goal (MDG) Carbon Finance Facility, which is designed to contribute directly to achieving the MDGs as an innovative carbon-finance mechanism featuring emission offsets derived from a pool of projects. The facility represents a fundamental building block in UNDP's poverty, environment and MDG delivery strategy as it bridges developing countries with the carbon market. Over the last few years, the facility has removed barriers and established efficient host-country procedures, such as supporting project development via the facility. The first MDG carbon facility project is the Manna Energy Limited which in 2011 registered the World's First United Nations Clean Water Carbon Credit Program for enabling the deployment of community scale water treatment in the rural areas of Rwanda.

Like the rest of Tanzania, the Kilimanjaro Region is heavily dependent on biomass for its energy needs. Energy security in the region is affected by several key development factors, such as population growth (1.8% annual average), accelerated urbanisation, economic development, and constant changes in prices of available energy sources [1]. Although the region's population growth rate is the twenty-fourth highest in the country, its density, at 124 people per square kilometre, is the eighth [2], implying that energy demand and consumption is concomitantly high. In fact, studies indicate that the consumption of wood-based biomass will increase in relative terms over the next 30 years as demographic growth continues to outstrip access to other modern fuels. The demand for cooking energy is projected to increase threefold [3] in urban Tanzania from 2002 to 2030, but the high price and low dependability of electricity will help maintain the demand for charcoal beyond that time.

In the region, increasing demand for wood-based biomass is visible in public institutions such as hospitals, prisons, boarding secondary schools, university campuses, and factories which consume large quantities of wood fuel, estimated at 35,661 tonnes of firewood and charcoal annually, thereby significantly contributing to deforestation and land degradation [4]. A recent study has shown that fuel wood is seldom the primary source of forest depletion. Other factors such as land clearing for agriculture and human settlement, commercial development and other land use changes are more important in causing forest depletion [5]. The fact that 35,661 tonnes of firewood and charcoal is consumed undoubtedly means that large patches of forests are being cleared per year. Besides, the Ministry of Natural Resources and Tourism estimates that 97.9% of total wood consumed in Tanzania is firewood and charcoal. To compound the problem, charcoal carbonization technologies that are presently being used for making charcoal are traditional and highly inefficient, with low conversion efficiency. The traditional methods include earth-mounds and pit kilns with conversion efficiencies ranging from 10% - 14%. This low efficiency often translates to raw material waste of up to 50%. This means that the current tonnage of charcoal could, theoretically, be produced by half the weight of wood used. This situation has to be addressed if the remaining natural forest resources are to be conserved.

The introduction of energy-efficient technologies such as improved stoves, improved charcoal kilns as well as renewable energy technologies including solar photovoltaic (PV) systems, solar water heating, and briquetted charcoal date as far back as 1975 when biogas was introduced by SIDO (Small Industries Development Organization) according to the history of biogas in Tanzania [6]. Their adoption, however, has remained disappointingly low because of high transaction and distribution costs, and incomplete lack of regulation and enforcement. As an attempt towards addressing this problem, the SLM project funded by the URT Global Environment Facility (GEF), and UNDP intends to design an energy-improvement strategy for the Kilimanjaro Region. It is expected that the implementation of this strategy will lead to carbon emissions reduction linked to a carbon-credit-earning scheme and generate revenue from the sale of certified emission reduction (CER) credits under the carbon finance pilots. The first step in designing such a strategy was to examine types of energy and stoves preferred by institutions, and the latter's willingness to switch energy sources.

2. Study Area

Kilimanjaro is one of the 30 regions of the United Republic of Tanzania (Tanzania Mainland and Zanzibar Island).

It is located in the north eastern part of Tanzania Mainland. It lies south of the equator between latitudes 2° and 4°. Longitudinally, the region lies between longitudes 36° and 38° east of the Greenwich. To the north and east, it is bordered by Kenya, to the south, southwest, and west by the Tanga, Manyara, and Arusha Regions respectively. It covers an area of 13,209 km² or 1.4% of the area of the entire Tanzania Mainland. The region comprises 7 administrative councils, namely Rombo, Hai, Moshi, Mwanga, Siha, Same and Moshi Municipality. In terms of institutions, major consumers of wood fuel, there are 1731 of according to the Kilimanjaro Regional Secretariat (2011) (**Table 1**).

3. Methods

A questionnaire was used to collect data because it is considered to be more appropriate than other techniques [7], for example the Rapid Rural Appraisal (PRA) and Participatory Rural Appraisal (RRA). Questionnaire design considered content, wording, and format, as recommended by another author [8]. To this end, a structured and open-ended questionnaire was formulated. The structured questions were used on issues where a specific range of known responses was expected. Such questions allow easier interpretation and analysis than open-ended questions. On the other hand, the open-ended questions were included so as to allow interviewees to construct their own accounts of experiences because the explanatory power of structured questions is limited. It comprised 22 variables that fell under five major sections that were considered as encompassing in as far as energy use and cooking devices are concerned. The sections included general information of the study area, institutional details, energy types used, reasons for those preferences, sources of biomass used, and carbon financing. Prior to the survey, a pilot study of 20 respondents was undertaken to check a number of questionnaire design aspects, such as clarity, appropriateness of the questions, and respondents' willingness to answer the questions.

A total of 300 institutions (17%) were targeted (75 boarding schools, 9 colleges, 49 hospitals, 54 hotels, and 45 industries). However, the enumerators managed to interview only 204 (12%) because of constraints such as time and distance to some of the institutions. These changes did not influence the final results as the difference between the target and achieved sample sizes is minor.

The judgment (purposive) sampling method was also employed [9], a non-probability sampling technique in which an experienced individual selects the sample based on his or her judgment about some appropriate characteristics required of the sample member. In this case, being an institution was the required characteristic. To ensure equal representation, we selected institutions to sample based on our knowledge of the establishments. Data analysis involved calculations of sums and percentages.

4. Results and Discussion

4.1. Main Types of Energy Used by Institutions

Firewood is the most widely used fuel in institutions followed by electricity, charcoal, liquefied petroleum gas (LPG) and others (**Figure 1**). Electricity is used mostly in industries, hotels and hospitals. Respondents revealed that they prefer firewood (**Figure 2**) to other sources not only because of availability, affordability and reliability, but also because the institutions have invested heavily in improved cook stoves (*jiko banifu*) that are designed for firewood use. That institutions have invested in energy-saving stoves is an indication that they are striving to contribute to the reduction in use of biomass energy.

Most of the institutions indicated that they would prefer to use LPG (37%), followed by firewood (32%) and biogas (12%) respectively. The main reasons given for preferring LPG and biogas were that these fuels are convenient, efficient and clean, *i.e.*, don't have some limiting factors like smoke during cooking. However, the major hindrance to the use of LPG is the high cost of cylinders and gas. For example the cost of an empty cylinder (38.0 kg) at Oryx Energies, one of suppliers of gas in Tanzania is TZS 110,000.00 or US $62.21 (exchange rate as of 15 Aug 2014 was US $1.00 to TZS 1661.49). Refilling such a cylinder costs TZS 120,000.00 or US $72.22, *i.e.*, approximately US $1.90/kg. Compared firewood or charcoal, this price is about 950 or 11 times more expensive respectively. A 5-tonne truck of firewood retails at TZS 150,000.00 or US $0.002/per kg, while a bag of charcoal (50 - 60 kg) goes for TZS 18,000 or US $0.18/kg. This high disparity in prices forces institutions to use wood fuel notwithstanding the limiting factors.

In the case of biogas, some educational institutions mentioned that school holidays posed a challenge in the

Table 1. Institutions in the Kilimanjaro Region, Tanzania.

Type of institution	Type of ownership		Total
	Public	Private	
Primary schools	887	46	933
Secondary schools	215	98	313
Teachers' colleges	3	4	7
Vocational training centres	6	71	77
University colleges	1	4	5
Hospitals	6	13	19
Health centres	22	14	36
Dispensaries	152	189	341
Total	**1292**	**439**	**1731**

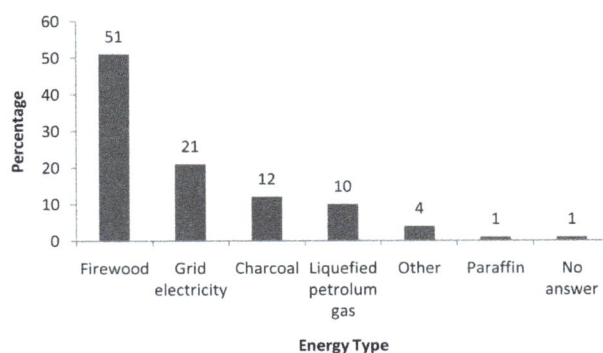

Figure 1. Main energy types used in institutions.

Prefered Energy Source in Kilimanjaro

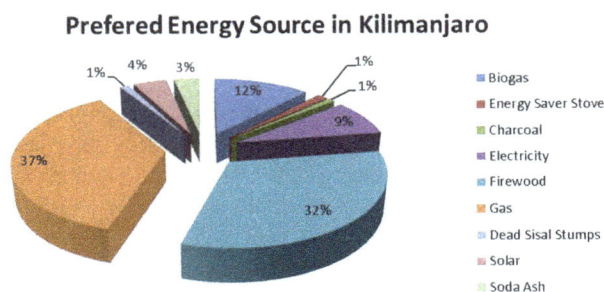

Figure 2. Preferred energy types by institutions.

feeding of bio-digesters. Once students are on holidays, there is a significant reduction in the amount of human waste to keep the biogas system operational.

A study by the WHO [10] states that the introduction of liquefied petroleum gas, despite initial investment costs, can result in a 7-fold return on investment. Government intervention is therefore required to lower the value added tax (VAT) of 16% imposed on LPG to make it more affordable than presently is the case. This recommendation applies to the whole range of renewable energy technologies to increase their affordability and adoption to spur growth of the industry in alternative energy sources.

4.2. Sources of Wood Fuel

Most institutions buy wood fuel from contracted suppliers who also organise delivery (**Figure 3**). In some instances, institutions also buy firewood from a number of local farmers in the neighborhood. However, public secondary schools in Moshi Municipal Council obtain firewood from suppliers through a centralised tendering system.

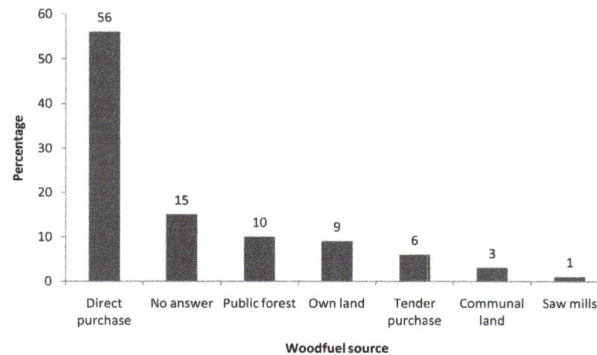

Figure 3. Sources of fuel wood for institutions.

Institutions like prisons, four of them in the region, and the police academy in Moshi Municipality obtain fire wood from government forests. The disadvantage is that their trucks have to cover increasingly long distances, up to 100 km, to reach the forests. Such long trips lead to unsustainably high fuel costs. The problem is exacerbated during the rainy season when a round a trip can take up to three days, yet a load of firewood (5 tonnes) is consumed within 10 days only. Such a consumption rate means that a truck has to make at least three trips a month. Although high costs of fuel, frequency of firewood collection trips, and labour ultimately raises the price firewood, the institutions still view it as less costly to use biomass energy than alternative sources because apart from fuel expenses, the rest are considered as hidden. Institutions use their own trucks and labour (student and inmates in the case of the police academy and prisons respectively).

4.3. Stoves Used in the Institutions

A majority of the institutions (56%) used energy saving stoves (*jiko banifu*) for cooking and heating purposes, particularly in secondary schools, colleges, and prisons (**Figure 4** and **Figure 5**). A broad range of models in various shapes and capacities have been installed. Unfortunately, most of the stoves are badly designed, poorly operated, and maintained. Firewood is not cut into pieces that fit into the stove fire chamber. Instead, logs are used meaning that fire doors are left open during operation. It is, therefore, not surprising that most kitchens were found to be smoky, hot and unhygienic making the working environment very uncomfortable for the kitchen staff (**Figure 4**). In addition, the activities of cooks are not closely monitored and supervised by senior staff. Records of fuel consumption are not properly kept making it difficult to know how much fuel is consumed over a given period of time, and the cost. Worse still, firewood is not properly prepared, stored and dried before it is used (**Figure 6**).

Some of the institutions, however, still use the traditional 3-stone fire (13%) and the charcoal stove (*jiko mkaa*), 3%. The use of the 3-stone fire, a traditional cooking system, is very inefficient. A high proportion of the fire wood burned goes to waste.

4.4. Willingness to Switch Energy Sources

Eighty eight percent of the institutions expressed willingness (**Figure 7**) to switch to biogas for cooking which can only mean that widespread adoption of this technology can take place once the issue of financing system installation costs, which is the main barrier, is addressed. The availability of funding under the SLM Project offers such an opportunity. This level of willingness is encouraging because if an energy-switch project were to be initiated, it would be well received and implemented. A further opportunity for institutions to participate in the carbon credits is the abundance of suppliers of technologies for alternative energy sources not only within the region, but Tanzania as a whole. A good proportion of these are supplied by the Centre for Agricultural Mechanization and Rural Technology (CARMATEC) based in Arusha which has been operational for more than 20 years. CAMARTEC's mission is to develop and disseminate improve technologies suitable for agricultural and rural development. The organisation promotes improved household stoves, community stoves, biogas and solar energy systems. Camartec also works with other organisations to promote and disseminate knowledge of improved wood and charcoal cook stoves.

Figure 4. Poorly desined cook stoves.

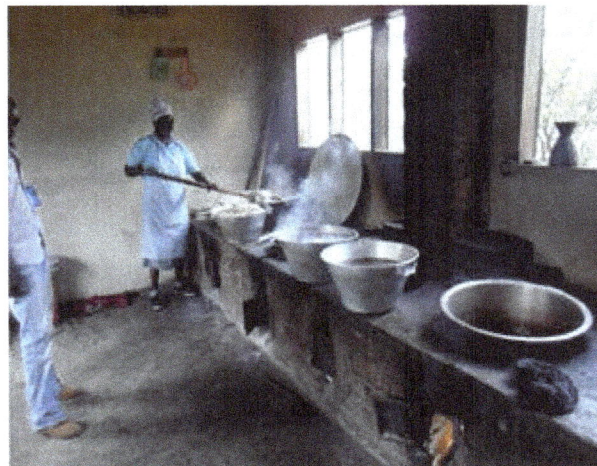

Figure 5. Cookstoves in a better organised kitchen.

Figure 6. Firewood in a stored in a shed.

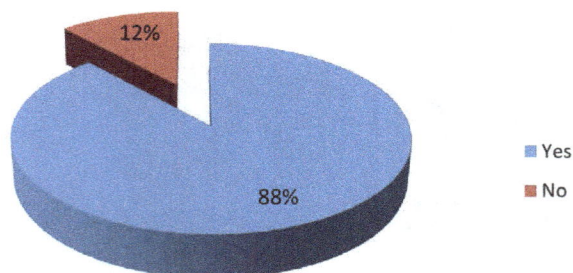

Figure 7. Willingness to participate in a carbon finance project.

Proliferation of different cook stove models is a good development in that it offers institutions various options based on capacity and price from the different suppliers. However, the stoves sector in Tanzania operates more or less in an environment that has no regulations or guidelines, making it difficult for local authorities to maintain order and discipline in the industry. Without regulations in place, manufacturers and vendors impose unreasonably high prices on the cook stoves, a factor that hinders wider adoption of this technology. For instance, a two burner gas cooker ranges from 2500.00 (US $28.40) to KSH (Kenyan Shillings) 5000.00 (US $56.72) in Kenya depending on the brand. In Tanzania, however, such types of stoves are about US $61.00 and US $77.00 a two and three-burner stove. Secondly, there are no standards specifying range and quality of materials that should be used, gauge, mode of construction and installation. Such standards would ensure that only those products meeting certain specifications qualify to be referred to as energy-saving stoves. Apart from models supplied by CARMATEC, TaTEDO (Tanzania Traditional Energy Development), SIDO, KIDT (Kilimanjaro Industrial Development Trust), and a few other reputable suppliers, most of the other models are not energy-saving. It can be safely deduced that these stoves are not having the intended impact of helping institutions reduce firewood consumption by 60% - 70% which is achievable with well designed, properly operated and maintained improved institutional cook stoves.

5. Conclusion

This study sets out to provide an understanding of type of energy used, cooking devices, and willingness on the part of institutions to switch energy sources. As such, it reveals that firewood is by far the most widely used fuel because of availability, affordability, and reliability. That 88% of the institutions are willing to switch to biogas implies that there is an opportunity for energy-switch. This level of willingness means that widespread adoption of this technology can take place once the issue of financing system installation costs, which is the main barrier, is addressed. That 56% of the institutions have energy saving stoves for cooking is a sign of commitment towards reduction in the use of biomass energy. A further opportunity for institutions to participate in the use of the energy-switch principle is the abundance of suppliers of technologies for alternative energy sources not only within the region, but also within Tanzania as a whole.

References

[1]   Kilimanjaro Regional Secretariat (2011) Kilimanjaro Region Strategic Plan 2011/12-2015/16.

[2]   United Republic of Tanzania Population and Housing Census (2013) Population Distribution by Administrative Areas. National Bureau of Statistics, Ministry of Finance, Dar es Salaam, and Office of Chief Government Statistician, President's Office, Finance, Economy and Development Planning, Zanzibar, 244 p.

[3]   Bauner, D., Sundell, M., Senyagawa, J. and Doyle, J. (2012) Sustainable Energy Markets in Tanzania; Report II: Analysis and Conclusions. Renewable Energy Technologies. Stockholm Institute, 36 p.

[4]   Mutimba, S. (2014) Scaling up Fuel Efficient Technologies for Domestic, Institutions and Industrial Use with Carbon Benefits in the Kilimanjaro Region, Final Report. Consultancy Report, Sustainable Land Management Project. Regional Administrative Secretary-Kilimanjaro/United Nations Development Programme, Tanzania, 77 p.

[5]   Arnold, J.E.M., Kohlin, G. and Persson, R. (2005) Fuelwoods, Livelihoods, and Policy Interventions: Changing Perspectives. *World Development*, **34**, 596-611. http://dx.doi.org/10.1016/j.worlddev.2005.08.008

[6]   Tanzania Domestic Biogas Programme (Undated) History of Biogas in Tanzania. http://www.biogas-tanzania.org/tdbp/about/category/history_of_biogas_in_tanzania

[7]   Parfitt, J. (1997) Questionnaire Design; Methods in Human Geography. A Guide for Students Doing a Research Project, 77-109.

[8]   Valentine, G. (1997) Tell Me about Using Interviews as a Research Methodology; Methods in Human Geography. A Guide for Students Doing a Research Project, 110-253.

[9]   Anon (Undated) Sampling Design and Procedures. http://www.cengage.com/marketing/book_content/1439080674_zikmund/book/ch16.pdf

[10] International Energy Agency and World Health Organization (2010) Energy Poverty: How to Make Modern Energy Access Universal? Special Early Excerpt of the World Energy Outlook 2010 for the 2010 UN MDG Review Summit.

18

Effects of Land Use Change on Land Degradation Reflected by Soil Properties along Mara River, Kenya and Tanzania

Ally-Said Matano[1,2], Canisius K. Kanangire[1], Douglas N. Anyona[2], Paul O. Abuom[2], Frank B. Gelder[3], Gabriel O. Dida[4*], Philip O. Owuor[5], Ayub V. O. Ofulla[4]

[1]Lake Victoria Basin Commission Secretariat, Kisumu, Kenya
[2]School of Environment and Earth Sciences, Maseno University, Maseno, Kenya
[3]Probe International, Inc., Auckland, New Zealand
[4]School of Public Health and Community Development, Maseno University, Maseno, Kenya
[5]Department of Chemistry, Maseno University, Maseno, Kenya
Email: *gdidah@gmail.com

Abstract

Human-induced changes to natural landscapes have been identified as some of the greatest threats to freshwater resources. The change from natural forest cover to agricultural and pastoral activities is rampant especially in the upper Mara River catchment (water tower), as well as along the course of the Mara River. The objective of this study was to determine the effect of land use change on the physico-chemical properties of soil (bulk density, carbon, nitrogen, phosphorus and pH) along the course of the Mara River. Five major land uses (agricultural lands, livestock/pastoral lands, forested lands, conservancy/game reserves, and natural wetland) were explored. Results revealed that the mean soil bulk density was 0.956 g/cm³ and differed significantly between sites (p < 0.001). Live biomass values differed significantly between sampling sites (land use types) within the Mara River Basin ($F_{(4, 147)}$ = 8.57, p < 0.001). The mean infiltration over a period of 150 minutes differed, not only among sampling sites, but also between different sides of the river (left and right) within the same sampling site. Soil pH was generally acidic across the five sites and varied significantly ($F_{(4, 63)}$ = 19.26, p < 0.0001) between sites along the Mara River Basin. The mean percentage soil nitrogen across all sampling blocks was 4.87%, with significant differences observed in percentage soil nitrogen ($F_{(4, 63)}$ = 3.26, p < 0.006) between sampling sites. The results indicated that the five land use types affected land degradation differently along the Mara River, while adjacent land degradation affected water physico-chemical properties. These results point to the need to have focused policies on integrated land and water resource management strategies in the Mara River Basin.

*Corresponding author.

Keywords

Catchment Area, land Degradation, Land Use Types, Mara River, Riparian Land

---

1. Introduction

Land use and land cover changes associated with human activities and natural factors compromise many ecosystems including watersheds of important rivers [1]. Land degradation resulting from human activities has been a major global challenge since the 20th century and will remain high on the international agenda in the 21st century [2]. According to Bai *et al.* [3], land degradation is increasing in severity and extent in many parts of the world, with more than 20% of all cultivated areas, 30% of forests and 10% of grasslands undergoing degradation. In recent centuries, an increasing amount of riparian lands have been developed and utilized for agriculture, human settlements and development of cities and towns [4]. This has significantly impacted on critical catchment areas, thus altering water quality in aquatic ecosystems.

Land degradation encompasses the whole environment including individual factors such as soils, water resources (surface and ground), forests (woodlands), grasslands (rangelands), croplands (rainfed and irrigated) and biodiversity (animals, vegetative cover and soil) [5]. Different studies have examined the effects of land use/ cover change on soil physico-chemical properties, and most concur that despite its consequences vary, land use change frequently leads to nutrient losses and reduction of organic matter inputs in the soil [6]-[8]. Conversion of natural forest to other forms of land uses such as farmlands and pasturelands can provoke soil erosion and lead to a reduction in soil nutrients and modification of soil structure [9]. Rai and Sharma [10] also concur that change in land use types negatively affects soil productivity characteristics such as soil bulk density and hydraulic conductivity. Cultivation of forests for instance can diminish soil carbon (C) within a few years of initial conversion [11] and substantially lower mineralizable nitrogen (N) [12]. Islam and Weil [13] reported an increase in bulk density and a reduction in porosity and aggregate stability following the conversion of forest land to crop land, with consequent degradation of adjacent aquatic system.

Soil erosion, salinity and absence of vegetation cover are early warning signs of land degradation, which are likely to influence adjacent aquatic systems through sediment loading. The relationship between land use and water quality has also been demonstrated over the last two decades [14] [15]. There is convincing evidence that watersheds dominated by agriculture and/or human settlement have significantly higher river and nutrient levels [15] [16]. Since rivers provide many ecological and social service functions, they are subject to increased human exploitation and pollution [17]. A river system degrades under severe interference by anthropogenic activities in the catchment [4]. Interference can either occur directly on the river itself or indirectly by degradation of adjacent riparian land which then impacts on the aquatic system. Numerous studies have demonstrated the effects of land use change on erosion and sediment loading patterns in aquatic ecosystems (Alin *et al.* [18]). The spatial relationship between land use and water quality has also been examined by many researchers (Tong and Chen [19], Ngoye and Machiwa [20] and King *et al.* [21]). More specifically, some investigators have compared land cover within certain distances from a stream or sampling site [22]. Burcher *et al.* [1] also observed that changes in land use and land cover interact with anthropogenic and natural drivers to impact negatively on the water quality of watersheds.

Three important spatial scales that influence physical, chemical and biological conditions in a river are basin-wide conditions, riparian (area adjacent to the stream) conditions, and in-stream conditions. The impact of land use change on the watershed environment has also been reported to vary across different spatial scales [23]. Land use changes therefore have both direct and indirect effects on freshwater ecosystems with the former having immediate ecological impacts (e.g. destruction of wildlife habitats), while the latter has impacts that are normally transmitted via altered flow or sediment transport patterns (e.g. lower productivity due to increasing turbidity). Sediment deposition in adjacent water bodies is driven by soil erosion, which is the most widely recognized and most common form of land degradation. Overgrazing of rangelands, poor cultivation of croplands, deforestation and urbanization are some of the land use practices that result in increased soil erosion and subsequent load of sediments and nutrients into aquatic systems [24].

Land use changes in the Lake Victoria Basin have transformed land cover to mainly farmlands, grazing lands, human settlements and urban centres from the previous natural vegetation cover [9]. Over the last 50 years, the

Mara River basin has undergone major changes in land use and land cover. Forests and savannah grasslands have been cleared and turned into land with the main purpose of expanding agricultural activities [25]. Many researchers, including Mugisha [26], Misana *et al.* [27] and Olson *et al.* [28], all concur that most of the changes observed in land use/cover in many parts of Africa are mainly associated with extension and intensification of agricultural activities to new areas. Livestock and wildlife grazing has also been cited as a source of soil degradation. Over cultivation and overgrazing have been linked to increased nutrient transfer and bulk density through nutrient loss and compaction of the soil leading to accelerated soil erosion [29]. High soil bulk density has been a serious land degradation problem in the entire Lake Victoria basin landscape including the Mara River Basin, due to unsustainable land uses [30]. Over the last 50 years, the Mara River Basin has undergone unprecedented changes in land use, just like many other river basins within the larger Lake Victoria Basin (LVB). Accelerated loss of vegetation at the upper Mara River basin has been reported in several studies, including Mati *et al.* [31] and WWF [32].

The Mara River Basin (MRB) plays a major ecological and socio-economic role in communities living in the basin. There is growing evidence of land degradation in the MRB due to improper land use practices, which directly impacts on adjacent aquatic ecosystems. According to Dwasi [25], forests and savannah grasslands in MRB have been cleared and turned into agricultural lands. Rapidly increasing population in the Mara River Basin puts an even greater pressure on the limited natural resources, resulting in increased pollution. Records from the Government of Kenya [33] indicate that over 7000 hectares of Mau forest, which is one of the major water towers in Kenya, were destroyed between 2000 and 2003. The records further show that the area under cultivation in the Amala sub catchment increased from less than 20% in 1960 to more than 51% in 1991 to give way to Olenguruone Settlement Scheme. Such changes in land use result in increased degradation of water quality, thus affecting aquatic biota (flora and fauna). The Lake Victoria Basin region has witnessed increased land use changes in recent times, some of which have led to accelerated land degradation reflected through diminishing vegetation cover, reduced biomass, increased bulk density, and reduced soil nutrients among others [31] [33]. A recent study by USAID EA [34] also linked land use practices in the Mara River Basin to environmental flows and water quality degradation along the Mara River. Over time, the Mara River Basin has witnessed intensified land use changes resulting in increased pollution of the Mara River waters, visible through the high turbidity, total dissolved and total suspended solids, which is an indication of off-site effects of soil erosion. This has resulted in changes in the physical, chemical and biological properties of water in the Mara River ecosystem. However, the exact link between land use types and soil properties along the Mara River Basin has not been well established. This study therefore set out to determine the effects of land use types on land degradation as reflected by soil properties (soil bulk density, soil organic carbon, nitrogen, and phosphorus) along the Mara River.

2. Materials and Methods

2.1. Study Area Description

The Mara River basin covers 13,750 km^2 and lies between Kenya and Tanzania. The Basin is located roughly between longitudes 33°47'E and 35°47'E and latitudes 0°38'S and 1°52'S, with the upper 65% area (8941 km^2) being in Kenya and the remaining 35% in Tanzania (**Figure 1**). The Mara River originates from the Napuiyapui swamp in the Mau escarpment in the highlands of Kenya, with altitudes ranging from 2932 m at its source to 1134 m at Musoma bay.

Rainfall varies inter-annually by a factor of about four between extreme wet and dry years [35]. The river which has for a long time been considered as one of the more pristine rivers draining into Lake Victoria [36], traverses through different land use types including forests, farmlands, open lands, urban centers, game reserves and conservancy before flowing through the Mara Swamp at Musoma Bay in the lower Mara and finally into Lake Victoria. Five distinct land use types along the Mara River were selected for this study. These were: Silibwet sampling site (forested land but with some human interference), Kapkimolwa (agropastoralism with subsistence and large scale farming), Bomet town sampling site (urban set-up with high population and economic activities), Ngerende sampling site (protected area/conservancy) and Kirumi bridge sampling site (a natural wetland with relatively low human interference).

2.2. Study Design Based on Land Degradation Surveillance Framework (LDSF)

A modified version of the LDSF was used in this study. The cross sectional field component of this study was

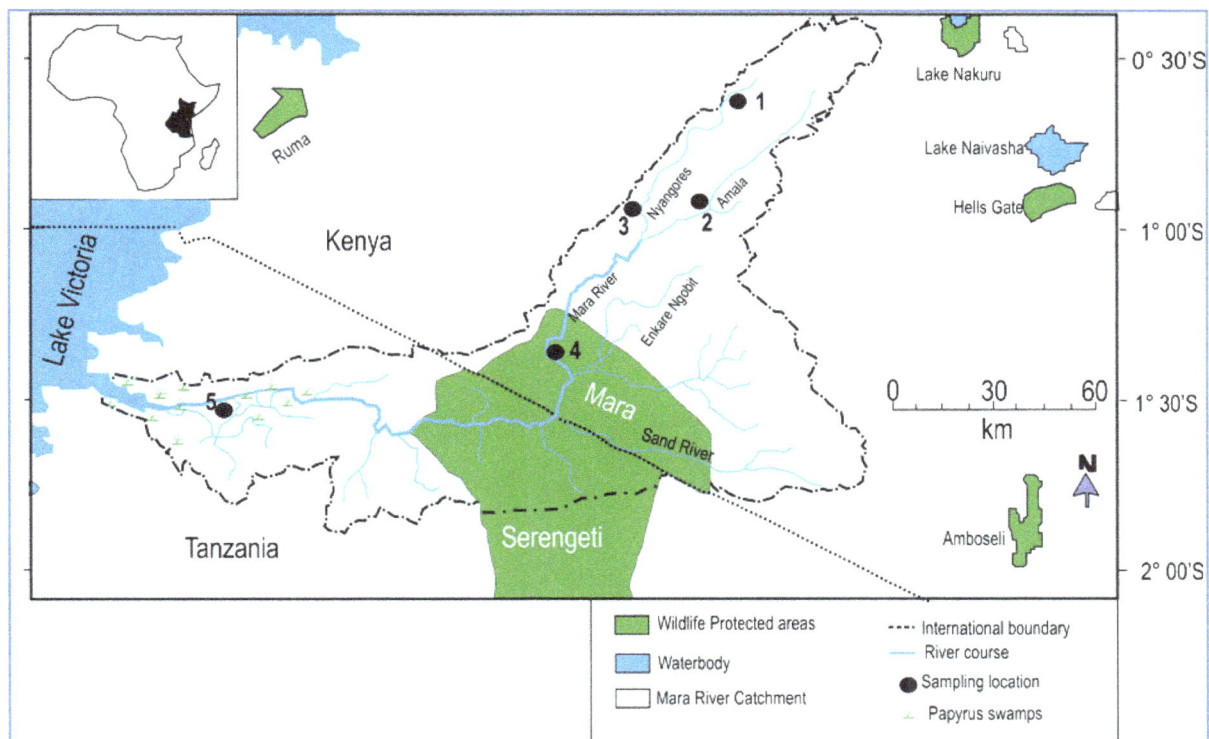

Figure 1. Map of the Mara River basin showing sampling blocks numbered 1 - 5.

carried out between July 2011 and September 2011 during which soil samples were collected for analysis in established laboratories at the International Centre for Research in Agroforestry (ICRAF), Kisumu, Kenya and the Kenya Forestry Research Institute (KEFRI) in Maseno, Kenya. In the study, the entire Mara River Basin represented a block with 5 sampling sites, namely, Silibwet, Kapkimolwa, Bomet, Ngerende and Kirumi, and purposively selected to represent different predominant land uses. In each of the sampling sites there were 8 plots (four on each side of the river) which were laid on a line transect on either side of the river. Each sampling site was also projected to within a 5 km radius from a designated central point within the river, out of which 4 plots were selected within an area lying between 22.5° and 45° degrees on either side of the river. The left and right sides of the river were determined with the researcher facing downstream (**Figure 2**).

Thereafter, each of these sites was laid out as a straight-line transect, with sampling distances for the 4 plots doubling from the central point up to a maximum of the 5 km limit. The same was repeated on the other side of the river. For instance a 3 km transect had the first plot (L1 or R1) 200 m from the central point, with the second (L2 or R2) being located 400 m from the first plot (600 m from the central point) and the third 800 m from the second plot (1.4 km from the central point). The final plot in such transect (L4 or R4) was 1.6 km from the third plot, and thus 3 km from the water sampling point. In all cases the sampling point was located at a bridge.

Infiltration rates were determined *in situ* at each sampling point, while soil samples were collected for determination of soil bulk density, soil carbon, % soil nitrogen, % soil phosphorus and soil pH in the laboratory. A total of 16 soil samples, 8 from the left and 8 from the right side of the river at each sampling site were collected making a total of 80 samples for soil pH, soil carbon, and percentage nitrogen and phosphorus determination. Modifications aside, all other LDSF protocols were observed.

2.3. Land Degradation Surveillance Framework Sampling Plan

In this study, plots measuring 1000 m², 4 subplots (SP) measuring 100 m² (**Figure 3**) and their equidistance were established from each of the 5 sampling sites. Each subplot yielded at least 1 top-soil (0 - 20 cm) and 1 sub-soil (20 - 50 cm) aggregate samples (yielding at least 8 soil sample replicates per plot). Bulk Density (BD) sample collection was done at SP 1 (centre) using a standard Bulk Density (BD) ring, and weight readings were

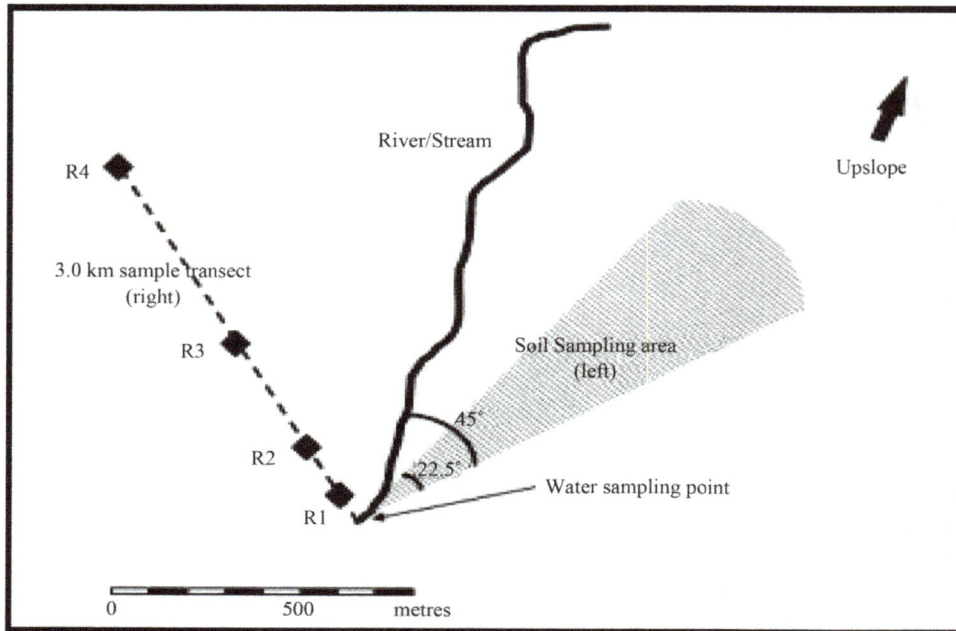

Figure 2. Sampling plot within a sampling site.

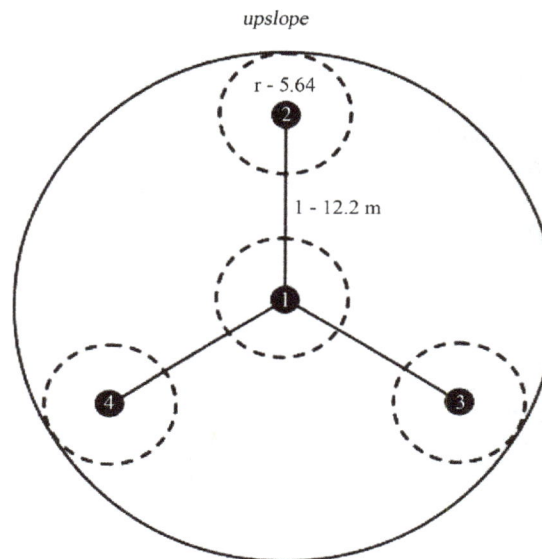

Figure 3. A 1000 m^2 LSDF plot layout showing the four 100 m^2 sub-plots and distances.

determined on-site. In addition, soil hydraulic conductivity tests (infiltration) were carried out at SP 1 (center) in 8 plots within the sampling site thus providing data for 40 soil conductivity samples.

2.4. Soil Sampling for Land Degradation Analysis

Soil sampling was done on both sides of the river within the area lying between the bearings of 22° - 45° upstream from the water sampling site as depicted in **Figure 2** (R1 to R4 and L1 to L4). Soil samples for the soil pH, soil organic carbon (SOC), soil nitrogen, soil phosphorus, and soil bulk density were collected from the river catchment sites using a Dutch auger. Care was taken to avoid mixing the top and sub soil samples, and also to avoid collapsing the hole during sampling. As such, augering was done by taking small and steady bites and by keeping the auger as vertical as possible. Once a significant amount of soil was collected, it was emptied into a suitably labeled bucket to avoid overfilling the auger. In each site, 4 plots of 1000 m^2 were identified systematically. From each plot four sub plots were identified from which top soil (0 - 20 cm) and subsoil (20 - 50 cm) were collected giving 8 different soil samples replicates (4 topsoil and 4 subsoil samples) packed in 8 different polythene bags sealed with rubber band for laboratory analysis at the Kenya Forestry Research Institute (KEFRI) laboratory at Maseno, Kenya. Each polythene bag was clearly labeled with catchment name, area, plot, position and depth.

2.5. Soil Sampling for Land Degradation Analysis

Soil samples for determining soil nitrogen and soil phosphorous were air dried by spreading the sample out as a thin layer into shallow trays. Maximum care was taken to ensure that no materials from the sample were lost or discarded. Drying time depended on the condition of samples and ambient conditions, but they were nevertheless dried thoroughly. The choice for air dried samples was preferred because of the convenience in handling as well as to reduce variation due to moisture [37]. The air dried samples were then ground gently with a wooden pestle and mortar and pressed through a 2 mm sieve [38]. In line with Ben-Dor and Banin [37], the screened sample was subjected to reflectance spectrometer which provided non-destructive rapid prediction of soil physical, chemical and biological properties in the laboratory at KEFRI, Maseno, Kenya.

2.6. Determination of Soil Bulk Density

Bulk Density (BD) samples were collected at each of Sub-Plot 1 using a standard Bulk Density (BD) ring. This involved a cylindrical metal sampler (52 mm high and 51 mm diameter) being driven into the soil until at the same level with ground. The sample was removed by digging around the ring with the trowel underneath it to prevent any loss of soil. Excess soil from the sample was removed with a flat bladed knife and the bottom of the sample made flat and even with the edges of the ring.

The total sample was put into a polythene bag for laboratory analysis and sealed with rubber band. Each polythene bag was identified by the sample site identification code. Top soil bulk density was used to characterize differences in soil compaction among different land use types in the landscape. A total of 32 soil sample replicates were collected per cluster (1 top soil sample × 4 sub-plots × 4 plots × 2 sides of the river) during the field work. Coarse organic matters were removed manually from bulk density soil samples. In the laboratory at International Centre for Research in Agroforestry (ICRAF), Kisumu, Kenya, the samples were oven dried completely at 105°C, weighed on sensitive weighing machine, and weight recorded in grams. The bulk density was then calculated based on mass/volume ratio of the bulk density sampling ring and values recorded.

2.7. Determination of Soil Infiltration Rate

All infiltration measurements were carried out at the mid sub-plots (SP 1) along the sampling transects. Eight (8) measurements were made per sampling site (1 mid subplot × 4 plots × 2 sides of the river). Surface saturated hydraulic conductivity was measured using single-ring cylinder, 16 cm inner diameter according to Reynolds and Elrick [39] taking into account ring radius, depth of ring insertion and depth of ponding. The test calculations were based on shape factors that suggest that field saturated hydraulic conductivity could be obtained with accuracy of about ±20% [39]. The ring was driven into a pre-wetted soil surface according to procedure reported by Reynolds and Elrick [39]. The ring was then filled with water up to a given initial depth and the initial water depth noted while a given time interval (in minutes) was set.

At each and every interval the water depth was noted and the ring refilled to the initial depth. Using a stopwatch, readings were taken first, at 5 minute intervals (where possible) for at least the first 30 minutes, then the interval were increased to 10 minutes (1 hour), and finally to 20 minutes (1 hour or until the readings had stabilized). After two and half to three hours of ponding water in the ring, steady infiltration (Q_s, in mm per hour) and steady ponding depth (W, in cm), the depth of ring insertion (d in cm), was recorded for the ring. The determination of field saturated hydraulic conductivity was based on mass balance equation of flow into soil. Using mass balance, the steady infiltration from a single ring (Q_s) in mm/hr was approximated by the formula: $Q_s = Q_p + Q_g$, where Q_p and Q_g are the steady water out flows from the ring due to hydrostatic and capillary pressure, and due to gravity, respectively.

3. Statistical Methods

The field data was first entered in to Excel spreadsheet and analyzed for descriptive statistics. Further analysis was done using various statistical packages, including Genstat and the Statistical Package for Social Sciences (SPSS) software version 11.0. One way analysis of variance (ANOVA) was performed to test for significant differences among different soil and water quality parameters in relation to the five different land use types along the Mara River. If the main effects were found to be significant at $p < 0.05$, a post hoc separation of means analysis was done by means of Duncan Multiple Range Test (DMRT) to further elucidate the specific differences, while correlations/regression and principal component analysis (PCA) were employed to establish the inter-relationships between and within land and water based variables.

4. Results

Five distinct land use types were identified along the Mara River based on their dominant land uses and characteristics. These were roughly forested but with human interference, agro-pastoralism, urban setting, protected/conservancy area and natural wetland. The link between degradation and its effect on land use is central to nearly all published definitions of land degradation. In the current study, several soil physical properties were measured including soil bulk density, soil infiltration rates among others.

4.1. Soil Texture and Soil Particle Size Grades

Four different soil texture particles, namely, clay (C), sandy loam (SL), sandy clay loam (SCL) and silt clay (SC), were recorded. The current results show a decreasing trend in the proportion of clays towards the lower Mara. Proportions of sandy loam and sandy clay loam, which were negligible in the upper Mara River basin were clearly manifested at Ngerende towards the lower parts of the Mara River basin. All soil samples from Silibwet and Bomet sites showed higher properties of pure clay (C) particles, while soils from Kapkimolwa and Ngerende sampling sites recorded 3% and 24% of silt and clay, respectively. On the contrary, soils from Kirumi sampling site had higher proportions of sandy clay loam, followed by sandy loam.

4.2. Land Degradation Based on Soil Infiltration Rates

The mean infiltration over a period of 150 minutes differed, not only among sampling sites but also between different sides of the river (left and right) within the same sampling site. For instance, while mean infiltration rate on the left side of the river at Silibwet sampling site was 22.6 mm/hr that recorded on the right side was 16.2 mm/hr. The highest infiltration rate of 66.7 cm/h was recorded on the left side of the river at Bomet sampling site and lowest 1.2 mm/hr) on the left side of the river at Ngerende sampling site (**Figure 4**).

4.3. Land Degradation Based on Live Biomass Levels at Different Sampling Sites

Live biomass values differed significantly between sampling sites (land use types) within the Mara River basin

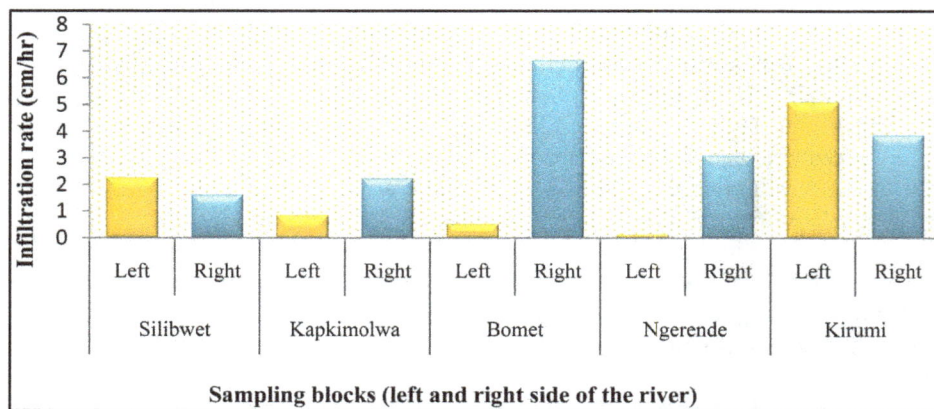

Figure 4. Infiltration rates on the right and left sides of the river per site.

(one-way ANOVA, $F_{(4, 147)}$ = 8.57, p < 0.001). Duncan Multiple Range Test (DMRT) further established that live biomass at Kirumi sampling site was significantly higher compared to all the other sites. Live biomass at Silibwet, Kapkimolwa and Bomet sampling sites did not however show any significant variation between them. Likewise, live biomass recorded at Bomet sampling site was not significantly different from that recorded at Ngerende, Silibwet and Kapkimolwa sampling sites. In addition, live biomass values recorded on the left and right sides of the river at all the sampling sites were different, with some sampling sites having higher values on one side of the river than others, as shown in **Table 1**.

4.4. Land Degradation Based on Soil Chemical Properties Due to Land Use Changes

Soil pH Levels

Soil pH was generally acidic across the five sites and varied significantly (One-way ANOVA, $F_{(4, 63)}$ = 19.26, p < 0.0001) between sites along the Mara Basin (**Table 2**). Duncan Multiple Range Test (DMRT) further established that soil pH levels at Silibwet and Bomet sampling sites differed significantly from the other three sites and among themselves. Lowest soil pH levels (5.53 ± 0.43) were recorded at Silibwet sampling site and highest (6.85 ± 0.33) at Ngerende sampling site (**Figure 5**).

4.5. Percentage Soil Nitrogen and Phosphorus Levels

The mean percentage soil nitrogen across all sampling blocks was 4.87%, with significant differences observed in percentage soil nitrogen (ANOVA, $F_{(4, 63)}$ = 3.26, p < 0.006) between sampling sites (**Table 2**). DMRT

Table 1. Live biomass levels recorded on the left and right sides of the river.

Sampling Sites	Side of River	Live Biomass (g/m²)	Overall Mean (g/m²)
Silibwet	Left	22.63	17.14[BC]
	Right	11.65	
Kapkimolwa	Left	20.35	21.83[B]
	Right	23.33	
Bomet	Left	19.63	22.96[B]
	Right	26.29	
Ngerende	Left	10.48	10.32[C]
	Right	10.16	
Kirumi	Left	93.14	75.44[A]
	Right	57.73	

[A, B, C, BC]Means with different superscripts in the same column are significantly different at p < 0.05 (data analyzed by Duncan's multiple range test).

Table 2. Mean (±SD) soil parameters at different sampling blocks along Mara River.

Sampling Blocks	Soil Bulk Density g/cm³	Live Biomass	Soil pH	% Soil Phosphorus	% Soil Nitrogen
Silibwet	0.867 ± 0.12[C]	22.96 ± 6.79[B]	5.53 ± 0.43[C]	1.24 ± 0.72[B]	2.07 ± 2.20[C]
Kapkimolwa	0.938 ± 0.10[BC]	21.83 ± 10.41[B]	6.55 ± 0.26[A]	1.58 ± 0.52[B]	5.93 ± 4.85[ABC]
Bomet	0.881 ± 0.13[C]	16.82 ± 7.77[BC]	5.94 ± 0.73[B]	1.40 ± 0.39[B]	3.49 ± 4.67[BC]
Ngerende	1.016 ± 0.18[B]	10.50 ± 6.08[C]	6.85 ± 0.33[A]	2.27 ± 0.72[A]	7.43 ± 6.22[AB]
Kirumi	1.214 ± 0.19[A]	41.53 ± 48.65[A]	6.69 ± 0.40[A]	2.41 ± 0.90[A]	8.18 ± 7.94[A]
Mean	0.956	21.204	6.24	1.65	4.87

[A, B, C, AB, BC, ABC]Means with different superscripts in the same column are significantly different at p < 0.05.

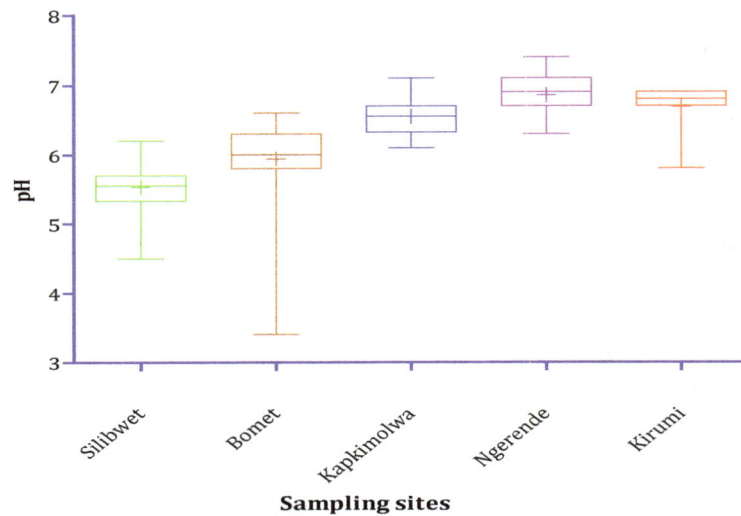

Figure 5. Box plot showing soil pH from various sampling sites along the Mara River.

showed that percentage soil nitrogen at Kirumi sampling site differed significantly from those recorded at Silibwet and Bomet sampling sites, but did not show any significant difference with those recorded at Kapkimolwa and Ngerende sampling sites.

Further, percentage nitrogen levels recorded at Ngerende, Bomet and Kapkimolwa sampling sites did not however show any significant differences. Silibwet sampling site recorded the lowest mean soil nitrogen, (2.07% ± 2.20%) followed by Bomet sampling site-an urban area that recorded 3.49% ± 4.67% nitrogen (**Figure 6**).

Percentage soil phosphorus increased downstream with the highest percentage soil phosphorus being recorded at Kirumi followed and lowest at Silibwet samplign site (**Figure 7**).

4.6. Soil pH, N and P as Influenced by Different Land Use Types and Characteristics

Soil pH was lowest in soils obtained from pastoral land (fenced) and highest in the agropastoral (combination of crop agriculture and livestock) land which also serves as human settlement area. Percentage nitrogen was highest in agricultural and pastoral land while it was lowest on land use for controlled pastoralism (fenced). Highest phosphorus percentage was recorded in land used for agropastoral land and lowest in agricultural land and human settlement (**Table 3**).

4.7. Land Degradation Status Based on Soil Bulk Density

The mean soil bulk density within the Mara River basin was 0.956 g/cm^3, while there were significant variations in soil bulk density between different land use types (sites) within the Mara River basin (one-way ANOVA, $F_{(4, 140)}$ = 19.03, p < 0.001). DMRT further established that mean soil bulk density recorded at Kirumi sampling site was significantly higher than all the other sites, with the highest soil bulk density (1.214 ± 0.19 g/cm^3) recorded at a grazing field at Kirumi sampling site and lowest (0.867 ± 0.12 g/cm^3) at atea farm within Silibwet sampling site. However, soil bulk density at Bomet, Kapkimolwa and Silibwet sampling sites did not vary significantly from each other, as was also the case for soil bulk density between Ngerende and Kapkimolwa sampling sites (**Table 2** and **Figure 8**).

5. Discussion

Characterization of land use along the Mara River profile provided a clear insight on the various land use types and rating. Mara River Basin exhibited five diverse land uses with forests (trees and shrubs) and mixed agriculture characterizing the upper Mara (Silibwet and Kapkimolwa sampling sites), agriculture, pastureland, and nat-

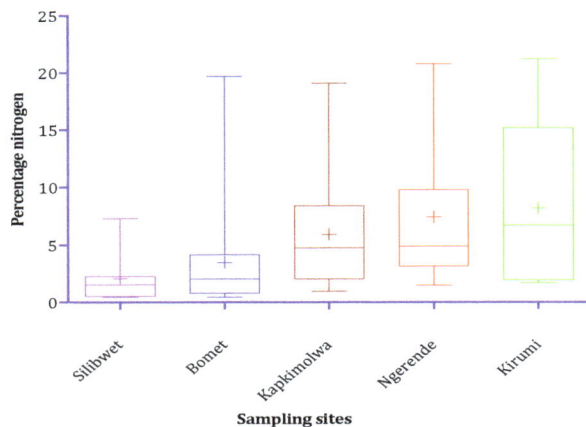

Figure 6. Box plot showing percentage soil nitrogen at various sites within blocks/clusters.

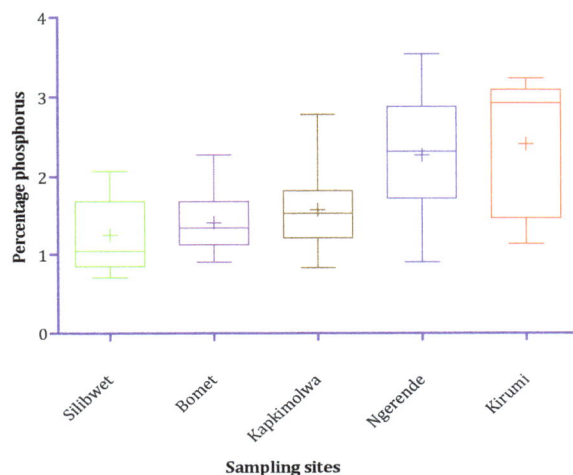

Figure 7. Box plot showing percentage soil phosphorus at different sites along the Mara River profile.

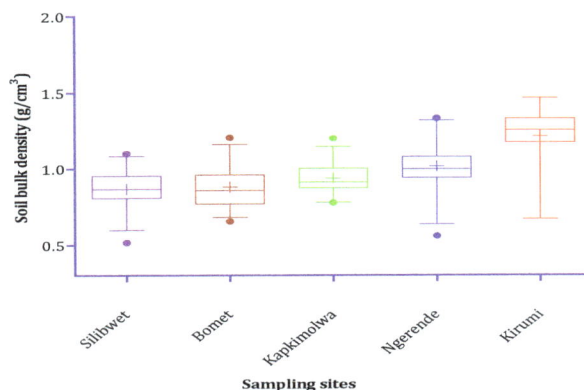

Figure 8. Box plot showing mean soil bulk density at different sites along the Mara River.

Table 3. Statistics of pH, % Nitrogen, and % Phosphorus by land use types.

Land Use Type	pH		% Nitrogen		% Phosphorus	
	Mean	s.e	Mean	s.e	Mean	s.e
Agricultural & pastoral land	5.94	0.18	0.099	1.17	0.013	0.10
Agricultural land	6.36	0.13	0.011	0.69	0.023	0.13
Agricultural land and settlement	6.43	0.13	0.018	0.64	0.010	0.66
Agropastoral land	5.33	0.09	0.005	0.09	0.024	0.27
Agropastoral & settlement land	7	0.17	0.016	0.99	0.013	0.03
Agropastoral land	6.12	0.13	0.015	0.86	0.014	0.14
Ujamaa village since 1967	5.8	0.12	0.008	1.47	0.023	0.55
Pastoral/settlement (school) land	6.4	0.19	0.004	0.74	0.011	0.11
Pastoral land (fenced)	5.42	0.31	0.014	0.03	0.012	0.08
Pastoral land (range land)	5.89	0.19	0.014	1.25	0.015	0.1

ural savanna characterizing the middle (Bomet and Ngerende sampling sites) and parts of the lower basin, and expansive wetlands characterizing the Lower Mara (Kirumi sampling site). Land-use change is primarily influenced by local needs, urbanization and remote economic forces. Land use change at the upper Mara River could have been triggered by the high human population growth, forcing encroachment of forest land for agricultural purposes, livestock grazing and human settlement (including urban development). Consistent with the current study findings, previous studies show that change in land use is highly dependent on the immediate needs of the inhabitants. Different land use types influence land degradation differently, with those used for agriculture and human settlements thought to cause more degradation than those used as pasture land [40]. According to Celik [41] the magnitude of land use changes vary with land use type and land management options applied. However, studies show that while hasty conversion of land use from one type to another may provide a quick remedy to the inhabitants, the long term effect on the land and by extension the aquatic ecosystem nearby may be profound.

Many researchers, including Mugisha [26], Misana et al. [27] and Olson et al. [28], all concur that most of the changes observed in land use in many parts of Africa, are mainly associated with extension and intensification of agricultural activities to new areas. The current findings showed that most plots within Silibwet, Kapkimolwa, Bomet and Ngerende sampling sites that had undergone changes in land use were initially forested before they were converted to farmlands, pasture lands and human settlements. One of the plots in Ngerende which was initially a forested area had also been cleared to create room for construction of a school.

Like in the current study, Maitima et al. [29] while studying the linkages between land use change and land degradation and biodiversity in East Africa, reported that land use changes in East Africa have transformed land cover to farmlands, grazing lands, human settlements and urban centers at the expense of natural vegetation. These changes are associated with deforestation, biodiversity loss and land degradation. Studies by Lambin et al. [42] showed that land conversion to agriculture in East Africa has outpaced the proportional human population growth in recent decades. Natural vegetation cover has given way not only to cropland but also to native or planted pasture [42]. A synthesis of results of long term research by an interdisciplinary team reveals the linkages between land use change, biodiversity loss and land degradation [29]. Maitima et al. [29] reported that land use changes in East Africa have transformed land cover to farmlands, grazing lands, human settlements and urban centers at the expense of natural vegetation. They further observed that land use changes are now associated with deforestation, biodiversity loss and land degradation.

Studies by EAC [43] showed that on average 90% of the LVB population depends on crop and livestock agriculture as the main land use activity, with farm size often less than 1 hectare. This overdependence on limited land use activities contributes substantially to land degradation. Livestock and wildlife grazing has been cited as a source of soil degradation [44]. However, grazing has also been linked to increased bulk density through compaction and exposure of the soil to the sun, but reduces most soil nutrients through feeding and subsequent erosion due to the reduced ground cover [29]. One potentially degrading effect on soil condition is that of soil compaction [45]. Because soil is a complex system of biotic and abiotic components, soil compaction influences several properties of soils that may in turn affect vegetation. These include changes in root growth, availability and movement of air and water, and microbial activity [46]. Established evidence links land use change to losses of soil nutrients among them soil organic carbon, soil nitrogen and soil phosphorus, all of which reduce the productivity of the land, while adjacent aquatic systems are diminished by increased nutrient and sediment load [19].

Studies show that in lands used for pastoral activities, soil compaction can be affected by stocking rates [47] [48], soil texture [47] [49], season of grazing [48], and water content and organic matter [50]. Overgrazing of rangeland; over-cultivation of cropland; water logging; deforestation; and pollution and industrial causes are the most frequently recognized land uses that cause of land degradation [51]. Among the key indicators of land degradation that this study comprehensively used was soil quality, characteristics and productivity. It is acknowledged that land degradation processes are not always induced by man but rather can take natural forms. In the natural state, the rate of water erosion under natural forest corresponds with the subsoil formation rate, hence there is always equilibrium. This boils to the point that accelerated land degradation commonly transpire after human intervention in the environment through different land uses.

Cultivated lands tend to have higher soil bulk densities than forested lands. However, in the current study, soil bulk density and other soil characteristics (N, P and pH) at Silibwet sampling block were the lowest compared to all other land use types within the Mara River basin. On the contrary, though they had relatively less human influence, Ngerende and Kirumi sampling blocks recorded significantly higher soil bulk density, % carbon, % phosphorus, % nitrogen as well as soil pH, compared to all other land use types (sites) probably due to wildlife and livestock trampling effects on the top soils which results in pore volumes reduction and thus high soil bulk density as well as accumulation of fertile soils from uplands to the lower reaches of the Mara River basin resulting in the high nutrient levels recorded at this section of the basin. Consistent with the current study findings, Christensen et al. [52] also demonstrated that total nitrogen and total phosphorus concentrations were highest at sampling sites with little to no agricultural activities such as Ngerende and Kirumi and lower at sampling sites that had a higher percentage of agricultural activities such as Kapkimolwa.

Soil Infiltration rates can be linked to two factors mainly the soil texture and bulk density [53]. The low infiltration rates recorded in some sampling sites such as Silibwet and Kapkimolwa could possibly be due to soil compaction from the perennial farming activities as well as due to cattle trampling effect among other factors. Studies show that soil texture controls the infiltration rate and the amount of water that can be stored in a given thickness of soil for plant use [53]. Clay soils for instance provide the highest surface area, but if clay content is great enough to restrict air and water movement, these critical variables may limit its productivity. Soils in the pure sand range on the other hand have high rates of water infiltration but are low in productivity because they do not retain water or nutrients [53]. The ideal substrate is therefore texturally balanced soil in the loam range [54].

Land use has significant effects on soil chemical properties. Different studies have examined the effects of land use on soil physico-chemical properties [6]-[8]. The levels of soil nutrients (P and N), soil bulk density and soil infiltration rates can be used to deduce the degree at which a given site is degraded. Majule [55] and Gachimbi [56] reported that soils in areas with continuous cultivation and without appropriate management practices have low fertility levels due to overutilization. Various researchers have however reported that agricultural intensification often includes a substantial increase soil nitrogen emanating from the increased application of nitrogen (N) fertilizer, which improves yields but has deleterious consequences on adjacent aquatic systems, where nutrient loading can lead to eutrophication [57] [58]. It is however possible that nutrient uptake by crops, leaching during heavy downpours or further removal during plant harvest time could have contributed to the relatively low soil nutrient levels at Silibwet (an agricultural area) compared to other terrestrial sampling blocks downstream, while the upper ridges and urban areas like Bomet town are also susceptible to erosive nature of landscapes especially during heavy down pours which could have facilitated the active removal of top-soil nutrients especially soil nitrogen and increased erosion through runoff leaving the uplands devoid of the fertile topsoil, which are instead swept into the stream.

In this study, soil phosphorus, soil nitrogen and soil pH were assessed, to determine variations among different land use types. Studies by Maitima et al. [29] reported that grazed sites were significantly higher in soil pH and lower in bulk density, nitrogen, moisture content, percent organic matter and organic carbon than un-grazed sites (p < 0.05). This is a further indication that livestock and wildlife grazing activities have significant effects on soil properties with subsequent effect on water quality. The current study findings showed that soil characteristics as measured by various soil parameters including soil nutrients and soil pH varied significantly under different land use types along the Mara River.

The mean soil pH recorded in this study was 6.24, implying that the soils within the Mara River basin were slightly acidic. The soil pH of soil samples from Silibwet sampling site-which is a relatively forested site and Bomet sampling site-which was characterized by urban land use, were relatively low probably as a result of washing out of solutes from these parts as was also reported in the central highlands of Eastern Ethiopia by Mohammed et al. [59]. The high acidity at highly agricultural lands could be attributed to the decomposition and formation of carboxylic acid during the farming process. Weak acids (corresponding to vinegar) are produced in the soil when plant residues and organic matter decompose. These weak acids react and readily combine with nutrients such as calcium, magnesium, potassium, and sodium as the soil solution (water) moves down through and below the root zone (leaching). During this process, if soil pH is less than 5.2, hydrogen or aluminium replaces basic cations causing the soil in the leached zone to become more acidic [60]. Tamirat [61] also attributed soil pH of an area to the nature of the parent material, climate of the region, organic matter and topographic situation.

The relatively acidic nature of the soils could also be attributed to the high rainfall resulting in the leaching of some basic cations especially calcium from the surface horizons of the soils [62]. A pH value of less than 5.5 like that recorded at Silibwet sampling site is considered problematic for most microbial activities, and this directly influences availability of nutrients to plant [63]. Consistent with the findings of Ahmed [64], continuous cultivation practices, excessive precipitation, steepness of the topography and application of inorganic fertilizer could also have resulted to the reduction of pH in the soil profile particularly at Silibwet sampling site where agricultural activities especially tea plantations were most dominant. Juo and Manu [65] also reported that growing vegetation tended to decrease soil pH, a phenomenon they related to cation uptake by plant, with subsequent release of H+ ions, organic matter decomposition into organic acids, increased carbon dioxide levels through root respiration and nitrification. Bobbink et al. [66], however, reported that low soil pH can decrease plant diversity in forests especially when the soil pH is less than 4.2 making aluminium potentially toxic.

The relatively high soil pH at Ngerende sampling site could be attributed to low organic matter input in the grazing fields and probably accumulation of bases resulting from a compromised hydraulic conductivity that results from minimum leaching of the soluble bases. Simmons [67] reported that as soil pH increases above 6.5, potassium mobility slows down, and as the soil pH reaches 7.0, mobility is severely hindered making it unavailable. This is important as three of the five sites sampled within the Mara River basin recorded pH levels greater than 6.5. It was however interesting to note that high pH on land did not necessarily correlate directly to high pH levels in water. This was clear at Ngerende sampling site whereby land pH was amongst the highest but the same site recorded the lowest water pH, probably implying that pH in water was not necessarily driven or impacted by adjacent terrestrial soil pH. A number of studies show that in cropped fields, the ash deposited by the common practice of slash-and-burn releases alkaline cations (Ca, Mg and K), causing high pH, and low exchangeable Aluminium values [65]. This could probably explain the relatively high pH values recorded at Kapkimolwa, Ngerende and Kirumi sampling sites, as most studies have reported that most agricultural land is regularly limed to undo the effects of natural acidification. Studies show that many acid soils "fix" or hold phosphorus, making it unavailable for plant growth. Soil acidity can also be a barrier to root development, limiting the plants' ability to reach moisture in the sub-soil. In the humid tropics, soil acidity and associated problems often lead to land abandonment and the perpetuation of slash-and-burn agriculture [68].

Although mean soil pH across most of the land use types in the current study were higher than the critical value of 5.5, mean soil pH at Silibwet block was much lower than the critical value. Consistent with our findings, studies show that soils from tea plantations as was the case at Silibwet tend to be strongly acidic, with some as less as 4.5 [69]. The low pH, at Silibwet sampling site corresponded with a high soil organic carbon at the same site. The long-term effect of fertilizer application could have been attributed to the low pH and high soilorganic carbon as was also observed by Wang et al. [70] in commercial tea plantations in China.

Nitrogen and phosphorus are essential nutrients for growth and development of crops, whose optimal levels in the soil is indicative of the productivity of the soil [71]. However, soils in areas with continuous cultivation and without appropriate management practices often have low soil fertility levels due to over utilization with studies further showing that major plant nutrients, e.g., potassium (K) and phosphorous (P) are the soil properties most affected by cultivation over time [55] [56]. The mean percentage soil nitrogen across all sampling blocks was 4.87%, with significant differences observed in percentage soil nitrogen ($F_{(4, 63)}$ = 3.26, p < 0.006) between sampling sites. The continuous conversion of vegetated areas to non-vegetated surfaces as was the case at Bomet sampling site could have resulted in the reduced soil nutrients through increased soil erosion, while the significantly low nitrogen and phosphorus proportions in soils from Silibwet sampling site could have been due to crop (mainly tea) uptake as well as additional loss through food crop harvests or when vegetation is uprooted during land preparation as was also observed by Elliot [72].

Studies show that though its consequences vary, land conversion frequently leads to nutrient losses through disruption of surface and mineral horizons (e.g. by mechanical disturbance) and reduction of organic matter inputs. Cultivation of forests, for example, diminishes soil carbon within a few years of initial conversion and substantially lowers mineralizable nitrogen [11]. Phosphorus is critical to biotic function and essential to the development and maintenance of ecosystems [73]-[75]. It is an essential element classified as a macronutrient because of the relatively important part it plays in the growth of plants. However its mismanagement can pose a threat to water quality of adjacent aquatic systems. In the current study, percentage phosphorus levels varied significantly between different sites/land use types (p < 0.0001) with DMRT further establishing that Ngerende and Kirumi sampling sites located at the lower Mara River Basin had significantly higher percentage phosphorus compared to the other three sites (Silibwet, Kapkimolwa and Ngerende sampling sites). Just like nitrogen, the relative low phosphorus percentages recorded at the sites located on the upper catchment particularly Silibwet, Bomet and Kapkimolwa where crop farming was dominant could have been due to removal by the crops' edible parts and plant residues. Mohammed *et al.* [59] linked the variability of phosphorus levels to land use, altitude, slope position and other characteristics, such as clay and calcium carbonate content. Further analysis established significant differences in percentage phosphorus level between different vegetation types.

Studies have also shown that soils devoted to crop production can lose far more phosphorus than soils that are covered by relatively undisturbed forest or natural grass land [76]. Phosphorus is an essential nutrient for plant growth hence its' active uptake by plants and subsequent removal through harvest can have an acidifying effect on the soils. Studies show that the amount of nutrients removed by cropping depends on the type of crop grown, part of the crop harvested, and the stage of growth at harvest [77]. Silibwet sampling site which is a relatively forested area characterized by small scale farms recorded the lowest phosphorus percentage compared to all the other sites, probably as a result of continuous cultivation that facilitated phosphorus uptake and removal by food crops during harvest. Livestock and wildlife herbivory can also cause shifts in plant species composition by replacing highly palatable grasses with unpalatable species [78] and therefore cause changes in soil nutrients indirectly. This is because as vegetation cover declines, soil nutrients are also depleted, while soil erosion increases, generating negative consequences on rangeland productivity [79]. Due to severe grazing, a reduction in plant biomass leads to depletion of existing nutrients among them nitrogen and phosphorus thus resulting in soil fertility reduction [79].

Soil bulk density is highly dependent on soil texture and the densities of soil mineral (sand, silt, and clay) and organic matter particles, as well as their packing arrangement [80]. Further, bulk density is influenced by crop and land management practices that affect soil cover, organic matter, soil structure, and/or porosity. In the current study, the mean soil bulk density within the Mara River basin was 0.956 g/cm^3, while there were significant variations in soil bulk density between different land use types (sites) within the Mara River basin ($F_{(4, 140)}$ = 19.03, p < 0.001). Most soil bulk densities fall between 1.0 g/cm^3 and 2.0 g/cm^3; while root penetration is severely impacted at bulk densities greater than 1.6 g/cm^3 [81]. As density increases, pore space decreases and the amount of air and water held in the soil also decreases [81]. In the current study, high soil bulk density observed at the lower Mara River watershed compared to other sites upstream could have been contributed by the high livestock and wild life population at those areas which together with farming activities might have had an impact on the soil structure thus increasing the bulk density and subsequently contributed to land degradation. Studies show that lower soil bulk density is desirable for plant growth, whether those plants are agricultural crops, trees, or turf grass [82]. This is because low bulk density soils have greater water infiltration rates which minimize runoff, improve water quality, and reduce storm water flow [80].

Activities such as plowing, timber harvesting or compaction of the soil during home construction, some of which were observed at the Mara River basin can however increase soil bulk density and reduce pore space. Compacted soils may result in little to no vegetation growth in some locales [83]. The high land degradation status at the lower Mara River was also manifested in the high turbidity and nutrient concentration (*i.e.* total nitrogen and total phosphorus) recorded in the adjacent waters of the Mara River at this point. There was also surface runoff and soil erosion increase because these soils had lost the ability to absorb rain water. Some guidelines to minimize the damage caused by compaction include confining traffic to designated paths; reducing livestock population per unit area among others can control erosion [84].

The significant differences in soil bulk density that existed between Nyangores and Amala and those observed at the lower Mara sub-catchments represents spatial processes occurring along the length of the river, with reduced human impact and high vegetation cover at the upland areas probably protecting soils from degradation thus leading to low bulk density. Likewise, several researchers have also reported that increased land degradation can be accelerated by the position of the land within a landscape, with lowland areas likely to suffer impacts of increased erosion, surface runoff and silt deposition, thus resulting in high bulk density [85] [86]. Silibwet sampling site which was relatively forested and with a relatively large area covered by tea plantation recorded significantly low soil bulk density, high soil conductivity and live biomass compared to all other land use types. However, bulk density recorded within Bomet and Kapkimolwa sampling sites, both located at the upper Mara River basin, were relatively higher compared to those at Silibwet.

In the current study, areas where land use types alternated between cultivated and grazing fields recorded intermediate mean soil bulk density between the agricultural crop lands and grazing fields. This is explained by the intermediate amount of litter inputs, vegetation cover and the moderate level of surface ground disturbance due to seasonality alteration of use. The ability of livestock to alter soil bulk densities is a function of stocking density and intensity of the same land use [43]. Studies show that livestock activities like simple grazing often results in soil compaction due to the weight of the animals and the mechanical forces that cattle apply when walking on land [87]. Soil compaction may in turn have negative consequences such as reduced rainfall infiltration, enhanced soil erosion [88] and degradation of the herbaceous vegetation cover [89] [90].

6. Conclusion

Based on the current study findings, it was presumed that land use changes in the low land areas of the Mara River Basin took place earlier than in the uplands, as exhibited by the differences in soil particle sizes in soil bulk densities recorded at different points along the river. This phenomenon can also explain relatively low soil bulk density at Silibwet, Bomet, and Kapkimolwa sampling sites as opposed to high soil bulk density and low soil conductivity recorded in Ngerende and Kirumi sampling sites. These results point to the need to have focused policies on integrated land and water resource management strategies in the Mara River Basin.

Acknowledgements

East Africa Community-Lake Victoria Basin Commission Secretariat (EAC-LVBC) provided funds for this study. We are grateful to the Kenya Forestry Research Institute (KEFRI), Maseno, Kenya for providing the time, material, and technical support.

References

[1]   Burcher, C.L., Vallet, H.M. and Benfield, E.F. (2007) The Land Cover Cascade. Relationship Coupling Land and Water. *Ecology*, **88**, 228-242. http://dx.doi.org/10.1890/0012-9658(2007)88[228:TLCRCL]2.0.CO;2

[2]   Eswaran, H., Lal, R. and Reich, P.F. (2001) Land Degradation: An Overview. In: Bridges, E.M., Hannam, I.D., Oldeman, L.R., Pening de Vries, F.W.T., Scherr, S.J. and Sompatpanit, S., Eds., *Responses to Land Degradation, Proceedings of 2nd International Conference on Land Degradation and Desertification*, Khon Kaen, Thailand. Oxford Press, New Dehli, India. Kaen, Thailand. Oxford Press, New Dehli, India.

[3]   Bai, Z.G., Dent, D.L., Olsson, L. and Schaepman, M.E. (2008) Proxy Global Assessment of Land Degradation. *Soil Use Management*, **24**, 223-234. http://dx.doi.org/10.1111/j.1475-2743.2008.00169.x

[4]   Jones, K.B., Slonecker, E.T., Nash, M.S., Neale, A.C., Wade, T.G. and Hamann, S. (2010) Riparian Habitat Changes across the Continental United States (1972-2003) and Potential Implications for Sustaining Ecosystem Services. *Landscape Ecology*, **25**, 1261-1275. http://dx.doi.org/10.1007/s10980-010-9510-1

[5]   Food and Agriculture Organization (FAO) (2005) Global Forest Resources Assessment 2005. Progress towards Sustainable Forest Management. Forestry Paper 147, Rome.

[6]   Emadi, M., Baghernejad, M., Fathi, H. and Saffari, M. (2008) Effect of Land Use Change on Selected Soil Physical

and Chemical Properties in North Highlands of Iran. *Journal of Applied Sciences*, **8**, 496-502. http://dx.doi.org/10.3923/jas.2008.496.502

[7] Agoume, V. and Birang, A.M. (2009) Impact of Land-Use Systems on Some Physical and Chemical Soil Properties of an Oxisol in the Humid Forest Zone of Southern Cameroon. *Tropicultura*, **27**, 15-20.

[8] Gol (2009) Effects of Land Use Change on Soil Properties and Organic Carbon at Dagdami River Catchment in Turkey. *Journal of Environmental Biology*, **30**, 825-830.

[9] Lake Victoria Basin Commission (LVBC) (2007) Regional Transboundary Diagnostic Analysis (RTDA) of Lake Victoria Basin. No. 4, Lake Victoria Basin Commission Publication, Kisumu.

[10] Rai, S.C. and Sharma. E. (1998) Comparative Assessment of Runoff Characteristics under Different Land Use Patterns within a Himalayan Watershed. *Hydrological Process*, **12**, 2235-2248. http://dx.doi.org/10.1002/(SICI)1099-1085(19981030)12:13/14<2235::AID-HYP732>3.0.CO;2-5

[11] Murty, D., Kirschbaum, M.F.U., Mcmurtrie, R.E. and Mcgilvray, H. (2002) Does Conversion of Forest to Agricultural Land Change Soil Carbon and Nitrogen? A Review of the Literature. *Global Change Biology*, **8**, 105-123. http://dx.doi.org/10.1046/j.1354-1013.2001.00459.x

[12] Richter, D.D., Markewitz, D., Heine, P.R., Jin, V., Raikes, J., Tian, K. and Wells, C.G. (2000) Legacies of Agriculture and Forest Regrowth in the Nitrogen of Old-Field Soils. *Forest Ecology and Management*, **138**, 233-248. http://dx.doi.org/10.1016/S0378-1127(00)00399-6

[13] Islam, K.R. and Weil, R.R. (2000) Land Use Effects on Soil Quality in a Tropical Forest Ecosystem of Bangladesh. *Agriculture, Ecosystems and Environment*, **79**, 9-16. http://dx.doi.org/10.1016/S0167-8809(99)00145-0

[14] Benoit, M. and Fizaine, G. (1999) Quality of Water in Forest Catchment Areas. *Revue Forestiere Francaise*, **50**, 162-172.

[15] Berka, C., Schreier, H. and Hall, K. (2001) Linking Water Quality with Agricultural Intensification in a Rural Watershed. *Water, Air, and Soil Pollution*, **127**, 389-401. http://dx.doi.org/10.1023/A:1005233005364

[16] Wang, X. (2001) Integrating Water Quality Management and Land Use Planning in a Watershed Context. *Journal of Environmental Management*, **61**, 25-36. http://dx.doi.org/10.1006/jema.2000.0395

[17] Liu, C.M. and Liu, X.Y. (2009) Healthy River and Its Indication, Criteria and Standards. *Journal of Geographical Sciences*, **19**, 3-11. http://dx.doi.org/10.1007/s11442-009-0003-6

[18] Alin, S.R., Eilly, O.R., Ohen, C.M.C., Ettman, A.S.D., Alacios, D.L.P., Fest, M.R. and Mckee, B.A. (2002) Effects of Land-Use Change on Aquatic Biodiversity: A View from the Paleorecord at Lake Tanganyika, East Africa. *Geology*, **30**, 1143-1146. http://dx.doi.org/10.1130/0091-7613(2002)030<1143:EOLUCO>2.0.CO;2

[19] Tong, S.T.Y. and Chen, W. (2002) Modeling the Relationship between Land Use and Surface Water Quality. *Journal of Environmental Management*, **66**, 377-393.

[20] Ngoye, E. and Machiwa, J.F. (2004) The Influence of Land-Use Patterns in the Ruvu River Watershed on Water Quantity in the River System. *Physics and Chemistry of the Earth*, **29**, 1161-1166. http://dx.doi.org/10.1016/j.pce.2004.09.002

[21] King, R.S., Baker, M.E., Whigham, D.F., Weller, D.E., Jordan, T.E., Kazyak, P.K. and Hurd, M.K. (2005) Spatial Considerations for Linking Watershed Land Cover to Ecological Indicators in Streams. *Ecological Applications*, **15**, 137-153. http://dx.doi.org/10.1890/04-0481

[22] Sponseller, R.A., Benfield, E.F. and Valett, H.M. (2001) Relationships between Land Use, Spatial Scale, and Stream Macroinvertebrate Communities. *Freshwater Biology*, **46**, 1409-1424. http://dx.doi.org/10.1046/j.1365-2427.2001.00758.x

[23] Houlahan, J.E. and Findlay, C.S. (2004) Estimating the "Critical" Distance at Which Adjacent Land-Use Degrades Wetland Water and Sediment Quality. *Landscape Ecology*, **19**, 677-690. http://dx.doi.org/10.1023/B:LAND.0000042912.87067.35

[24] Seitzinger, S.P., Mayorga, E., Bouwman, A.F., Kroeze, C., Beusen, A.H.W., Billen, G., Van Drecht, G., Dumont, E., Fekete, B.M., Garnier, J., Harrison, J., Wisser, D. and Wollheim, W.M. (2010) Global River Nutrient Export: A Scenario Analysis of Past and Future Trends. *Global Biogeochemical Cycles*, **24**, 1-16. http://dx.doi.org/10.1029/2009GB003587

[25] Dwasi, J.A. (2002) Trans-Boundary Environmental Issues in East Africa: An Assessment of the Environmental and Socio-Economic Impacts of Kenya's Forestry Policy. Nairobi.

[26] Mugisha, S. (2002) Root Causes of Land Cover/Use Change in Uganda: An Account of the Past 100 Years. LUCID Working Paper No.14, International Livestock Research Institute, Nairobi. WWW.Lucideastafrica.org

[27] Misana, S.B., Majule, A.E. and Lyaruu, H.V. (2003) Linkages between Changes in Land Use, Biodiversity and Land Degradation on the Slopes of Mount Kilimanjaro, Tanzania. LUCID Working Paper No. 38, International Livestock

Research Institute, Nairobi.

[28] Olson, J.M., Misana, S.B., Campbell, D.J., Mbonile, M.J. and Mugisha, S. (2004) The Spatial Pattern and Root Causes of Land Use Change in East Africa. LUCID Working Paper No. 4, International Livestock Research Institute, Nairobi. WWW.Lucideastafrica.org

[29] Maitima, J.M., Mugatha, S.M., Reid, R.S., Gachimbi, L.N., Majule, A., Lyaruu, H., Pomery, D., Mathai, S. and Mugisha, S. (2009) The Linkages between Land Use Change, Land Degradation and Biodiversity across East Africa. *African Journal of Environmental Science and Technology*, **3**, 310-325.

[30] ICRAF (2002) Improved Land Management in the Lake Victoria Catchment. Linking Land and Lake, Research and Extension, Catchment and Lake.

[31] Mati, B.M., Mutie, S., Home, P., Mtalo, F. and Gadain, H. (2005) Land Use Changes in the Trans-Boundary Mara Basin: A Threat to Pristine Wildlife Sanctuaries in East Africa. *Proceedings of the 8th International River Symposium*, Brisbane, 6-9 September 2005.

[32] WWF (2009) Mara River Basin Trans-Boundary Water Resources Management Programme Progress Report. Nairobi.

[33] Government of Kenya (2009) The Mau Taskforce Report. Prime Minister's Office, Republic of Kenya, Nairobi.

[34] USAID EA (2010) Assessing Environmental Flows for the Mara River. Nairobi.

[35] Mutie, S.M., Mati, B., Home, P., Gadain, H. and Gathenya, J. (2006) Evaluating Land Use Change Effects on River Flow Using USGS Geospatial Stream Flow Model in Mara River Basin, Kenya. Center for Remote Sensing of Land Surfaces, Bonn, 28-30 September 2006.

[36] Nile Equatorial Lakes Subsidiary African Program (NELSAP) (2002) Management of the Water Resources of the Mara River Basin. Project Identification No.3, 1-13.

[37] Ben-Dor, E., Irons, J.R. and Epema, G.F. (1999) Soil Reflectance. In: Rencz, N., Ed., *Remote Sensing for the Earth Sciences*: *Manual of Remote Sensing*, John Wiley & Sons, New York, 111-188.

[38] Hunt, G.R. and Salisbury, J.W. (1970) Visible and Near-Infrared Spectra of Minerals and Rocks. I. Silicate Minerals. *Modern Geology*, **1**, 283-300.

[39] Reynolds, W.D. and Elrick, D.E. (1990) Ponded Infiltration from a Single Ring: I. Analysis of Steady Flow. *Soil Science Society of America Journal*, **54**, 1233-1241. http://dx.doi.org/10.2136/sssaj1990.03615995005400050006x

[40] Greijn, H. (1994) A Missed Opportunity. *Our Planet*, **6**, 23-24.

[41] Celik, I. (2005) Land Use Effects on Organic Matter and Physical Properties of Soil in a Southern Mediterranean Highland of Turkey. *Soil & Tillage*, **83**, 270-277. http://dx.doi.org/10.1016/j.still.2004.08.001

[42] Lambin, E.F., Geist, H.J. and Lepers, E. (2003) Dynamics of Land Use and Land Cover Change in Tropical Regions. *Annual Review of Environment and Resources*, **28**, 206-241. http://dx.doi.org/10.1146/annurev.energy.28.050302.105459

[43] East African Community (2005) Potentials and Constraints of Promoting Lake Victoria Basin as a Regional Economic Growth Zone. EAC, Arusha.

[44] Dregne, H., Kassas, M. and Rozanov, B. (1991) A New Assessment of the World Status of Desertification. *Desertification Control Bulletin*, **20**, 6-18.

[45] Soane, B.D. and Van Ouwerkerk, C., Eds. (1994) Soil Compaction in Crop Production. Developments in Agricultural Engineering Series, Volume 11. Elsevier Science, Amsterdam, 662.

[46] Lal, R. (2005) Forest Soils and Carbon Sequestration. *Forest Ecology and Management*, **220**, 242-258. http://dx.doi.org/10.1016/j.foreco.2005.08.015

[47] Van Haveren, B.P. (1983) Soil Bulk Density as Influenced by Grazing Intensity and Soil Type on a Short Grass Prairie Site. *Journal of Range Management*, **36**, 586-588. http://dx.doi.org/10.2307/3898346

[48] Naeth, M.A., Pluth, D.J., Chnnasyk, D.S., Bailey, A.W. and Fedkenheuer, A.W. (1990) Soil Compacting Impacts of Grazing in Mixed Prairie and Fescue Grassland Ecosystems of Alberta. *Canadian Journal of Soil Science*, **70**, 157-167. http://dx.doi.org/10.4141/cjss90-018

[49] Orr, H.K. (1960) Soil Porosity and Bulk Density on Grazed and Protected Kentucky Bluegrass Range in the Black Hills. *Journal of Range Management*, **13**, 80-86. http://dx.doi.org/10.2307/3895129

[50] Howard, R.F., Singer, M.J. and Frantz, G.A. (1981) Effects of Soil Properties, Water Content, and Compactive Effort on the Compaction of Selected California Forest and Range Soils. *Soil Science Society of America Journal*, **45**, 231-236. http://dx.doi.org/10.2136/sssaj1981.03615995004500020001x

[51] Stocking, M. and Murnaghan, N. (2000) Land Degradation—Guideline for Field Assessment. Overseas Development Group, University of East Anglia, Norwich.

[52] Christensen, V.G., Lee, K.E., Sanocki, C.A., Mohring, E.H. and Kiesling, R.L. (2009) Water Quality and Biological

Responses to Agricultural Land Retirement in Streams of the Minnesota River Basin, 2006-2008. US Geological Survey Scientific Investigations Report 2009-5215, USGS, Reston, VA.

[53] USDA-NRCS (2001) Rangeland Soil Quality: Water Erosion. Soil Quality Information Sheet. http://soils.usda.gov/sqi

[54] Tiedemann, A.R. and Lopez, C.F. (2004) Assessing Soil Factors in Wildland Improvement Programs. In: Monsen, S.B., Stevens, R. and Shaw, N.L. (Compilers), *Restoring Western Ranges and Wildlands*, USDA Forest Service, Rocky Mountain Research Station, Fort Collins, General Technical Report RMRS-GTR-136-Vol. 1, 39-56.

[55] Majule, A.E. (2003) A Study on Land Use Types, Soils and Linkages between Soils and Biodiversity along the Slopes of Mt. Kilimanjaro, Tanzania.

[56] Gachimbi, L.N. (2002) Technical Report of Soil Survey and Sampling Results: Embu-Mbeere Districts, Kenya. LUCID Working Paper Series No. 9, KARI-Kabete, Nairobi.

[57] Beman, J.M., Arrigo, K.R. and Matson, P.A. (2005) Agricultural Runoff Fuels Large Phytoplankton Blooms in Vulnerable Areas of the Ocean. *Nature*, **434**, 211-214. http://dx.doi.org/10.1038/nature03370

[58] Turner, R.E. and Rabalais, N.N. (1991) Changes in Mississippi River Water Quality This Century. *BioScience*, **41**, 140-147. http://dx.doi.org/10.2307/1311453

[59] Mohammed, A., Leroux, P.A.L., Barker, C.H. and Heluf, G. (2005) Soil of Jelo Micro-Catchment in the Central Highlands of Eastern Ethiopia. I. Morphological and Physiochemical Properties. *Ethiopian Journal of Natural Resource*, **7**, 55-81.

[60] Spies, C.D. and Harms, C.L. (2004) Soil Acidity and Liming of Indiana Soils. Purdue University, West Lafayette. www.agry.purdue.edu/ext/forages/publications/ay267.htm

[61] Tamirat, T. (1992) Vertisol of Central Highlands of Ethiopia: Characterization and Evaluation of Phosphorus Statues. Master's Thesis, Alemaya University, Dire Dawa.

[62] Iwara, A.I. (2011) Soil Erosion and Nutrient Loss Dynamics in Successional Fallow Communities in a Part of the Rainforest Belt, South-Southern Nigeria. Ph.D. Proposal Presented at the Staff/Postgraduate Seminar, University of Ibadan, Ibadan.

[63] Solomon, D. (2008) Presentation on the Relationships Existing in Minerals Soil between pH on the One Hand and the Activity of Microorganisms and the Availability of Plant Nutrients on the Other. Bahir Dar University, Bahir Dar.

[64] Ahmed, H. (2002) Assessment of Spacial Variability of Some Physicochemical Property of Soil under Different Elevation and Land Use Systems in the Western Slopes of Mount Chilalo, Arisi. Master's Thesis, Alemaya University, Dire Dawa.

[65] Juo, A.S.R. and Manu, A. (1996) Chemical Dynamics in Slash-and-Burn Agriculture. *Agriculture, Ecosystems Environment*, **58**, 49-60.

[66] Bobbink, R., Hornung, M. and Roelofs, J.G.M. (1998) The Effects of Air-Borne Nitrogen Pollutants on Species Diversity in Natural and Semi-Natural European Vegetation. *Journal of Ecology*, **86**, 717-738. http://dx.doi.org/10.1046/j.1365-2745.1998.8650717.x

[67] Simmons, J. (1998) Balancing Soil Nutrition. Rutgers Turf Management Program.

[68] TropSoils (1991) Technical Report for 1988-1989. TropSoils Management Entity, Raleigh, 357 p.

[69] Brady, N. (1990) The Nature and Properties of Soils. 13th Edition, Macmillan, New York.

[70] Wang, R., Shi, X., Wei, Y., Yang, X. and Uoti, J. (2006) Yield and Quality Responses of Citrus (*Citrus reticulate*) and Tea (*Podocarpus fleuryi* Hickel.) to Compound Fertilizers. *Journal of Zhejiang University Science*, **7**, 696-701. http://dx.doi.org/10.1631/jzus.2006.B0696

[71] Zhao, Q., Zeng, D.H., Fan, Z.P. and Lee, D.K. (2008) Effect of Land Cover Change on Soil Phosphorus Fractions in Southeastern Horqin Sandy Land, Northern China. *Pedosphere*, **18**, 741-748. http://dx.doi.org/10.1016/S1002-0160(08)60069-7

[72] Elliot, W.J. (2003) Soil Erosion in Forest Ecosystems and Carbon Dynamics. In: Kimble, J.M., Heath, L.S., Birdsey, R.A. and Lal, R., Eds., *The Potential of US Forest Soils to Sequester Carbon and Mitigate the Greenhouse Effect*, CRC Press, Boca Raton, 175-190.

[73] Richter, D.D., Allen, H.L., Li, J.W., Markewitz, D. and Raikes, J. (2006) Bioavailability of Slowly Cycling Soil Phosphorus: Major Restructuring of Soil P Fractions over Four Decades in an Aggrading Forest. *Oecologia*, **150**, 259-271. http://dx.doi.org/10.1007/s00442-006-0510-4

[74] Crews, T.E., Kitayama, K., Fownes, J.H., Riley, R.H., Herbert, D.A., Muellerdombois, D. and Vitousek, P.M. (1995) Changes in Soil Phosphorus Fractions and Ecosystem Dynamics across a Long Chronosequence in Hawaii. *Ecology*, **76**, 1407-1424. http://dx.doi.org/10.2307/1938144

[75] Turner, B.L. and Engelbrecht, B.M.J. (2011) Soil Organic Phosphorus in Lowland Tropical Rain Forests. *Biogeoche-*

mistry, **103**, 297-315. http://dx.doi.org/10.1007/s10533-010-9466-x

[76] Brady, N.C. and Weil, R.R. (1996) The Nature and Properties of Soil. 11th Edition, Prentice-Hall, Inc., Englewood Cliff, 740 p.

[77] Whibread, A., Blair, G., Konboon, Y., Lefroy, R. and Naklang, K. (2003) Managing Crop Residues, Fertilizers and Leaf Litters to Improve Soil C, Nutrient Balances, and the Grain Yield of Rice and Wheat Cropping Systems in Thailand and Australia. *Agriculture, Ecosystems & Environment*, **100**, 251-263. http://dx.doi.org/10.1016/S0167-8809(03)00189-0

[78] Owen-Smith, N. (1999) The Animal Factor in Veld Management: Implications of Selective Patters of Grazing. In: Tainton, N.D., Ed., *Veld Management in South Africa*, University of Natal Press, Pietermaritzburg, 129-130.

[79] Morgan, R.P.C. (1995) Soil Erosion and Conservation. 2nd Edition, Longman Group, Essex.

[80] Sakin, E., Deliboran, A. and Tutar, E. (2011) Bulk Density of Harran Plain Soils in Relation to Other Soil Properties. *African Journal of Agricultural Research*, **6**, 1750-1757.

[81] Froese, K. (2004) Bulk Density, Soil Strength, and Soil Disturbance Impacts from a Cut-to-Length Harvest Operation in North Central Idaho. Master's Thesis, University of Idaho, Moscow, ID, 72 p.

[82] Catherine, S. and Rock, P. (2007) Organic Carbon, Organic Matter and Bulk Density Relationships in Boreal Forest Soils. *Canadian Journal of Soil Science*, **88**, 315-325.

[83] Gomez, A., Powers, R.F., Singer, M.J. and Horwath, W.R. (2002) Soil Compaction Effects on Growth of Young Ponderosa Pine Following Litter Removal in California's Sierra Nevada. *Soil Science Society of America Journal*, **66**, 1334-1343. http://dx.doi.org/10.2136/sssaj2002.1334

[84] Hesselbarth, W. and Vachowski, B. (2000) Trail Construction and Maintenance Notebook. 2000 Edition, 0023 2839P, USDA Forest Service, Missoula Technology and Development Center, Missoula.

[85] Lasanta, T., Garcia-Ruiz, J.M., Perez-Rontome, C. and Sancho-Marcen, C. (2000) Run-Off and Sediment Yield in a Semi-Arid Environment: The Effect of Land Management after Farm Land Abandonment. *Catena*, **38**, 265-278. http://dx.doi.org/10.1016/S0341-8162(99)00079-X

[86] Jiang, P., Anderson, S.H., Kitchen, N.R., Sadler, E.J. and Sudduth, K.A. (2007) Landscape and Conservation Management Effects on Hydraulic Properties of a Claypan-Soil Toposequence. *Soil Science Society of America Journal*, **71**, 803-811. http://dx.doi.org/10.2136/sssaj2006.0236

[87] Steffens, M., Kölbl, A., Totsche, K.U. and Kögel-Knabner, I. (2008) Grazing Effects on Soil Chemical and Physical Properties in a Semiarid Steppe of Inner Mongolia (P.R. China). *Geoderma*, **143**, 63-72. http://dx.doi.org/10.1016/j.geoderma.2007.09.004

[88] Russell, J.R., Betteridge, K., Costall, D.A. and Mackay, A.D. (2001) Cattle Treading Effects on Sediment Loss and Water Infiltration. *Journal of Range Management*, **54**, 184-190. http://dx.doi.org/10.2307/4003181

[89] Bouman, B.A.M., Nieuwenhyse, A. and Ibrahim, M. (1999) Pasture Degradation and Its Restoration by Legumes in Humid Tropical Costa Rica. *Tropical Grasslands*, **33**, 142-165.

[90] Alados, L., Ahmed, E.L., Aich, A., Papanastasis, V.P., Ozbek, H., Navarro, T., Freitas, H., Vrahnakis, M., Larrosi, D. and Cabezudo, B. (2004) Change in Plant Spatial Patterns and Diversity along the Successional Gradient of Mediterranean Grazing Ecosystems. *Ecological Modelling*, **180**, 523-535. http://dx.doi.org/10.1016/j.ecolmodel.2003.10.034

19

X-Ray Computed Tomography for Root Quantification

Bente Foereid

Environment and Climate Division, NIBIO, Ås, Norway
Email: bente.foereid@bioforsk.no

Abstract

Soil cores from a field growing barley and barley mutants without root hairs under conventional and minimum tillage were sampled. They were X-ray scanned to produce a 3D image and then the roots were washed out and weight and length were determined by conventional means. Root volume and surface area were then calculated from the 3D images using state of the art software and methodology, and the measured and calculated measures were correlated. The only strong and significant correlation was between measured weight and calculated volume for mutants without root hairs. It is concluded that the software cannot segment out very small roots, but segmentation accuracy also depends on root structure in some unknown way. Any study using X-ray computed tomography to quantify roots as they grow *in situ* should start with a calibration for the conditions in question.

Keywords

Roots, 3D Image, X-Ray Computed Tomography

1. Introduction

Whilst aboveground plant development and productivity can usually be easily observed and quantified, at least on the plot scale, quantification of roots still pose significant challenges. None of the methods available can quantify root biomass or turnover reliably. Furthermore, most methods are destructive, and there are few methods available to observe roots as they grow.

X-ray CT was first developed for medical applications, and was later used for a variety of industrial applications. The development scanners that could scan a variety of size classes, including small ones (micro-tomography) made it useful also for geological applications, including soils [1]. X-ray CT has had a variety of useful applications in studies of soil physics (e.g. [2]-[6]), but studies of biological properties have turned out to be difficult as organic matter has a similar X-ray attenuation as air [7].

Although imaging roots using X-ray CT can be challenging because roots are difficult to distinguish from pores, progress has been made [8]. The power of the technology lies in that the method is non-destructive, so that roots can be observed repeatedly over time as they grow. Some progress has been made in segmenting soil from pore space [9]-[11]. Segmentation of roots has also been attempted, but has so far succeeded for very young roots or small parts of roots [12] [13].

In 3D scans of roots in soil the roots can be segmented out and volume and surface area can be calculated using dedicated software. However, it is not known how accurate these measures are. The purpose of this paper is to compare surface and volume calculated by software on 3D scans to weight the length of roots measured by traditional destructive techniques on roots washed out from the samples after scanning.

2. Materials and Methods

2.1. Field Experiment

The field experiment was established in Invergowrie, Dundee, Scotland, UK (56°27'N, 3°W) in 2003 (see [14] for details) to compare tillage treatments, among them conventional and minimum tillage. The field was situated in a mainly agricultural area in eastern Scotland close to the sea. The soil was Dystric-Fluvic Cambisol (FAO) with a sandy-loam texture. It had a pH (1 part soil to 2.5 parts 1M $CaCl_2$) of 5.7, was freely drained and underlain by colluvial sand at 60 cm depth. In 2012 the fields were planted to barley with different rooting pattern, among them wild type and a mutant lacking root hairs. The mutants and their origin are described in [15].

2.2. Sampling and Direct Root Measurements

Soil cores were sampled in plastic rings (4 cm height, 4 cm diameter). Before sampling, aboveground plant material and top soil were removed in the area to be sampled, so that the sampling depth was 4 - 8 cm. To make sure all samples were fresh when scanned and processed, each replicate (n = 3) were sampled on different days. Sampling days were 21, 23, 25 May. The samples were stored in a cold store (4°C) until scanning (after 1 day) and further sample processing (after 2 days). The treatments sampled were barley mutants without root hairs and the wild type (with root hairs) at minimum and conventional tillage.

Roots were washed out, and total roots length was determined by scanning the roots and using the software WinRhizo. The roots were then dried at 70°C overnight and dry weight determined.

2.2.1. Scanning Specifications and Root Segmentation Method

3D volumetric images in this study were obtained using a Metris X-Tek HMX CT scanner with a Varian Paxscan 2520 V detector and a 225 kV X-ray source (Nikon Metrology X-Tek Systems Ltd, Tring, UK) giving a resolution of up to 5 μm. Samples were scanned at 160 kV and 201 mA using a 0.1 mm Al filter to obtain 3003 angular projections (based on a 360° rotation).

VGStudio MAX 2.2 (Volume Graphics, Germany) was used for root segmentation. Roots were identified by eye, and segmented using "region grower". This was repeated several times in each sample until no more roots could be found. Erode/dilate (radius 2) was then used. The total volume and surface of the root region was calculated by the software. The procedure was repeated twice for each sample, to assess repeatability of the procedure.

2.2.2. Statistics

Minitab v15 was used. Correlation analyses between root weight (measured by weighing) and root volume (calculated by VGStudioMAX) and between root length (measured by WinRhizo) and root surface area (calculated by VGStudioMAX) and between root surface area measured by WinRhizo and calculated by VGstudioMax were performed. Varieties with and without root hairs were correlated separately.

3. Results and Discussion

It was expected that weight and volume of roots would be strongly correlated, and that testing if the volume measured in the 3D image correlates with the weight should be a good test of the calculation from the 3D image. Root length and surface area were also expected to be correlated, but this correlation was expected to be weaker.

Overall there was little correlation between measurements done on the roots and the parameters calculated from the 3D image (**Table 1**; **Figure 1**). The only strong and significant correlation was between measured weight and calculated volume for the varieties without root hairs showed (**Table 1**; **Figure 1**). Tillage treatment did not affect the relationship between measured and calculated parameters, but variety (with or without root hairs) did (**Figure 1**). How well roots can be quantified from 3D images therefore depends on the type or structure of the roots. Hirano et al. [16] found that detection frequency of roots below 1 cm in diameter was poor using root penetrating radar. Also in this study it was noted that very thin roots were not picked up by the software even when they were seen by the experimenter. Large roots or high resolution would help, but that means that with current technology only a small volume of soil can be examined. Any use of 3D imagery in following root development should start with a calibration like this for the type of roots and resolution to be used. Using a homogenous soil may also help, but the results here suggest that the quality of the roots are more important. It is not known why the roots without root hairs were easier to quantify. Although root hair are too small to be seen on a 3D image, they are also lost in root washing, so they would also not be included in the weight measurements. It is possible that varieties without root hairs compensate by being thicker or in other ways more distinct, and therefore easier for the software to follow.

4. Conclusion

Larger roots can be reliably quantified in 3D X-ray images, but the quantification is less reliable for smaller roots. However, reliability also depends on root structure in a way that is not fully understood, and any study of *in situ* root growth using X-ray tomography should start with a calibration of reliability for the roots and resolution to be used.

Table 1. Pearson's correlation coefficient and p-values for correlations between various parameters measured and calculated from the X-ray CT scans.

	Overall		Varieties with root hair		Varieties without root hair	
	R	p	R	p	R	p
Weight vs. volume	0.454	0.138	0.362	0.481	0.823	0.044
Surface vs. length	0.159	0.622	0.052	0.923	0.388	0.447
Surface, measured vs. calculated	0.210	0.512	-0.442	0.380	0.441	0.381

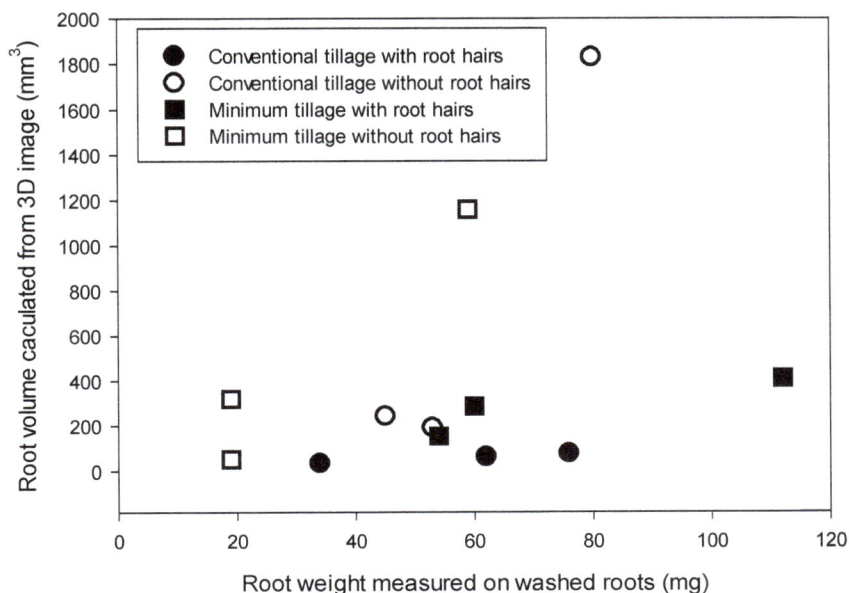

Figure 1. Root weight measured plotted against root volume calculated from the 3D image for the two tillage treatments and two varieties.

Acknowledgements

The author wishes to thank Dr. Tim George at James Hutton Institute in Dundee, UK for support at the field site VGstudioMAX support team in Heidelberg, Germany for support on the use of the software. This work was supported by the University of Abertay Dundee, UK.

References

[1]   Ketcham, R.A. and Carlson, W.D. (2001) Acquisition, Optimization and Interpretation of X-Ray Computed Tomographic Imagery: Applications to the Geosciences. *Computers and Geosciences*, **27**, 381-400. http://dx.doi.org/10.1016/S0098-3004(00)00116-3

[2]   De Gryze, S., Jassogne, L., Six, J., Bossuyt, H., Wevers, M. and Merckx, R. (2006) Pore Structure Changes during Decomposition of Fresh Residue: X-Ray Tomography Analyses. *Geoderma*, **134**, 82-96. http://dx.doi.org/10.1016/j.geoderma.2005.09.002

[3]   Nunan, N., Ritz, K., Rivers, M., Feeney, D.S. and Young, I.M. (2006) Investigating Microbial Micro-Habitat Structure Using X-Ray Computed Tomography. *Geoderma*, **133**, 398-407. http://dx.doi.org/10.1016/j.geoderma.2005.08.004

[4]   Sleutel, S., Cnudde, V., Masschaele, B., Vlassenbroek, J., Dierick, M., Van Hoorebeke, L., Jacobs, P. and De Neve, S. (2008) Comparison of Different Nano- and Micro-Focus X-Ray Computed Tomography Set-Ups for the Visualization of the Soil Microstructure and Soil Organic Matter. *Computers and Geosciences*, **34**, 931-938. http://dx.doi.org/10.1016/j.cageo.2007.10.006

[5]   Iassonov, P., Gebrenegus, T. and Tuller, M. (2009) Segmentation of X-Ray Computed Tomography Images of Porous Materials: A Crucial Step for Characterization and Quantitative Analysis of Pore Structures. *Water Resources Research*, **45**, W09415. http://dx.doi.org/10.1029/2009WR008087

[6]   Elyeznasni, N., Sellami, F., Pot, V., Benoit, P., Vieublé-Gonod, L., Young, I. and Peth, S. (2012) Exploration of Soil Micromorphology to Identify Coarse-Sized OM Assemblages in X-Ray CT Images of Undisturbed Cultivated Soil Cores. *Geoderma*, **179-180**, 38-45. http://dx.doi.org/10.1016/j.geoderma.2012.02.023

[7]   Taina, I.A., Heckl, R.J. and Elliot, T.R. (2007) Application of X-Ray Computed Tomography to Soil Science: A Literature Review. *Canadian Journal of Soil Science*, **88**, 1-19. http://dx.doi.org/10.4141/CJSS06027

[8]   Mooney, S.J. Pridmore, T.P., Helliwell, J. and Bennett, M.J. (2012) Developing X-Ray Computed Tomography to Non-Invasively Image 3-D Root Systems Architecture in Soil. *Plant and Soil*, **352**, 1-22. http://dx.doi.org/10.1007/s11104-011-1039-9

[9]   Hapca, S., Houston, A.N., Otten, W. and Baveye. P. (2013) New Objective Segmentation Method Based on Minimizing Locally the Intra-Class Variance of Grayscale Images. *Vadoze Zone*, **12**, 13 p. http://dx.doi.org/10.2136/vzj2012.0172

[10] Houston, A.N., Otten, W., Baveye, P. and Hapca, S. (2013) Thresholding of Computed Tomography Images of Heterogeneous Porous Media by Adaptive-Window Indicator Kriging. *Computers and Geosciences*, **54**, 239-248. http://dx.doi.org/10.1016/j.cageo.2012.11.016

[11] Houston, A.N., Schmidt, S., Otten, W., Baveye, P. and Hapca, S. (2013) Effect of Scanning and Image Reconstruction Settings in X-Ray Computed Tomography on Soil Image Quality and Segmentation Performance. *Geoderma*, **207-208**, 154-165. http://dx.doi.org/10.1016/j.geoderma.2013.05.017

[12] Schmidt, S. Bengough, A.G., Gregory, P.J., Grinev, D.V. and Otten, W. (2012) Estimating Root-Soil Contact from 3-D X-Ray Microtomography. *European Journal of Soil Science*, **63**, 776-786. http://dx.doi.org/10.1111/j.1365-2389.2012.01487.x

[13] Haling, R.E., Tighe, M.K., Flavel, R.J. and Young, I.M. (2013) Application of X-Ray Computed Tomography to Quantify Fresh Root Decomposition *in Situ*. *Plant and Soil*, **372**, 619-627. http://dx.doi.org/10.1007/s11104-013-1777-y

[14] Sun, B., Hallett, P.D., Caul, S., Daniell, T.J. and Hopkins, D.W. (2011) Distribution of Soil Carbon and Microbial Biomass in Arable Soils under Different Tillage Regimes. *Plant and Soil*, **338**, 17-25. http://dx.doi.org/10.1007/s11104-010-0459-2

[15] Brown, L.K., George, T.S., Barrett, G.E., Hubbard, S.F. and White, P.J. (2013) Interactions between Root Hair Length and Arbuscular Mycorrhizal Colonization in Phosphorus Deficient Barley (*Hordeum vulgare*). *Plant and Soil*, **372**, 195-205. http://dx.doi.org/10.1007/s11104-013-1718-9

[16] Hirano, Y., Yamamoto, R., Dannoura, M., Aono, K., Igarashi, T., Ishii, M., Yamase, K., Makita, N. and Kanazawa, Y. (2012) Detection Frequency of *Pinus thunbergii* Roots by Ground-Penetrating Radar Is Related to Root Biomass. *Plant and Soil*, **360**, 363-373. http://dx.doi.org/10.1007/s11104-012-1252-1

20

Analytical Framework Model for Capacity Needs Assessment and Strategic Capacity Development within the Local Government Structure in Tanzania

John F. Kessy[1], Abiud Kaswamila[2]

[1]Department of Forest Economics, Sokoine University of Agriculture, Morogoro, Tanzania
[2]Department of Geography, University of Dodoma, Dodoma, Tanzania
Email: jfkessy2012@gmail.com, abagore.kaswamila6@gmail.com

Abstract

This is a methodological paper prepared by senior academicians, researchers and consultants from renowned universities in Tanzania. The paper provides insights as to how best development agents can approach the challenge of capacity needs assessment and development of capacity building programs in the context of the local government structure in Tanzania. The paper is of original nature and is based on author's accumulated knowledge and practice in conducting capacity assessments and developing capacity building programs in Tanzania. The paper describes what can be considered to be best practices in conducting participatory capacity assessment through consultative processes which involves most of the key actors who would be engaged in implementing proposed interventions. The paper puts forward an analytical model for capacity assessment and program development in the Tanzanian context. The main features of the model can be summarized as participatory capacity assessment, strategic capacity building program development and complementarity through synergy building with like-minded stakeholders. The operational modality for utilizing the model in developing capacity building programs which among other components has monitoring and evaluation aspects is included. It is recommended that practitioners and development agents should test the model in their working environments to realize its potential benefits including program ownership by stakeholders.

Keywords

Capacity Assessment, Local Government Structure, Tanzania

1. Introduction

Conservation and sustainable land management is a challenge to both developed and developing countries. However, the problem is more critical in developing countries where existing capacities in terms of human, financial and other resources to deal with the challenge are very limited. To address the situation, governments and development agencies have been implementing among others capacity building interventions. A review of literature indicates that there is a considerable body of work available on capacity development [1]-[3] though very little has been published on approaches to assessing capacity needs in the Tanzanian context. Several development partners have developed capacity assessment approaches. However, most of the documented approaches do lack specificity and may not adequately capture the uniqueness of administrative structures in various countries. For example UNDP has recommended some guidelines and approaches which may assist countries in dealing with the capacity assessment challenges [4]-[6]. The UNDP guidelines introduce a note of realism by suggesting that putting the concept into practice is not a simple process and that it requires some degree of flexibility. The guidelines further make a case by arguing that the assessment should be taken within the broader socio-economic environment, as well as evaluated for specific organizations and individuals in different countries. In other words, the assessments might be undertaken in different situations and needs to be re-designed from time to time to ensure ownership, sustainability and ultimate success by the users. This makes it necessary to develop country specific models like the one presented in this paper.

A range of tools and approaches have been proposed for assessing capacity at different levels taking into consideration the enabling environment, the organizational level and individual level [1]. However, these approaches in most cases are general. Identifying the most appropriate entry point for assessing capacity and in particular the methodological part of it is critical to the realization of successful outcomes. It has been proposed [7] that the assessment should deploy a multi-dimensional approach through which the scope includes institutional issues, leadership capacities, accountability and dialogue. In this context knowledge gap assessment becomes one element of the comprehensive assessment. Logically assessment should start with the big picture at the level of enabling environment and then proceed to the lower levels. In reality, there may be many reasons why this is not only impractical but also impossible. In view of these challenges, this article tries to develop a general model that can guide capacity building interventions for conservation and development projects in the context of local government in Tanzania. The model has been tested in Kilimanjaro and Manyara regions in Tanzania and has proved to work.

2. Structure of the Local Government in Tanzania and SLM Interventions

There are two types of local authorities in Tanzania namely rural authorities normally referred to as *district councils* and urban authorities that include city, municipal and town councils. Hierarchically, a district council goes down through a ward, under which exists village governments and finally the 10-house cell system. Moreover, an urban council runs down through a municipality (if the top structure is a city) under which exists the ward, then the mtaa/street government and, finally, the 10-house cell system. There are no village government structures in urban authorities [8]-[10]. The local government is administratively divided into regions. The regions are divided into districts (urban and rural authorities), which are then sub-divided into divisions, wards and villages in case of rural authorities. Urban authorities consist of city councils, municipal councils and town councils, whereas included in the rural authorities are the district councils with township council and village council authorities. The district and urban councils have autonomy in their geographic area. The former coordinates activities of wards, divisions and villages. The village and townships councils have the responsibility for formulating plans for their areas [8]-[10].

In both village and townships there are a number of democratic bodies at lower levels to debate and oversee the implementation of local development needs. In rural system, the hamlet, the smallest unit of a village, is composed of an elected chairperson who appoints a secretary and three further members all of whom serve on an advisory committee. In urban areas the street, is the smallest unit within the ward of an urban authority. Unlike the hamlets, the street committees have a fully elected membership comprising of a chairperson, six members and an executive officer. The basic functions of the local government are [8]-[10] maintenance of law, order and good governance; promotion of economic and social welfare of the people within their areas of jurisdiction; and ensuring effective and equitable delivery of qualitative and quantitative services to the people within their areas of jurisdiction.

The head of the paid service is the district executive director in the district authorities and the town/municipal/ city director in the urban authorities. Below the director there are a number of heads of department. The departments (for provision of extension services) include personnel and administration, planning and finance, engineering/works, education and culture, trade and economic affairs; health and social welfare, cooperative and agriculture development, environment, and natural resources [10]. From the analysis of the basic structure and functions of local government authorities in Tanzania, it is apparent that most development and conservation interventions in the country are housed within the local government structures. For sustainability reasons it is strongly encouraged that all development interventions should form an integral part of the local government structure through systematic mainstreaming of project interventions into routine activities in these structures. That is why the model presented in this paper puts much emphasis on integration of SLM interventions into existing structures.

The most important, intended links between the local government and the local communities are the villages in rural areas and urban street committees in urban areas, which are designed to mobilize community participation in local development and conservation issues. Priorities for local service delivery and development projects are brought to village and/or street committees for discussion before being forwarded to the Ward Development Committee (WDC). In the rural system proposals reach the WDC via the village council.

3. The Strategic Capacity Development Model (SCDM)

As stated earlier on, the presented analytical model in this paper was tested in two projects financed by UNDP and Farm Africa in Kilimanjaro and Manyara regions respectively and proved to work well in the Tanzanian context [11] [12]. The model tries to describe the best practices in conducting participatory capacity assessment through consultative processes which involves most of the key actors who would be engaged in implementing proposed interventions and strategic development of capacity building interventions. The model provides a roadmap which can assist planners, practitioners and development partners aiming to organize and facilitate capacity building interventions in SLM practices in Tanzania and beyond. The model has three main components all being processes. These include a participatory assessment phase, the strategic program development phase and operational phase. These phases are described in this chapter.

3.1. Participatory Assessment Phase

The algorithm to follow involves four main steps namely conducting participatory consultative process; discussion with key informants in the region, districts, division, ward and level; collection of bio-data of employees and deficiencies at different levels; and holding village/community level discussions. It is envisaged that after all these steps, capacity needs including staffing and training needs for example will be determined. As put forward in literature [2] while conducting capacity needs assessment there is a challenge of developing a process which is detailed enough to allow a logical regression through the assessment but also flexible enough to respond to the wide variations in local level demands. The participatory assessment phase takes care of these propositions.

The assessment involves organizing participatory consultative meetings with local government officials and other stakeholders to develop a common understanding of the current situation, capacity gaps and needs for capacity building. The officials could include project staff, regional administration, district technical teams, collaborating NGOs/CBOs, and community level stakeholders. The process should be guided by a pre-designed checklist of issues upon which capacity gaps shall be identified. The checklist among others should solicit information regarding each stakeholders understanding of their roles in implementation of SLM interventions, staffing levels and knowledge deficiencies, status of other required capacities for effective implementation of interventions and potential partners in capacity development. The consultative process has to involve both individual and roundtable focused group discussions aimed to establish the capacity building needs at various levels within the local government structure. The assessment should pay special attention to issues of sustaining SLM interventions beyond the existing project lifespan. The assessment of local government staff capacities involves collection of bio-data of employees and their deficiencies should be undertaken at the regional, district and ward offices within the local government structure. This shall be guided by a pre-designed form so as to capture the entire list of existing staff in the region and districts, their capacities and deficiencies. The bio-data form should include data such as name of the employee, present designation, salary scale, date birth, qualification, responsibility and training needs.

The assessment at village and/community level should capture community needs and perceptions in relation to capacity building needs. Village level discussions should be held before discussions with regional and district teams. This is meant to obtain villager's opinions first. Such discussions could be organized by the development agencies and/or consultants in collaboration with district facilitating teams. In each district representative communities/villages should be selected taking into consideration a range of criteria that ensures representation of all categories of beneficiaries in the strict sense. The cross section of representatives is meant to ensure that even the marginalized segments of the society should be involved in proposing capacity building options for their respective areas.

3.2. Strategic Program Development Phase

This phase is summarized in the conceptual framework model presented in this section. According to the elements presented in **Figure 1**, the model considers three main aspects in strategic capacity development program design. These include facilitation of capacity building, the capacity building activities and expected outcomes from the whole process. Additionally, the framework elaborates the contribution of the model to monitoring and evaluation of capacity building interventions.

According to this framework the design process assumes that facilitation of trainings and other capacity building interventions needs to be supported by the financiers of the project such district councils (DCs), municipal councils (MCs), donors and/or government institutions. The design further emphasizes that collaborating partners and service providers (e.g. consultants) need to be mobilized for the capacity building interventions as technical facilitators. Specific training activities involving project staff, district level stakeholders, ward and village level stakeholders shall form part of the strategic interventions for capacity building. It is anticipated that the capacity building for project staff shall be on cross-cutting issues which should improve their capacity to provide leadership and monitoring roles for the trainings that will be given to the other stakeholders. For both district and lower levels the strategy emphasize that the trainings should be designed for both technical officers (extension staff) to concentrate more on technical issues as well as politicians who are aimed for awareness of project results and political support. In these aspects the model takes into consideration suggestions in literature [7] that capacity development processes should be seen as collaborative learning processes linked to concrete pilot field level actions as well as evolving international policy instruments and market opportunities.

The model puts forward the anticipated outcomes from the capacity building processes. The major purpose of providing capacity building includes the realization of successful implementation of all planned interventions in the project document as well as instituting elements of sustainable implementation of introduced interventions in the respective districts. These are expected to be the major measurable outcomes of the capacity building interventions. The model draws the link between implementation of capacity building activities and project monitoring and evaluation as suggested by other scientists [7] [13]. It is envisaged that the facilitation process by different actors will use input. It is therefore expected that input indicators shall be used on the M&E section of the project to monitor and evaluate the efficiency and effectiveness of resource utilization by the project.

The model realize further that the planned capacity building activities for various actors shall involve a number of processes and yield some short term deliverables as outputs. The responsibility of the M&E section of the project in this aspect shall be to make use of process and output indicators to measure efficiency and effectiveness of the project in realizing tangible results from the processes. Of equal importance is the need for M&E section of the project to measure the magnitude at which the target population has benefited from the capacity building efforts. This can only be measured through outcome mapping as indicated in the conceptual framework of the model (**Figure 1**).

3.3. Operational Phase

The operational phase involves developing a capacity development program to bridge the identified gaps in terms of knowledge, skills and other capacities. The program should have both short term and long term capacity building interventions. In the process of developing the program, there developers need to translate the concepts and processed summarized the conceptual model (**Figure 1**) into operational objectives and capacity building activities. In the text box bellow (**Figure 2**) an example of the transformation of the conceptual model to operational objectives and activities is provided.

In cases where capacity building involves trainings, key considerations in the capacity building program

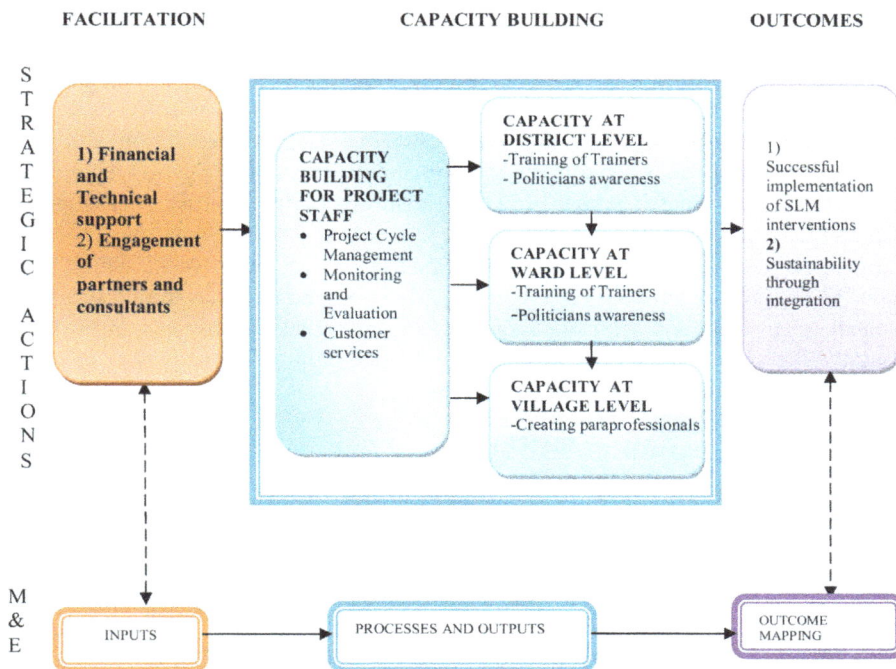

Figure 1. Conceptual framework of the strategic capacity development model (SCDM).

Text Box 1: An example of strategic capacity building objectives and activities derived from the conceptual model for purposes of developing capacity building programs

Strategic Capacity Building Objectives (SCBO) and activities:

SCBO-1. Resources for facilitating capacity building acquired

Strategic activities:

➤ Acquire financial resources
➤ Acquire Human resources
➤ Acquire Other resources

SCBO-2. Capacity building to specified stakeholders conducted

Strategic activities:

➤ Capacity building for Project staff conducted
➤ Capacity building for district staff conducted
➤ Capacity building for Ward and village levels conducted

SCBO-3: The contribution of capacity building to Project M&E Framework established
Strategic activities:

➤ Inputs and processes monitored
➤ Outputs monitored
➤ Outcome mapping conducted

Figure 2. Strategic objectives and strategies derived from the model.

should include identification of thematic training areas, targeted trainees, duration and the proposed time of the year when it should be executed; and whether to start trainings at the beginning of the project implementation phase in order not to frustrate implementation of project activities. As suggested in literature [14] the capacity development program should aim to strike a balance between various key aspects such as technical analysis, institutional development, policy reviews/formation and conflict management.

The possibility of collaborating with other actors in pursuing SLM capacity building interventions need to be examined. The other actors in this context are like-minded organizations such as NGOs, FBOs, the private sector and other government departments whose activities complement SLM interventions in the area. Therefore, the assessment need to be done in every district based on thematic areas (e.g. irrigation, energy, agriculture etc.) and the findings summarized in a matrix which shows the identified partner, partners activities and areas where collaboration can be focused.

The operational phase should also examined additional capacities at all levels which have a bearing on implementation of SLM capacity building interventions. It is important to assess these capacities in order to inform the project management team and sponsors about factors other than capacity building which can affect performance. Among the factors which need to be examined include issues like availability of reliable transport, computers facilities and accessories, and various equipment.

4. Conclusions and Recommendations

The paper has demonstrated that within the context of the local government structure in Tanzania it is possible to plan and execute capacity building interventions guided by a locally developed framework. The main features of the model can be summarized as participatory capacity assessment, strategic capacity building program development and complementarity during the implementation processes through synergy building with like-minded stakeholders.

The model is based on the philosophy of sustainability through integration into existing governance structures. As such the model largely depends on the existing local government staff within the established structures. Where capacities are low the model proposes capacity building for local government staff to be part of the package. This is because without the required capacities for example at the district and ward levels service delivery is likely to be ineffective.

The paper recommends that development agents, practitioners, local government authorities and other capacity builders should test the presented model in their local situations to generate more lessons of experience which are necessary in realizing the benefits of the model.

Acknowledgements

A word of appreciation goes to UNDP Tanzania Office and Farm Africa for giving us the opportunity to test the model in Kilimanjaro and Manyara Regions in Tanzania respectively. We are thankful to local government staff from the two regions for the cooperation which was rendered to the team during the assessments. The participation of community representatives in the assessments is also highly appreciated.

References

[1]   Kay, M., Franks, T. and Tato, S. (2004) Capacity Needs Assessment Methodology and Processes. FAO.

[2]   Stephen, P. and Triraganon, R. (2009) Strengthening Voices for Better Choices: A Capacity Needs Assessment Process. IUCN, Gland, Switzerland.

[3]   UNDP (1997) Capacity Development. Technical Advisory Paper 2, Management, Development and Governance Division, UNDP, New York.

[4]   Colville, J. (2008) Capacity Building Note. UNDP, USA.

[5]   Wignaraja, K. (2009) Capacity Development: A UNDP Primer. UNDP, USA.

[6]   UNDP (2007) Capacity Assessment Methodology. Users Guide. Capacity Development Group. Bureau for Development Policy, UNDP.

[7]   Karen, E., Kahana, L. and Kajembe, G. (2012) REDD+ Capacity Needs Assessment in Tanzania: A Policy Brief. Ministry of Natural Resources and Tourism and UN-REDD Program in Tanzania, Dar es Salaam.

[8]   Njunwa, M. (2005) Local Government Structures for Strengthening Society Harmony in Tanzania: Some Lessons for Reflection. *Proceedings of NAPSIPAG Annual Conference*, Beijing, 5-7 December 2005.

[9]   The Local Government System in Tanzania.
       http://www.tampere.fi/tiedostot/5nCY6QHaV/kuntajarjestelma_tansania_.pdf

[10] URT (1998) Policy Paper on Local Government Reform. Local Government Reform Program. Ministry of Regional Administration and Local Government. United Republic of Tanzania, Dar es Salaam.

[11] Kessy, J.F. (2013) Assessment of Training and Staffing Needs for Mainstreaming SLM Interventions in Kilimanjaro Region, Tanzania. Report Submitted to UNDP, Dar es Salaam.

[12] Kessy, J.F. and Kaswamila, A. (2013) Training Needs Assessment, Strategy and Program for Sustainable Nou Forest Ecosystem Management Project. Report Submitted to Farm Africa. Dar es Salaam and Manyara, Tanzania.

[13] Inna, B. (2009) Monitoring and Evaluation Capacity Building Needs Assessment. European Union and DMI Associates Consortium, Ukraine.

[14] Naya, S.P., Ojha, H. and Rana, S. (2010) Capacity Building Needs Assessment and Training Strategies for Grassroots REDD Stakeholders in Nepal. The Centre for People and Forests, Bangok.

21

The Effect of Clay Content and Land Use on Dispersion Ratio at Different Locations in Sulaimani Governorate—Kurdistan Region—Iraq

Saman M. K. Rasheed

Department of Soil and Water Sciences, Faculty of Agricultural Sciences, Sulaimani University, Sulaimani, Iraq
Email: saman.karim@univsul.edu.iq

Abstract

Land use changes from natural ecosystems into managed ecosystems resulted in negative impact on soil structure and quality. The purpose of this study was to determine the influences of different land-use types on physical and chemical properties of soils in Sulaimani governorate. Land use systems including natural forest, pastureland and agriculture were identified. Ten of soil samples were collected from the 0 - 30 cm depth, and some soil physical and chemical properties of soil were determined. The land use alters from forest to agriculture resulting in significant decrease in organic matter, calcium carbonate and soil surface area and with this change, dispersion ratio affected on the physical property. The value of DR was highest in the Zrguezi Gawra cultivated with Cucumber and the lowest value in Dukan is 13%, and correlation coefficient between dispersion DR with sand, silt and bulk density is positive, value is (0.4979, 0.0126 and 0.7536) respectively, and with clay and specific surface area (SSA) the correlation coefficient value is (−0.7281 and −0.4466).

Keywords

Particle Size Distribution, Dispersion Ratio, Bulk Density, Organic Matter

1. Introduction

The lack of success of dams and embankments as a result of the use of dispersive soil has been recognized by engineers and geologists in South Africa and internationally for many years. A serious problem, however, yet lies in the early identification of dispersive soils [1]. In spite of all the knowledge obtained over the years, there is still no fast, simple and dependable mean of conclusively recognizing dispersive soils. Many methods have been suggested including the pinhole, double hydrometer, crumb and chemical tests, individually or in group. These, but, have not always been totally credible and it looks like that the cause sets in the factual testing procedures. The standard testing procedures for the Soil Conservation Service (SCS) double hydrometer test, the pinhole test, crumb test and chemical analyses have lately been studied and problems/abnormalities identified. Although no discussion concerning these anomalies has been discovered in the literature, the recent testing proposes that many of these deficiencies may have been looked out during past routine investigations. This paper summarizes a proportional study including the testing of three specimen utilizing one standard laboratory test, namely the SCS double hydrometer test and discusses some potential means of conquering the problems identified. The other tests have been discussed separately [2]-[5].

SCS double hydrometer or dispersion soil test has been identified as one of the most suitable tests for classifying dispersive soils. The test assesses the dispersibility of a soil by measuring the natural propensity of the clay fraction to go into annotate in water. The procedure involves the estimation of the percentage of particles in the soil that are finer than 0.005 mm utilizing the standard hydrometer test. A parallel test is also carried out, in which no chemical dispersant is added and the solution is not mechanically excited. The quantity of particles finer than 0.005 mm in the parallel test is explicit as a percentage of this fraction determined in the standard test, which is defined as the dispersion ratio or dispersivity of the soil [6]. Dispersion ratios greater than 50% are observed extremely dispersive, between 30% and 50% are moderately dispersive, between 15% and 30% are a little dispersive and less than 15% are non-dispersive [7]. Similar systems with various limits were used by [1]-[19], and the dispersion test was first described by Volk (1937) [19] as a means of determining the potential dispersive of soils. The test has since been utilized extensively in this concern with minor modifications. Volk's test contrasts the weight of soil grains, 0.005 mm or smaller that slaked free when air-dried lumps of soil were soaked in silent distilled water with that of the complete soil. This was shown as the percentage dispersion. Measurements of the clay in the soil-water suspension were made by the pipette method [19].

The pipette method itself was announced by Middleton in 1930 as a means of determining the erosion potential of a soil. The difference in the methods was that the samples collected in the pipette depend on particles of a maximum diameter of 0.05 mm [16]. Volk's method was, however, chosen because studies in the southwest of the United States point out that the dispersion of the clay fraction (<0.005 mm diameter) was more significant in assessing the piping potential of soils [20]. The SCS involvement in earth dam construction in the United States reproduces in the 1940's and early 1950's. The dispersion test was used as a routine procedure for all samples submitted to SCS Laboratories. It was during this time that the test procedure was appropriate to utilize a hydrometer to instate a pipette [20].

2. Materials Methods

This study was performed in Sulaimani Governorate to determine the effect of ten land use and clay content on dispersion ratio. The soil samples from 0 - 30 cm depth were collected from different locations. All soil samples were air dried, sieved through a 2 mm physical properties were measured such as particle size distribution by the pipette method, bulk density by Clod method. A single point using water under isothermal condition was measured the (SSA) according to [17]. Chemical properties were measured like soil organic matter which determined by the Walkley-Black method (1934) [21] and calcium carbonate by titration method.

3. Micro Aggregate Stability

This was calculated from the amount of silt and clay in calgon-dispersed as well as water-dispersed samples using Bouyoucos hydrometer method of particle size described by [22]

The clay-dispersion indices were calculated as follows:

$$\text{Dispersion ratio}(\text{DR}) = \frac{(\text{silt} + \text{clay})\ \% \text{ for the soil dispersed in water}}{(\text{Silt} + \text{clay})\% \text{ for the soil dispersed by using Mechanical and chemical dispersion}}$$

Clay dispersion index:

$$(CDI)\frac{\% \text{ Clay H}_2\text{O}}{\% \text{ Clay}(\text{Calgon})} \times 100$$

4. Statistical Analysis

Correlation and regression were used to analyze the relationship between land use and some soil parameters of the samples (**Figure 1**). Analysis of variance (ANOVA) for complete randomized design (CRD) was used to compare the influence of the land usage types on the measured soil properties. DUNCAN at P < 0.5 level was used to separate the mean where applicable.

5. Results and Discussions

The soil texture of the study area was significantly affected by land use. Data obtained in **Table 1** showed the variation in sand, silt and clay fraction of the soil. The result shows that the textural class across all the land use types of the study area is different soil texture, indicating the different of parent material. The result is in agreement with [23]. However, over a very long period of time, paedogenesis processes such as erosion, deposition, eluviations, and weathering can change the soil texture. Data in **Table 1** relationship between clay dispersion ratio and dispersion ratio the highest value of clay dispersion ratio is 83.33% in Zrguezi Gawra (cultivated with Cucumber) and the lowest value is 11.29% in Azmar (Forest soil) and dispersion ratio the highest value is 40% in Zrguezi Gawra (cultivated with Cucumber) and the lowest value is 13% in Dukan (Bare soil).

Data in **Table 2** bulk density was highly impact by land use, the highest value in Zrguezi Gawra (cultivated with cucumber) is 1.73 g/cm^3 and lowest value in Azmar (forest soil) is 1.29 gm/cm^3. This was in agreement with works of [15] [18], who supposed that the significant difference is caused differences in the land management and land use history, and other reason may be due to the high content of OM which reduced bulk density. Loss of organic matter by transformation of the natural forest in to pasture and cultivated land caused a higher bulk density in cultivated soils. The greatest soil bulk density was observed in cultivated and followed by pasture and forest land, similar result were reported by [12].

Organic matter was highly affected by land use. The highest value in Azmar is 3.2% in this case deepened on land use (Forest soil) and the lowest value in Dukan, depended on land use (bare soil), may be as a result of

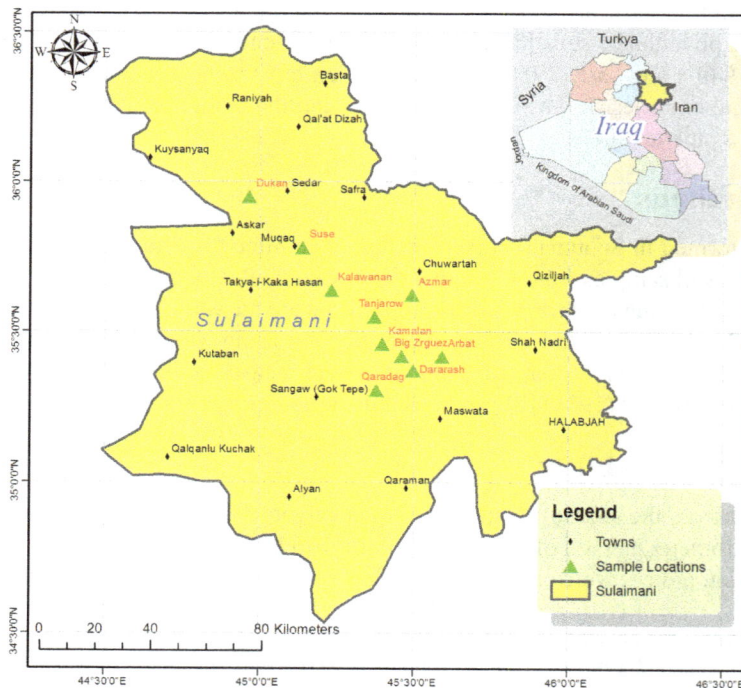

Figure 1. Site of soil samples.

Table 1. Effect of Land Use on particles size distribution of Soil, Clay dispersion index and dispersion ratio in Sulaimani Governorate.

Land Use Type	Sand (%)	Silt (%)	Clay (%)	Soil texture	Clay dispersion index (CDI) %	Dispersion ratio (%)
Arbat (Cultivated with Celery)	7	41	52	Silty Clay	23.07	24
Azmar (Forest Soil)	7	31	62	Clay	11.29	16
Zrguezi Gawra (Cultivated with Cucumber))	55	27	18	Sandy Loam	83.33	40
Dararash (Cultivated with Melon)	47	24	29	Sandy Clay Loam	34.48	38
Dukan (Bare Soil)	11	49	40	Silty Clay	12.5	13
Kalawanan Cultivated with Water Melon)	31	38	31	Clay Loam	25.8	25
Kamalan (Cultivated with Wheat)	55	10	35	Sandy clay	14.28	18
Qaradag (Cultivated with Barley)	17	50	33	Silty Clay Loam	33.33	28
Suse Cultivated with Tomato)	25	49	26	Loam	34.61	35
Tanjarow (Cultivated with Eggplant)	20	54	26	Silt Loam	46.15	30

Table 2. Effect of Land Use on bulk density, organic matter, calcium carbonate and specific surface area in Sulaimani governorate.

Land Use Type	bulk density (gm/cm^3)	Organic matter (%)	Calcium carbonate (%)	Specific Surface Area (m^2/gm)
Arbat (Cultivated with Celery)	1.33	2.3	24.39	130.80
Azmar (Forest Soil)	1.29	3.2	13.91	150.33
Zrguezi Gawra (Cultivated with Cucumber))	1.73	1.2	21.35	75.26
Dararash (Cultivated with Melon)	1.45	1.8	27.35	98.32
Dukan (Bare Soil)	1.35	1.1	18.35	73.55
Kalawanan Cultivated with Water Melon)	1.63	1.5	32.35	100.18
Kamalan (Cultivated with Wheat)	1.38	2.5	29.35	142.23
Qaradag (Cultivated with Barley)	1.53	1.9	20.35	122.07
Suse Cultivated with Tomato)	1.67	1.3	27.35	95.75
Tanjarow (Cultivated with Eggplant)	1.55	1.7	35.39	97.22

plant litter sufficient returned to the soil in these environments.

Calcium carbonate was highly affected by land use, the highest value in Tanjarow (35.39%) and the lowest value in Azmar (13.91), This was in agreement with works of [13].

Soil specific surface area was highly affected by land use, the highest value in Azmar (forest soil) (150.33) and the lowest value in Dukan (Bare soil) (73.55).

Table 3 shows the relationship between dispersion ratio and selected soil physical and chemical properties correlation model determination. There is a positive correlation between dispersion ratios with clay, the result is in disagreement with [9] [10] [14], with bulk density and organic matter was positively correlated, the result is in agreement with [11]. A negative correlation also estimates between dispersion ratio with sand, silt, specific surface area and calcium carbonate.

In this table showed the clay, soil bulk density and organic matter have a significantly on the dispersion ratio value is (0.0170, 0.0118 and 0.0130) respectively.

Figures 2-9 showed the relationship between dispersion ratio and selected soil physical and chemical properties parameter, regression model and coefficient of determinations. There is a positive correlation between dispersion ratio with sand, bulk density, CaCO^3 and CDI coefficient of determination value is (0.247, 0.568, 0.114

Table 3. Correlation between dispersion ratio with soil physical and chemical properties.

Variable	By Variable	Correlation Coefficient	Significant Probability
Dispersion ratio	Sand	0.4979	0.1430
Dispersion ratio	Silt	0.0126	0.9724
Dispersion ratio	Clay	−0.7281	0.0170[*]
Dispersion ratio	Soil bulk density	0.7536	0.0118[*]
Dispersion ratio	Soil Specific Surface Area	−0.4466	0.1957
Dispersion ratio	Organic Matter	−0.7472	0.0130[*]
Dispersion ratio	Calcium Carbonate	0.3381	0.3393

[*]Significant at $p \leq 0.05$.

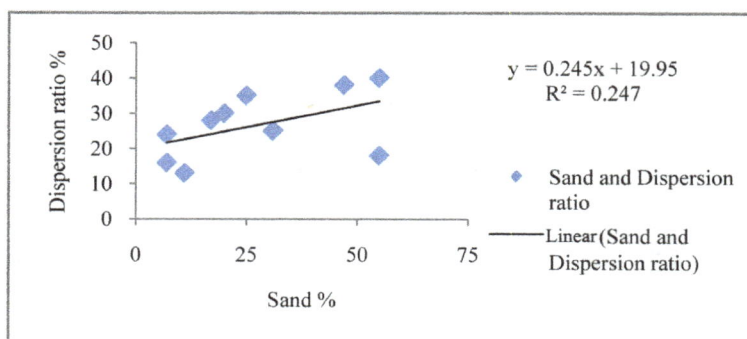

Figure 2. Effect of sand particle on dispersion ratio.

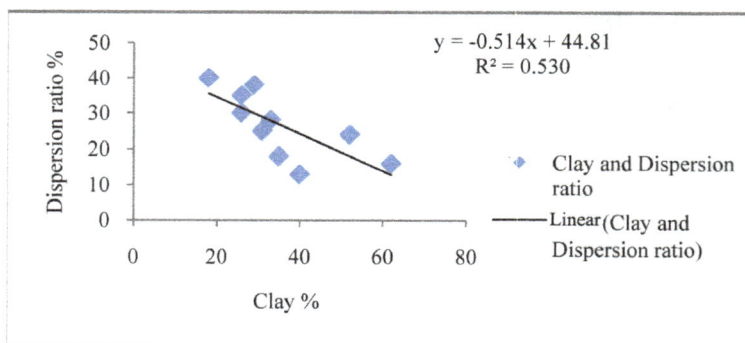

Figure 3. Effect of clay particle on dispersion ratio.

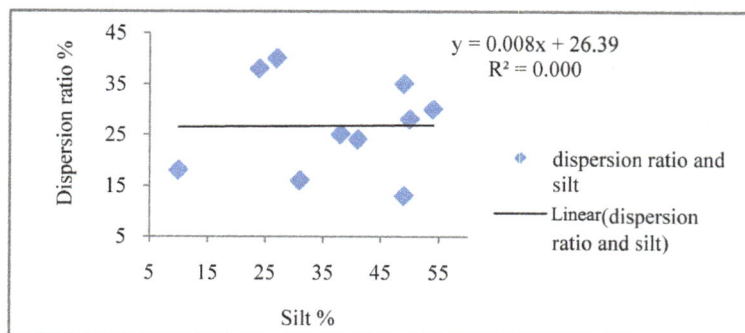

Figure 4. Effect of silt particle on dispersion ratio.

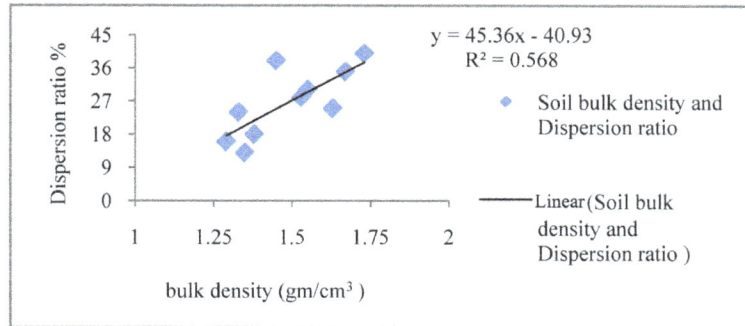

Figure 5. Effect of bulk density on dispersion ratio.

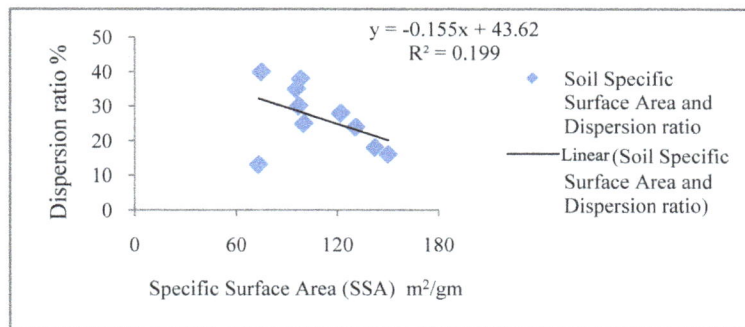

Figure 6. Effect of specific surface area on dispersion ratio.

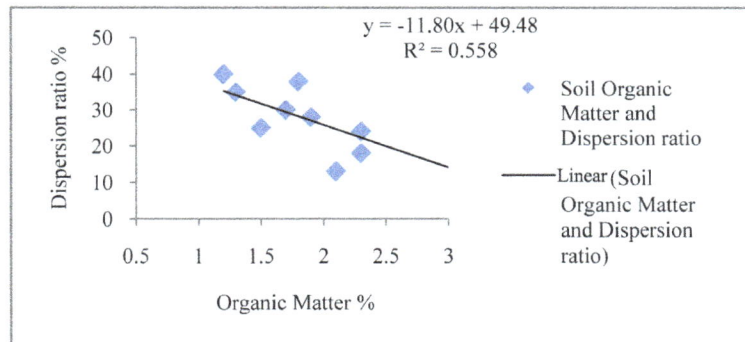

Figure 7. Effect of soil organic matter on dispersion ratio.

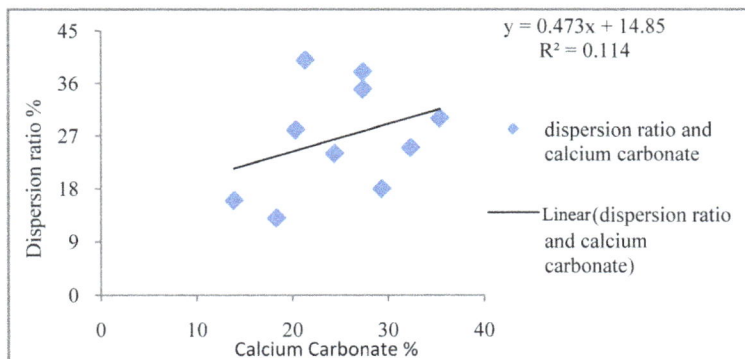

Figure 8. Effect of calcium carbonate on dispersion ratio.

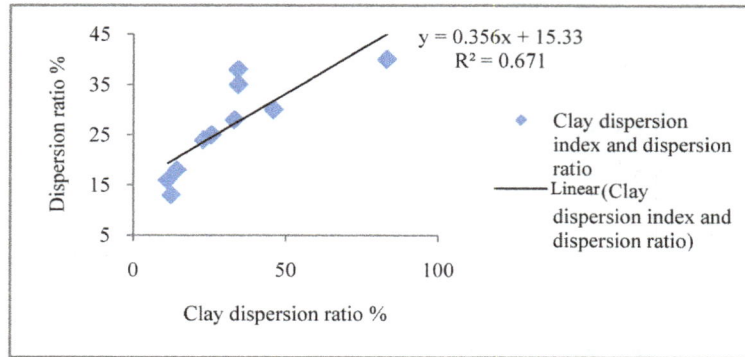

Figure 9. Effect of clay dispersion ratio on dispersion ratio.

and 0.671) respectively, and dispersion ratio negatively regression with clay, SSA and organic matter coefficient of determination value is (0.530, 0.199 and 0.558).

6. Conclusion

Variations in soil quality indicators with regard to land use were investigated in Sulaimani city, Kurdistan Region, Iraq. The most limiting soil properties impact quality and have significant effects on land use in the study area like soil organic matter, soil bulk density, soil texture and calcium carbonate. Among soil physical quality indicators, specific surface area and bulk density changed significantly with Land use and also among soil chemical quality indicators, soil organic matter and calcium carbonate varied significantly with land use.

References

[1]   Paige-Green, P. (2008) Dispersive and Erodible Soils—Fundamental Differences. *SAIEG/SAICE Problem Soils Conference*, Midrand, November 2008, 59-67.

[2]   Maharaj, A and Paige-Green, P. (2010) The Impact of Inconsistencies in the Interpretation of Soil Test Results on the Repeatable Identification of Dispersive Soils. *Proceedings IAEG2010*, Auckland, September 2010, 389-392.

[3]   Maharaj, A. (2010) Preliminary Observations of Shortcomings Identified in Standard Tests for Dispersive Soils. *Proceedings IAEG2010*, Auckland, September 2010.

[4]   Maharaj, A. (2011) the Use of the Crumb Test as a Preliminary Indicator of Dispersive Soils. *Proceedings of the 15th African Regional Conference on Soil Mechanics and Geotechnical Engineering*, Maputo.

[5]   Maharaj, A. (2012) Problems Associated with the Chemical Analysis If Dispersive Soils. *Proceedings of the 2nd European Conference on Unsaturated Soils (E-UNSAT 2012)*, Naples, 1-3 June 2012, 433-437.

[6]   Walker, D.J.H. (1997) Dispersive Soils in KwaZulu-Natal. Unpublished M.Sc. Thesis, University of Natal, Durban.

[7]   Elges, H.F.W.K. (1985) Problem Soils in South Africa—State of the Art. *The Civil Engineer in South Africa*, **27**, 347-353.

[8]   Gerber, F.A. and Harmse, H.J. von M. (1987) Proposed procedure for identification of dispersive soils by chemical testing. *The Civil Engineer in South Africa*, **29**, 397-399.

[9]   Gu, B. and Doner, H.E. (1993) Dispersion and Aggregation of Soils as Influence by Organic and Inorganic Polymers. *Soil Science Society of America Journal*, **57**, 709-716. http://dx.doi.org/10.2136/sssaj1993.03615995005700030014x

[10] Heil, D. and Sposito, G. (1993) Organic Matter Role in Illitic Soil Colloids Flocculation: I. Counter Ions and pH. *Soil Science Society of America Journal*, **57**, 1241-1246. http://dx.doi.org/10.2136/sssaj1993.03615995005700050014x

[11] Igwe, C.A. (2001) Clay Dispersion of Selected Aeolian Soils of Northern Nigeria in Relation to Sodicity and Organic Carbon. *Arid Land Research and Management*, **15**, 147-155. http://dx.doi.org/10.1080/15324980151062788

[12] Islam, K.R. and Weil, R.R. (2000) Land Use Effects on Soil Quality in a Tropical Forest Ecosystem of Bangladesh. *Agriculture, Ecosystem & Environment*, **79**, 9-16. http://dx.doi.org/10.1016/S0167-8809(99)00145-0

[13] Lahore Pour, Sh. (2004) The Effect of Grazing Intensity on the Physical and Chemical Properties of Soil in Charand-summer Pastures in Kurdistan. M.Sc. Thesis, Range Management, Tarbiat Modares University, Tehran, 68.

[14] Lebron, I. and Suarez, D.L. (1992) Variation in Soil Stability within and among Soil Types. *Soil Science Society of America Journal*, **56**, 1412-1421. http://dx.doi.org/10.2136/sssaj1992.03615995005600050014x

[15] Lemenih, M., Karltun, E. and Olsson, M. (2005) Assessing Soil Chemical and Physical Property Responses to Deforestation and Subsequent Cultivation in Small Holders Farming System in Ethiopia. *Agriculture, Ecosystems & Environment*, **105**, 373-386. http://dx.doi.org/10.1016/j.agee.2004.01.046

[16] Middleton, H.E. (1930) The Properties of Soils Which Influence Erosion. *US Department of Agriculture Technical Bulletin*, **178**, 1-16. http://dx.doi.org/10.2136/sssaj1930.036159950b1120010021x

[17] Puri, B.R. and Murarri, K. (1963) Studiers in Surface Area Measurement of Soil. Surface Area from a Single Point on the Water Isotherm. *Soil Science Journal*, **96**, 341-343.

[18] Sintayehu, M. (2006) Land Use Dynamics and its Impact on Selected Physicochemical Properties of Soils in Yabello-Woreda of Borana Lowlands, Southern Ethiopia. M.S. Thesis, Haromaya University, Haromaya.

[19] Volk, G.M. (1937) Method of Determination of Degree of Dispersion of the Clay Fraction of Soils. *Proceedings of Soil Science Society of America*, **2**, 561-567. http://dx.doi.org/10.2136/sssaj1938.036159950002000C0088x

[20] Decker, R.S. and Dunnigan, L.P. (1977) Development and Use of the Soil Conservation Service Dispersion Test. In: Sherard, J.L. and Decker, R.S., Eds., *Proceedings of Symposium on Dispersive Clays, Related Piping and Erosion in Geotechnical Projects*, ASTM Special Publication, Vol. 623, 94-109. http://dx.doi.org/10.1520/stp26982s

[21] Walkley, A. and Black, C.A. (1934) An Examination of the Degtjareff Method of Determining Soil Organic Matter and a Proposed Modification of the Chronic Acid Titration Method. *Soil Science*, 37-38.

[22] Gee, G.N. and Bauder, J.W. (1986) Particle Size Distribution. In: Klute, A., Ed., *Methods of Soil Analysis Part* 1. *Physical and Mineralogical Methods*, 2nd Edition, Agronomy Society of America/Soil Science Society of America, Madison, Wisconsin, 383-411.

[23] Abera, Y. and Belachew, T. (2011) Effects of Landuse on Soil Organic Carbon and Nitrogen in Soils of Bale, Southeastern Ethiopia. *Tropical and Subtropical Agroecosystems*, **14**, 229-235.

22

Tile Drainage and Nitrogen Fertilizer Management Influences on Nitrogen Availability, Losses, and Crop Yields

Rakesh Awale[1], Amitava Chatterjee[1], Hans Kandel[2], Joel K. Ransom[2]

[1]Soil Science, North Dakota State University, Fargo, ND, USA
[2]Plant Sciences, North Dakota State University, Fargo, ND, USA
Email: rakesh.awale@ndsu.edu

Abstract

Installing tile drainage facilitates early planting and field operations, and tiling has tremendously increased in the Red River Valley (RRV) due to recent wet cycles. This experiment studied tile drainage and N-fertilizer management effects on N availability, N losses, and yields of corn (*Zea mays* L.) and sugarbeet (*Beta vulgaris* L.) in a naturally poorly-drained Fargo soil during the 2012-2013 growing seasons. Regardless of drainage, applying 146 kg N ha^{-1} with nitrapyrin resulted similar soil N availability to 180 kg N ha^{-1} without nitrapyrin in sugarbeet in both years. In corn, application of nitrapyrin resulted either higher or similar soil N levels to split-N application. In 2013, application of urea alone increased soil N availability during the early corn growing season under the undrained condition, whereas nitrapyrin delayed N release in the tile-drained soils. Corn and sugarbeet yields averaged 7.4 and 47.0 Mg·ha^{-1} in 2012, and averaged 8.3 and 38.3 Mg·ha^{-1} in 2013, respectively, with no significant differences among N-sources. However, corn yield increased on an average by 27.6% with N-fertilizer application over unfertilized control in 2013. In 2012, sugarbeet root impurity (% sucrose loss to molasses) increased by 13.8% and 17.2% with 146 kg N ha^{-1} plus nitrapyrin and 180 kg N ha^{-1} treatments, respectively, compared to unfertilized control. Besides, higher N rates were usually associated with greater daily soil N$_2$O emissions, with the maximum flux of 105 g N$_2$O-N ha^{-1}·d^{-1} recorded under corn. Addition of fertilizer-N increased NH$_3$ volatilization losses up to 1.9% and 0.5% of the applied-N in corn and sugarbeet, respectively. Tile drainage influenced soil N availability more than crop yield during two years of study. Nitrogen management can have pronounced effects on N availability and losses. A long-term study is needed to investigate the fertilizer-N use efficiency of crops under tile drainage condition.

Keywords

Subsurface Drainage, Nitrogen, Nitrous Oxide, Ammonia Volatilization, Nitrate Leaching

1. Introduction

About 1.8 million hectares of soils in the RRV of North Dakota and northern Minnesota are naturally poorly drained [1], which has led to increased adoption of subsurface tile drainage [2]. Tile drainage can provide agronomic benefits through the gravimetric water removal, improved trafficability, and timeliness of field operations [3]. However, shifting water regimes influences below ground N dynamics through changes in soil moisture and temperature [4].

Soil water content has significant influence on N mineralization rates [5]. Tile drainage can accelerate soil N mineralization by improving soil aeration [6]. Optimum N rates for crop production can vary greatly among years based on N mineralization rates and possible N losses [7]. Fertilizer-N recommendations are usually based on the amount of N available at the beginning of growing season. However, in soils with high organic matter, a substantial amount of soil N mineralization is deemed possible during the crop growing season, which is less considered by the growers [8]. Therefore, monitoring mineral N contents in soils during the growing season can improve our knowledge on N availability to crops.

Denitrification losses of applied N increase with soil water content because of root-zone oxygen reduction through displacement of soil air by water [9]. Installing tile drainage can eliminate the chance of water logging by lowering the water table, and thereby, the denitrification N losses [10]. Apart from soil aeration, the availability of inorganic N substrates is also an important factor that controls the dynamics of soil nitrous oxide (N_2O) emissions [11]. Denitrification loss of N_2O from agricultural soils increases with N-application rates [12]. Large pulses of N_2O productions at times immediately following fertilizer N application and/or heavy rainfall events are not uncommon [9]-[12].

Tile drainage has potential to reduce denitrification loss of N_2O. However, nitrate (NO_3^-) leaching could be a major concern in tile-drained soils [13]. Annual rainfall, crop yield variations, and soil and nutrient management, in part or combination, govern NO_3^- leaching loss [14]. Greater N losses to subsurface drains are observed in years following a drought due to greater residual N in the soil profile [3] [13].

Tile drainage can also influence ammonia (NH_3) volatilization by changing soil water and temperature regimes, and/or influencing N substrate availability; however, the extent of control mechanism is still not understood [10]. Ammonia volatilization is a major N loss pathways associated with the use of surface applied NH_4^+ based fertilizers [12]. Soil water regimes have a strong impact on the NH_4^+ transport through profile and subsequently on the availability of NH_4^+ substrate for NH_3 volatilization [15]. Using tile drainage to lower soil water table provides surface soil water with more volume to percolate down the soil profile that can incorporate fertilizer N into the soil, resulting in low NH_3 volatilization [16]. Conversely, tile drainage may increase NH_3 volatilization by maintaining the soil water content at field capacity [17].

Soil N supply is essential for crop yield and quality [18]. Insufficient N limits crop yield, but excessive amounts of N can deteriorate air and water quality [12]-[15]. Application of fertilizer-N has generally shown to increase crop yields but it also increases N losses via N_2O emissions, NO_3 leaching, and NH_3 volatilization, especially when fertilizer-N is applied in excess of crop requirements [12]. Application of commercial nitrification inhibitor compounds such as nitrapyrin (NP) delay NH_4^+ oxidation to NO_3^-, and increase N availability to the crops during the periods of rapid crop N uptake [19]. Application of N fertilizers in split doses could be another approach to co-ordinate maximum crop N uptake [18]. Reducing the availability of inorganic N in soils when crop N requirement is small, N-inhibitors and split N application have the potential to increase N use efficiency and yields of crops, while minimizing potential N losses [8] [12] [14] [20].

Balancing the amount of N required for optimum crop growth while minimizing potential N losses under tile drainage conditions is a major challenge in the RRV. Simultaneous measurements of crop yields, N availability, and N losses as influenced by interaction of N management and tile drainage have not been intensively studied in the RRV. We hypothesize that crop yields would be increased under tile drainage condition due to increase in the N availability under favorable soil water level. Main objectives of this field experiment were to determine tile drainage and fertilizer-N management practices effect on (i) corn and sugarbeet yields, (ii) changes in the soil N availability (iii) denitrification loss of N in the form of N_2O, (iv) soil solution NO_3^- concentration at 60 cm soil depth, and (v) NH_3 volatilization loss of N in a poorly drained Fargo silty clay soil.

2. Materials and Methods

2.1. Description of Experimental Site

Field experiments were located at North Dakota State University research site (46.93°N, 96.85°W) near Fargo, North Dakota, USA. Broadly, the site has Fargo silty clay soil and is classified as Fine, smectitic, frigid Typic Epiaquerts with 0% - 1% slope [21]. Subsurface drainage tiles were installed at the 2.5 ha experimental area in 2008. The area was divided into eight units of 61 m long by 54 m wide, each unitconsisted of seven lateral sub-surface drainage tile lines. Tile lines of 10 cm in diameter were installed at a depth of 90 cm with a spacing of 7.6 m, and with a drainage coefficient of 7.5 mm·d^{-1}.

Each unit was controlled via a water table control structure (Agri-Drain Corp, Adair, IA, USA). Four of the units had the control structures open to represent subsurface drainage and the remaining four units had the control structures closed to represent undrained field conditions.

2.2. Field Experiments and Experimental Design

At the experimental site, corn and sugarbeet were grown during the 2012-2013 growing seasons. A randomized complete block design was used with four replicates in a split-plot arrangement with drainage (undrained and tile-drained) as the main plot factors and N fertilizer management as the sub plot factors in each crop for both years. For corn, the N treatments in 2012 were (i) 180 kg N ha^{-1} applied at preplant as urea (Urea180), (ii) 224 kg N ha^{-1} at preplant as urea (Urea224), (iii) 224 kg N ha^{-1} at preplant as urea plus nitrapyrin (Urea224 + NP), (iv) 112 kg N ha^{-1} at preplant as urea plus 112 kg N ha^{-1} at 6 leaf stage as urea ammonium nitrate (UAN) [Split (Urea112 + UAN112)]. The N-rates of 180 kg N ha^{-1} and 224 kg N ha^{-1} were selected on the basis of corn yield goals of 8.4 and 10.4 Mg·ha^{-1}, respectively, within this region [22]. In 2013, the N-rate was lowered to 134 kg N ha^{-1} considering residual soil N from fall soil test results, soybean N credits and corn N requirement for the yield goal of 10.4 Mg·ha^{-1}. The N treatments for corn in 2013 were (i) control (0 N), (ii) 134 kg N ha^{-1} at preplant as urea (Urea224), (iii) 134 kg N ha^{-1} at preplant as urea plus nitrapyrin (Urea134 + NP), (iv) 67 kg N ha^{-1} at preplant as urea plus 67 kg N ha^{-1} at 6 leaf stage as UAN [Split (Urea67 + UAN67)] in 2013. For sugarbeet, the N management included (i) control (0 N), (ii) 146 kg N ha^{-1} at preplant as urea (Urea146), (iii) 146 kg N ha^{-1} at preplant as urea plus nitrapyrin (Urea146 + NP), and (iv) 180 kg N ha^{-1} at preplant as urea (Urea180) in both years. The N-rate of 146 kg·ha^{-1} is considered as the recommended N-rate for sugarbeet production within this region [22].

2.3. Field Operations

During 2012 and 2013 growing seasons, corn was planted in the experimental field sections that were under soybean in the previous years. The plots were cultivated to a depth of 7.6 cm once in the fall following the harvest of soybean, and again in the spring before planting corn with a one-pass field cultivator. The preplant urea fertilizers were uniformly broadcasted by hand and incorporated immediately with the field cultivator on April 26 in 2012 and on May 15 in 2013. The nitrification inhibitor nitrapyrin (trade name Instinct, 17.6% by weight active ingredient solution, DOW AgroSciences LLC, IN, USA) was mixed with urea and applied to the soil, based on an area basis at the rate of 2.5 L·ha^{-1} (450 g a.i. ha^{-1}). Corn hybrid PH-8640 RIB (DuPont Pioneer, IA, USA) was planted on April 30 in 2012 and on May 15 in 2013 at the rate of 88000 seeds ha^{-1} with a 1010 John Deere seed planter (John Deere, Moline, IL, USA). The seeds were placed 3.8 cm deep with 15 cm in-row (seed to seed) spacing and 76 cm between-rows (row to row) spacing, such that four corn rows were included in an individual corn plot size of 6.1 m long by 3.0 m wide. For the split N treatments, the UAN was side-dress applied on June 4 in 2012 and on June 18 in 2013. Crops were grown under rainfed condition without any irrigation water inputs. Weeds were controlled using herbicides—applied twice during each corn growing seasons. The first application consisted of amixture of glyphosate (Monsanto, St. Louis, MO, USA) at 1.5 L·ha^{-1} + Status (dicamba and diflufenzopyr, and isoxadifen safener) (BASF, Research Triangle Park, NC, USA) at 13.1 g of a.i. ha^{-1} at V3 (three collar leaf) stage. The second application consisted of glyphosate at 1.5 L·ha^{-1} + Luadis (atrazine tembotrione and isoxadifen safener) (Bayer CropSciene LP, Research Triangle Park, NC, USA) at 96 ml of a.i. ha^{-1} at V7 (seven collar leaf) stage. All of the corn rows were machine harvested after physiological maturity on Oct. 9, and Oct. 24, respectively, in 2012 and 2013. Corn grain yields were determined at the moisture contents of 15.0%.

In 2012, sugarbeets were grown in the field sections previously under corn whereas in 2013, wheat preceded sugarbeets. The required rates of urea fertilizers were uniformly broadcasted by hand on May 10 and May 29, respectively in 2012 and 2013. Nitrapyrin was mixed with urea and applied to the soil, based on an area basis at the rate of 2.5 L·ha^{-1}. The fertilizers were then incorporated using a Triple K field cultivator with rolling basket. On the same day, sugarbeet variety Crystal 985 Roundup Ready (American Crystal Sugar Company, MN, USA) was planted with a John Deere Max Emerge II planter to an individual sub-plot size of 6.1 m long by 3.4 m wide. The seeds were placed 3.2 cm deep with 56 cm row spacing and 7.6 cm in-row spacing. The plots were thinned manually to maintain a plant population of 156500 plants ha^{-1} for the first year only. The beets were grown under rainfed condition without any irrigation water inputs. Glyphosate herbicide was applied on June 22 in 2012, and on July 6 in 2013 at 3.5 L·ha^{-1}. Two middle rows from each plot were machine harvested on Sept. 17 in 2012, whereas the beets were harvested manually (3.1 m long each from two middle rows) on Oct. 24 in 2013. The beets were weighed instantly (gross sugarbeet root yield) and subsamples of the sugarbeet roots were sent to American Crystal Sugar Quality Tare Lab, East Grand Forks, Minnesota, USA for yield determinations and quality analyses. From this data, net sugarbeet root yield was calculated after subtracting the external root impurities (tare %) from the gross sugarbeet root yield. The net sugarbeet root yield hereafter is referred to as sugarbeet (root) yield.

2.4. Basic Soil Properties

Before planting sugarbeets, three soil cores—up to a depth of 120 cm with depth intervals of 0 - 30, 30 - 60, and 60 - 120 cm—were collected using a truck mounted probe (3.6 cm internal diameter) and composited per replicate unit in order to determine initial soil inorganic N levels [23]. Also, separate soil cores (2 cm diameter) were taken from the upper 30 cm soil surface from each individual sugarbeet plots to determine bulk density [24], organic matter [25], soil pH and EC [26], cation exchange capacity [27], particle size analysis [28], Olsen-P [29], and available K [30]. The basic soil physical and chemical properties are presented in **Table 1**.

2.5. Growing Season Soil Inorganic N Content

After planting, soil samples were collected by hand using a soil probe (2 cm internal diameter) from the upper 30 cm soil profile—with 15 cm increments for both growing season in each crop. In 2012, soil samples were

Table 1. Basic soil characteristics in the surface 0 - 30 cm depth at the experimental site measured in sugarbeet plot during the 2012 and 2013 growing seasons.

Soil properties[†]	2012	2013
Sand (g·kg^{-1})	17 ± 3[‡]	17 ± 5
Silt (g·kg^{-1})	359 ± 47	374 ± 61
Clay (g·kg^{-1})	624 ± 89	609 ± 77
Bulk density (Mg·m^{-3})	1.22 ± 0.11	1.23 ± 0.06
pH	8.24 ± 0.27	8.32 ± 0.38
EC (dS·m^{-1})	1.59 ± 0.56	1.54 ± 0.62
Organic matter (g·kg^{-1})	71.8 ± 1.7	70.7 ± 2.3
NO$_3$-N, kg·ha^{-1} (0 - 30 cm)	36 ± 9	130 ± 31
NO$_3$-N, kg·ha^{-1} (30 - 60 cm)	13 ± 8	59 ± 22
NO$_3$-N, kg·ha^{-1} (60 - 120 cm)	44 ± 9	83 ± 28
Olsen-P (mg·kg^{-1})	14.5 ± 8.5	26.5 ± 10.2
Available-K (mg·kg^{-1})	333 ± 47	447 ± 45
Cation exchange capacity (cmol$_c$·kg^{-1})	NA[§]	37.1 ± 2.3

[†]Soil properties measured for 0 - 30 cm soil depth, unless stated; [‡]Values are mean ± standard deviations; [§]Not analyzed.

collected at a monthly interval, whereas soil samples were taken at a bi-weekly interval for the first two months and then at a monthly interval until harvest in 2013. Four soil cores were collected from between crop rows and composited for each sub-plot. The samples were transferred to the laboratory in a cooler and stored at −4°C until analyzed. In the laboratory, soil inorganic N (NH_4^+ and NO_3^-) contents were determined [23]. Field moist soil (6.5 g) was extracted with 25 mL of 2M KCl after shaking the mixture for 30 min in a reciprocal shaker. The soil suspension was then centrifuged for 5 min and filtered through a Whatman no. 2 filter paper. The extracts were then analyzed for inorganic N using an Automated Timberline TL2800 Ammonia Analyzer (Timberline Instruments, CO, USA). Three additional soil cores (2 cm internal diameter) with depth intervals of 0 - 15, and 15 - 30 were also collected using a soil probe per each replicate from each crop at the first soil sampling events in both years to determine bulk density [24]. The average bulk densities from each crop were used to calculate their respective growing season soil inorganic N contents. Soil NH_4^+ and NO_3^- concentrations (mg·kg^{-1}) in the 0 - 15 cm and 15 - 30 cm depth were multiplied by respective (depth-wise) bulk densities to express them into area basis (kg·ha^{-1}). The inorganic N contents at the two depth intervals were summed to obtain NH_4^+ and NO_3^- contents for the upper 30 cm soil profile. Finally, both the NH_4^+ and NO_3^- contents were added together to obtain total inorganic N contents for 0 - 30 cm soil depth.

2.6. Measurement of N₂O Emission

During the 2012 growing season, N$_2$O flux measurements were conducted from the four replicate units (two undrained and two tile-drained units) in both crops. The N$_2$O emission rates from surface soil was measured using semi-permanent vented static PVC chamber (25.4 cm internal diameter and 10 cm height) method following the GRACEnet project protocol [31]. A polyvinyl chloride (PVC) anchor ring with beveled edge was inserted into the soil between crop rows in each sub-plot. The germinating crops, if any, inside the PVC rings were plucked out during anchor installation. Gas samples were collected at four instances for both corn (52, 78, 87, and 100 d after treatment application) and sugarbeet (35, 42, 54, and 73 d after treatment application). Gas samples were taken in between 0800 h and 1200 h of the day assuming to represent the average flux of the day. On the observation day, the height of the anchor ring above the soil surface was recorded, in order to calculate the headspace volume after chamber enclosure. A chamber was placed on the anchor and gas samples (30 mL) were collected from the chamber headspace at 0, 15, and 30 min with a graduated polypropylene syringe. The samples were then transferred to 12 mL pre-evacuated glass serum vials and transported to the laboratory for analysis. In addition, soil temperature and volumetric soil water content at the 6 cm depth, adjacent to each gas chamber, were also measured by using GS3 soil moisture-temperature sensor (Decagon Devices, Inc., Pullman, WA 99163). All the gas samples were analyzed within 24 h, using a DGA 42-Master gas chromatograph (Dani Instruments, Milan, Italy), fitted with an electron capture detector (ECD). The ECD was operated at 300°C, He carrier gas at 10 mL·min^{-1}, Hayesep N 80/100 mesh (0.32 cm diameter × 50 cm length) and Porapak D 80/100 mesh (0.32 cm diameter × 200 cm length) columns in an oven operated at 80°C. Assuming a linear increase in gas concentration, flux was calculated using the following equation:

$$ F = kd \left(\frac{273}{T} \right) \left(\frac{V}{A} \right) \left(\frac{\Delta C}{\Delta t} \right) $$

where, F is the rate of gas emission (mass ha^{-1}·d^{-1}), k is unit conversion, d is gas density (g·cm^{-3}) at 273 °K, T is the air temperature (°K), V is the chamber volume (cm^3), A is soil area covered by chamber (cm^2) and $\Delta C/\Delta t$ is the rate of change of concentration over 15 and 30 min intervals [11].

2.7. Measurement of NH₃ Volatilization Loss

In 2013, NH$_3$ volatilization losses from the N fertilizers were measured from both corn and sugarbeet plots using semi-static open chambers [32]. A chamber was installed in the middle of each sub-plot in between the crop rows. The chambers were secured in an upright position on the soil surface using wire stakes, surrounded by rubber bands. Ammonia volatilization measurements were taken six times (5, 9, 19, 27, 33, and 40 d after N application), and five times (19, 22, 25, 32, and 60 d after N application) from corn and sugarbeet, respectively. On the day of measurement, the foam strips and the acid solution were collected, stored in 0.5 L mason jars containing 125 mL of 2 M KCl solution, and new traps were replaced. The sampled traps were transferred to the laboratory, where they were immediately extracted with 250 mL of 2 M KCl solution. The extracts were analyzed for NH_3 concentration using the ammonia analyzer as described above. Cumulative NH_3 volatilization loss from each sub-plot was obtained by adding NH_3 produced at individual days within the sub-plot.

2.8. Measurement of Soil Water NO_3^- Concentration

Samples of the soil water at 60 cm soil depth were collected from the middle of each sugarbeet plot between rows using suction lysimeters (68 cm in length and 2.2 cm diameter; Irrometer Company, Inc., CA, USA) during the 2013 sugarbeet growing season. The suction lysimeters were installed on June 13, 2013 (15 d after treatment application) and were allowed to equilibrate for a week such that the first water sample collected (22 d after treatment application) was discarded, and not used for data analysis. Then onwards sampling was conducted for a total of 9 times during the growing season (26, 29, 33, 36, 40, 47, 54, 62, 71 d) after treatment application). Using a hand pump, a vacuum of -60 kPa was applied to the tubes and maintained for a period until the time of water sampling. Water samples inside the lysimeters were extracted using a polypropylene syringe, collected into polypropylene conical tubes, and transferred to the laboratory for analyses. In the laboratory, NO_3^- concentrations in the water samples were analyzed using the ammonia analyzer. The lysimeters were devoid of water samples in all of the tile-drained plots on 26 d, as well as, in all of the plots (undrained and tile-drained) on 47, 62, and 71 d.

2.9. Statistical Analyses

Data were analyzed separately for each year per crop using a RCBD in a split-plot arrangement with drainage as main factor and N fertilizer management as sub factor for the analysis of variance as calculated by SAS PROC GLM (version 9.3, SAS Institute Inc., Cary, NC, USA). Because of limited number of replicates, the N_2O data were pooled across the drainage management and analyzed using a RCBD with N fertilizer management as the main factor in both crops. The growing season soil inorganic N contents, N_2O fluxes, soil water NO_3^- concentration were tested separately for each sampling date. Mean separations were tested using Fisher's protected least significant difference at $P \leq 0.05$.

3. Results and Discussion

3.1. Weather Conditions

Daily precipitation and mean air temperatures during the 2012 and 2013 growing seasons recorded at the research site by the Fargo NDAWN station are presented in **Figure 1** [33]. In 2012, the growing season precipitation totaled 53% of the long-term (1981-2010) normal precipitation (461 mm). Consequently, the crops were under visible drought stress during the growing period. In contrast, the 2013 growing season was relatively wet compared to the normal years, but the distribution was uneven. More than half of the total season precipitation fell within the first two months of the growing season in 2013. As a result, the entire plot area was intermittently flooded during May and June of 2013. July and August were relatively dry and the crops were under drought stress. The last part of the 2013 growing season had above normal rainfall. In 2012, the growing season air temperatures were between 7°C and 25°C, which is slightly higher than the normal years. In 2013, air temperatures during the growing season ranged similar to that of the normal years, and were between 7°C and 22°C.

3.2. Soil N Availability during the Crop Growing Season

Changes in soil inorganic N availability as influenced by N-management and drainage under corn in 2012 and 2013 are presented in **Figure 2**. In 2012, management of N application had no influence on soil N availability throughout the growing season, regardless of drainage probably due to large inherent soil N mineralization and/or residual mineral N [6]. In fact, soil inorganic N measured during the growing season even exceeded the actual amount supplied through the N sources. In 2013, the pattern of N release in response to N management varied with drainage conditions under corn. Under the undrained condition in 2013, application of Urea134 led to a rapid buildup of inorganic N in soils during the early corn growing season (2 d after treatment application) compared to the application of urea + NP, despite both the treatments were applied at the similar N-rate. In sharp contrast under the tile-drained condition in 2013, the Urea + NP treatment delayed N release in soil until 37 d after treatment application, suggesting the potential of NP to hinder nitrification activity [9]. Corn N uptake is

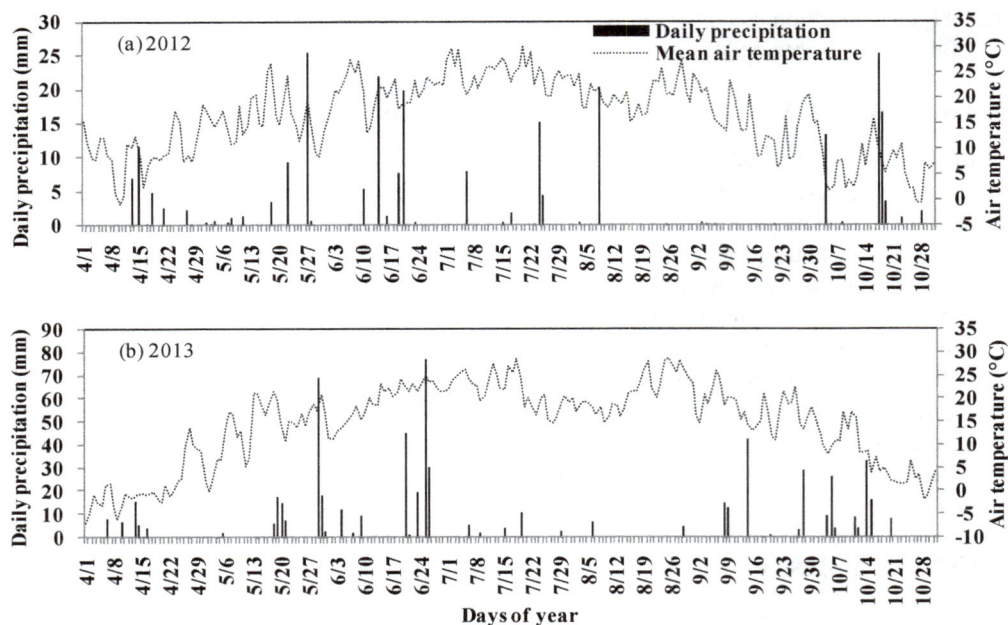

Figure 1. Daily precipitation and mean air temperature for the (a) 2012, and (b) 2013 growing seasons at the research site recorded by Fargo NDAWN station [33].

usually low during the early growth period [8]. And, given that the early 2013 growing season was relatively wet with a greater probability of N loss through denitrification, applying NP to urea was efficient in conserving N in soils for crop uptake during rapid crop growth. Similar to our results, previous studies have also shown that the application of NP to urea fertilizers delayed the rapid buildup of mineral N in soils compared to without NP in silty clay [9] and in silt loam [19]. Similarly, during the 2013 corn growing season, applying N fertilizer in split doses (half at planting and half at V6 corn growth stages) either resulted in lower soil N levels at early growing period (2 to 22 d) or released proportionate amount of mineral N afterwards compared to applying the entire N fertilizer at planting. Corn grain yields between these two N treatments (Urea134 and split N treatment) were not different under both drainage conditions (discussed below). Therefore, applying N in split doses could have also provided the benefit in the reduction of N loss during early growing season via denitrification, while maintaining crop yield similar to entire preplant N application [18].

Under sugarbeet, the application of N fertilizers increased soil N levels compared to unfertilized control on 48 d under the tile-drained condition in 2012, and on 7 d under both drainage conditions in 2013 (**Figure 3**). The application of Urea180 extended this increment until 15 d, particularly under the tile-drained condition in 2013. The higher inorganic N levels in soils with N-fertilization over unfertilized control during the crop growing season were expected. However, application of recommended N-rate (Urea146), with or without NP, resulted in similar soil N levels as higher N rate (Urea180), regardless of drainage management. The lack of differences in soil N availabilities among the fertilizer N management practices could be attributed to the presence of large residual soil NO_3-N levels (**Table 1**). There were no appreciable differences in sugarbeet root yields and net sucrose concentrations among these N treatments in either year (discussed below). These results indicate that excess application of N-fertilizers should be avoided not only to attain higher gross revenue but also to reduce the risk of losses to the environment [12] [14].

3.3. Soil N_2O Emissions

In corn, the elevated N_2O emissions were recorded on 54 d after N application (**Table 2**) in response to a total of 28 mm of precipitation that occurred on June 13 - 14 (52 - 53 d after N application) (**Figure 1**), with the Urea224 treatment exhibiting the largest flux of 105 g N_2O-N $ha^{-1} \cdot d^{-1}$. Cumulative precipitation for the following 3 wk period totaled only 9 mm. Consequently, the emission rates measured on 78 d after N application did not exceed 10 g N_2O-N $ha^{-1} \cdot d^{-1}$ in any of the N treatments. The N_2O flux rates increased to about 27 g N_2O-N

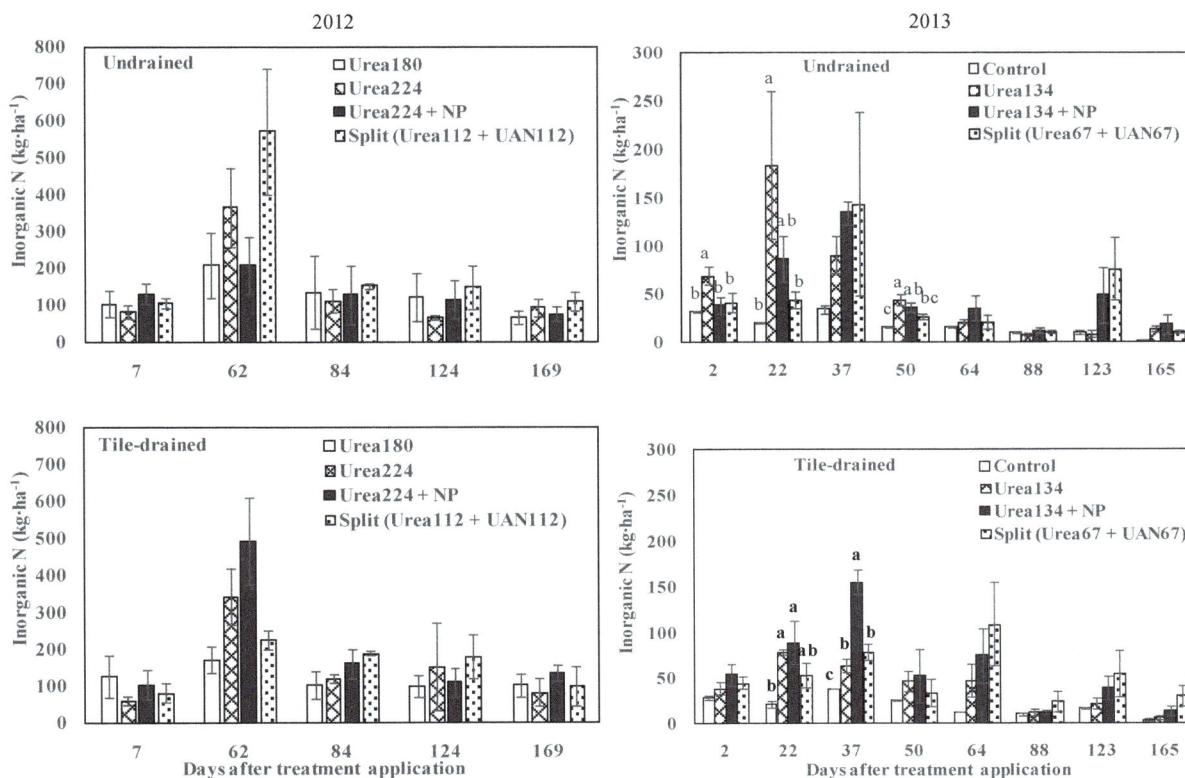

Figure 2. Inorganic N (NH_4^+ + NO_3^-) dynamics at 0 - 30 cm soil depth for N treatments under undrained and tile-drained conditions over a corn growing season in 2012 and 2013. Bars represent standard errors (n = 4). Different lower case letters within a day indicate significant difference at 0.05 level of significance.

Table 2. Soil N_2O fluxes as influenced by N management during the 2012 growing season in corn and sugarbeet.

N fertilizer	Nitrous oxide flux (g N_2O-N $ha^{-1} \cdot d^{-1}$)			
Corn	52 d	78 d	87 d	100 d
Urea180	74 ± 12	4 ± 1	25 ± 4	7 ± 1
Urea224	105 ± 69	7 ± 4	26 ± 12	11 ± 2
Urea224 + NP	74 ± 18	10 ± 2	28 ± 12	5 ± 1
Split (Urea112 + UAN112)	73 ± 17	7 ± 4	21 ± 10	6 ± 1
LSD (P ≤ 0.05)[‡]	NS[§]	NS	NS	NS
Sugarbeet	35 d	42 d	54 d	73 d
Control	10 ± 3	13 ± 4	13 ± 3	13 ± 4
Urea146	20 ± 4	15 ± 9	23 ± 5	18 ± 4
Urea146 + NP	17 ± 5	16 ± 4	24 ± 5	28 ± 13
Urea180	24 ± 7	17 ± 1	36 ± 17	29 ± 5
LSD (P ≤ 0.05)[†]	NS[§]	NS	NS	NS

[†]Values are means ± standard errors (n = 4); [‡]Least significant difference (LSD) values provided for P ≤ 0.05; [§]NS, non-significant.

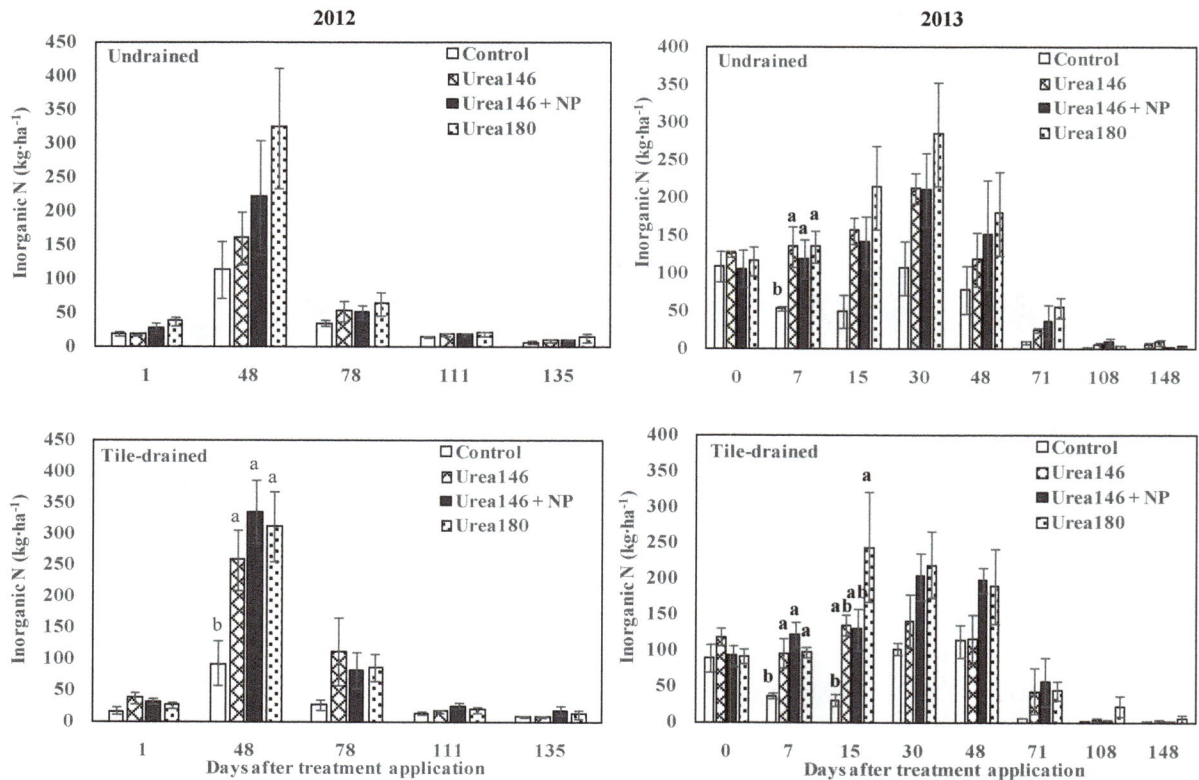

Figure 3. Inorganic N (NH_4^+ + NO_3^-) dynamics at 0 - 30 cm soil depth for N treatments under undrained and tile-drained conditions over a sugarbeet growing season in 2012 and 2013. Bars represent standard errors (n = 4). Different lower case letters within a day indicate significant difference at 0.05 level of significance.

$ha^{-1} \cdot d^{-1}$ on average across N treatments on 87 d, owing to 20 mm of total precipitation received on July 24 - 25 (86 - 87 d after N application). By 100 d, the N_2O rates averaged 8 g N_2O-N $ha^{-1} \cdot d^{-1}$ on average across N treatments. Similar to corn, N_2O emissions from the sugarbeet plots were also characterized by the total amount of precipitation received during the measurement periods, with the flux values ranging between 10 to 36 g N_2O-N $ha^{-1} \cdot d^{-1}$ across N treatments (**Table 2**).

Larger N_2O flux rates in response to the precipitation events—as a consequence of increased soil saturation—are commonly reported in the literature [10] [12] [20]. However, except for one of the flux measurement days in corn, soil N_2O fluxes remained mostly below 36 g N_2O-N $ha^{-1} \cdot d^{-1}$ across the crops, throughout the measurement period. These results are comparable to those reported for silt loam soils in Mandan, North Dakota, where peak values varied from 19 to 27 g N_2O-N $ha^{-1} \cdot d^{-1}$ in spring wheat—fallow and spring wheat—safflower (*Carthamus tinctorius* L.)—rye (*Secale cereale* L.) cropping systems [34]. In contrast, daily N_2O fluxes in the present study are comparatively lower than the reported peak values of 740 g N_2O-N $ha^{-1} \cdot d^{-1}$ for silt loam soils in Indiana [19]. Poorly drained soils usually favor denitrification because of O_2 inhibition in these soils upon soil saturation [10]. In general, N_2O emissions are lower from soils with less than 60% water filled pore space (WFPS), increases slowly between 60 and 80% WFPS, and then increase more rapidly above 80% WFPS [35]. The 2012 growing season was abnormally dry (**Figure 1**), with the highest daily precipitation of 25 mm received for the growing season. Therefore, the dry growing period in 2012 could have restricted N_2O emissions in our study as opposed to others that have commonly reported larger spikes in N_2O production following rainfall and/or irrigation events greater than 70 mm [19]. Moreover, lower gas diffusivity and/or higher cation exchange capacity associated with fine-textured soils in the study site may have also limited N_2O production [36].

Soil N_2O fluxes did not differ among the N management treatments on any sampling date in both corn and sugarbeet (**Table 2**). Studies have shown that application of mineral N fertilizers tends to cause positive and linear N_2O emissions because of greater inorganic N substrate availability [12]. Soil N availability measured at 0 -

30 cm soil depth in sugarbeet was higher with the N-fertilized treatments as compared to the control plots during the 2012 mid-growing season. However, soil N levels did not differ among the N-fertilized treatments under both corn and sugarbeet. Due to inconsistency in temporal release of mineral N from the applied fertilizers as well as soil inherent N mineralization, and due to episodic nature of N_2O emission induced by rainfall events, N_2O emissions did not vary among different N fertilizer sources during the dry growing season in 2012 [10] [37]. Besides, N_2O fluxes measured during the mid-growing season in 2012 were highly variable within the N treatments in both crops, and consequently no significant differences were observed among the treatments [20].

3.4. Soil NH_3 Volatilization

Cumulative losses of NH_3 did not respond to drainage management under both corn and sugarbeet in 2013 (**Table 3**). Approximately, 46 mm of precipitation fell within the next 5 d after N application under corn, and about 70 mm of precipitation fell on the day after N application under sugarbeet, which could have considerably incorporated the N fertilizers, regardless of drainage (**Figure 1**). Moreover, the N fertilizers were incorporated manually at 15 cm soil depth after N application. Therefore, soil incorporation of N fertilizer by high rainfall and/or manually after N application could have limited NH_4^+ substrate availability [38]. And considering pro-portionate amounts of NH_4^+ transport from the soil surface down to soil profile, NH_3 emissions from subsur-face drained and undrained plots are unlikely to be different [16].

In both corn and sugarbeet, NH_3 losses increased with N application over the unfertilized control (**Table 3**). Higher NH_3 losses from N-fertilized treatments are expected due to greater availability of N in the fertilized plots (**Figure 2** and **Figure 3**). However, there were no significant differences in NH_3 losses between N sources (with vs. without NP treatments in corn) and N rates (higher vs. recommended rates in sugarbeet). In corn, across drainage, about 1.9% of applied N was lost as NH_3 from the Urea134 + NP treatment, and NH_3 volatiliza-tion from the Urea134 alone was about 1.2% of applied N. Similarly, in sugarbeet, about 0.4 and 0.5% of ap-plied N was lost as NH_3 from Urea146 and Urea180, respectively, with no significant differences between the N-rate treatments. Studies have shown that the application of nitrification inhibitor to urea usually enhances NH_3 loss compared to untreated urea due to retention of fertilizer N in the NH_4^+ form for a prolonged period of time by the inhibitor [39]. And, NH_3 volatilization usually increases with increasing N application rate as a conse-quence of greater N availability [12]. However, in our study, soil N levels were similar between the N-rates throughout the sugarbeet growing season probably due to crop N uptake. And, under corn, although the availa-bility of N tended to be greater with the application of NP during the early growing period, it apparently was not

Table 3. Cumulative NH_3 volatilization losses from N fertilizers under undrained and tile-drained conditions in corn and su-garbeet during the 2013 growing season.

N fertilizer	Cumulative NH_3 loss[†]		Emission factor	
	Undrained	Tile-drained	Undrained	Tile-drained
	----------------kg·ha^{-1}--------------		-----------%-----------	
Corn				
Control	1.22 ± 0.13 b	1.26 ± 0.12 b	-	-
Urea134	2.51 ± 0.39 ab	3.23 ± 0.20 a	0.96	1.47
Urea134 + NP	3.69 ± 1.27 a	3.95 ± 1.19 a	1.84	2.01
LSD (P ≤ 0.05)[‡]	1.79	1.63	-	-
Sugarbeet				
Control	0.89 ± 0.11 b	0.77 ± 0.02 b	-	-
Urea146	1.44 ± 0.09 a	1.43 ± 0.14 a	0.38	0.45
Urea180	1.51 ± 0.15 a	1.75 ± 0.14 a	0.41	0.54
LSD (P ≤ 0.05)[‡]	0.35	0.46	-	-

[†]Values are means ± standard error (n = 4); Different lowercase letters within a column indicate significant difference at 0.05 level of significance.
[‡]Least significant difference (LSD) value provided for P ≤ 0.05.

enough to significantly influence NH_3 volatilization under wet early growing season, when most of the NH_4^+ would have already been stabilized to the soil exchange complex [32]. Nevertheless, the amounts of NH_3 volatilization losses in our study are comparable to the previously reported values of 2.3% for surface applied urea for clay loam soil at a simulated irrigation of 16 to 19 mm of precipitation applied after 1 d of urea application [32] and 2.8% for sandy loam soils at 21.4 mm of simulated rain applied after 1 d of N application [40]. As with these studies, high precipitation (46 and 70 mm in corn and sugarbeet, respectively) received shortly after N application considerably incorporated fertilizer N into soil and restricted NH_3 volatilization [38]. Higher CEC associated with high clay soil at the research site could have also limited the NH_4^+ substrate required for volatilization [41]. Furthermore, the urea were incorporated into soils shortly after their application in our study, which would reduce NH_3 loss because of increased contact of urea with soil exchange complex, and where it is converted to the stable NH_4^+ form [16].

3.5. Soil Water NO_3^- Concentration

Soil water NO_3^- concentration (mg·L^{-1}) measured at the 60 cm soil depth during the 2013 sugarbeet growing season for the N treatments under undrained and subsurface drained conditions are presented in **Figure 4**. The concentrations of NO_3^- at 60 cm depth was highly variable. However, the concentrations of NO_3^- were slightly lower under subsurface drained than undrained condition across all the N treatments throughout the measurement period. Under the undrained condition, peak concentrations of NO_3^- in Urea146, Urea146 + NP, and

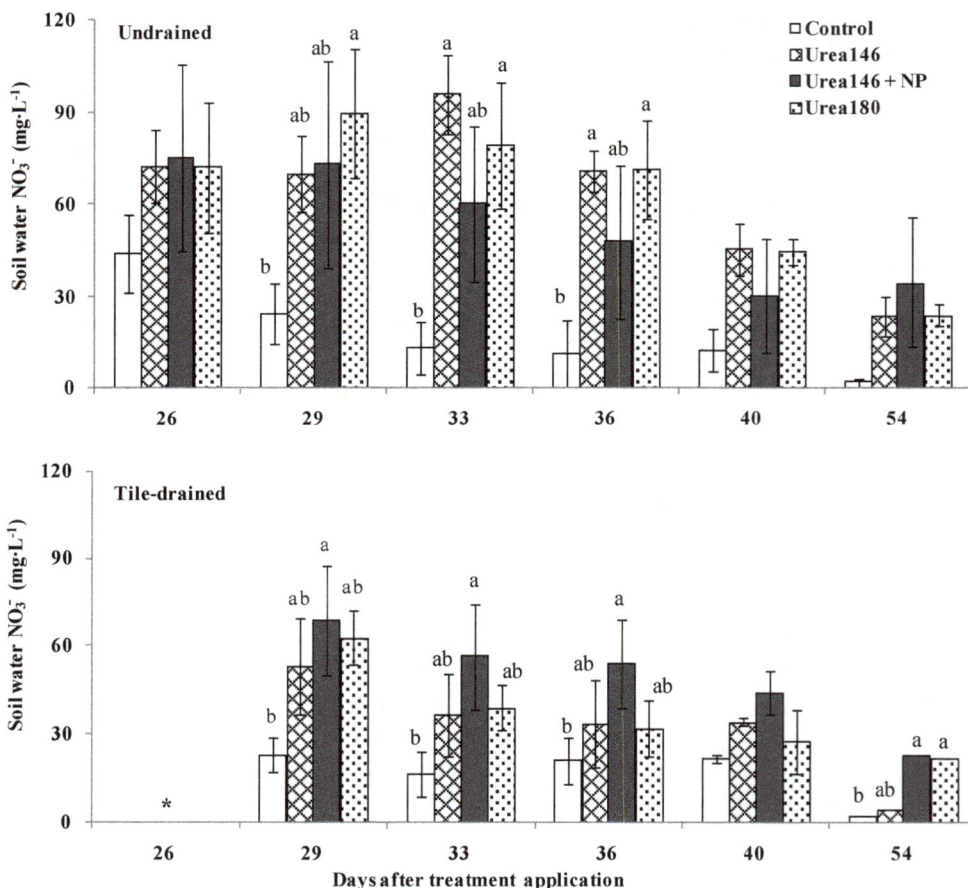

Figure 4. Soil water NO_3^- concentration (mg·L^{-1}) measured during the 2013 sugarbeet growing season as influenced by N management under undrained and drained conditions. Bars represent standard errors (n = 4). Different lower case letter within a day indicate significant difference at 0.05 level of significance. *Not available for the day.

Urea180 were 96, 75, and 85 mg·L^{-1}, respectively, whereas the corresponding peak concentrations for these treatments under the tile-drained condition were 53, 69, and 63 mg·L^{-1}, respectively. A consistently lower NO_3^- concentration with tile drainage could in part be due to greater N uptake by sugarbeet crops under tile drained condition. Sugarbeet N demand is usually high from the early growth period until canopy growth phase for the development of above-ground plant parts, during which maximum N is assimilated [42]. And, the measurement of leaf chlorophyll conducted at 54 d revealed that, across the N treatments, the spad values were slightly higher for the tile drained sugarbeet plots compared to undrained plots (data not presented).

Soil NO_3^- concentrations varied among the N treatments with drainage. Nitrate concentration generally increased with N fertilization over the control irrespective of drainage management, however. The lower NO_3^- levels with the control treatments were expected. Under the tile-drained condition, application of NP accumulated significantly greater soil water NO_3^- levels than the control, while the urea treatments without NP were only slightly greater than control. Under the undrained condition, the soil water NO_3^- concentration appeared not to depend on NP but N fertilization. In the present study, the extraction of soil water for the determination of NO_3^- concentration began only after 29 d following N application due to untrafficable soil condition at the study site. Despite the delayed measurements, under the tile-drained condition, the NO_3^- concentration tended to be slightly higher with the application of Urea146 + NP than without NP treatments (Urea146 and Urea180). Since nearly a third of the total growing season precipitation fell within the initial 30 d (**Figure 1**), considerable losses of N were deemed possible through denitrification without NP application during these initial 4 wk period [9] [19]. These results suggest the efficacy of nitrification inhibitor NP to conserve more NO_3^- in soils under the tile-drained condition, which otherwise could have potentially lost through denitrification without the application of NP [10].

3.6. Crop Yields

Tile drainage had no influence on crop yields and quality parameters during the 2012 and 2013 growing seasons (**Table 4** and **Table 5**). Several studies have indicated significant crop yield improvements under tile drainage compared with undrained conditions due to improvement in soil environment and/or N availability for crops with drainage [3] [4]. However, crop yield response to drainage management can vary with the amount and the pattern of precipitation received during the growing season [10]. And, studies have confirmed that water stress

Table 4. Corn grain yield as affected by N management under undrained and tile-drained conditions in 2012 and 2013.

N fertilizer	Grain yield[†]	
	Undrained	Tile-drained
	--------------Mg·ha^{-1}--------------	
2012		
Urea180	7.46 ± 0.81 a	8.10 ± 0.82 a
Urea224	6.84 ± 0.72 a	7.53 ± 0.92 a
Urea224 + NP	7.56 ± 1.09 a	6.42 ± 0.49 a
Split (Urea112 + UAN112)	7.71 ± 1.01 a	7.38 ± 0.37 a
LSD (P ≤ 0.05)[‡]	NS[§]	NS
2013		
Control	5.97 ± 0.41 b	6.01 ± 0.27 b
Urea134	8.79 ± 0.37 a	8.33 ± 0.57 a
Urea134 + NP	8.65 ± 0.59 a	8.59 ± 0.68 a
Split (Urea67 + UAN67)	7.42 ± 0.19 a	7.84 ± 0.57 a
LSD (P ≤ 0.05)[‡]	1.43	0.86

[†]Values are means ± standard error (n = 4); Different lowercase letters within a column within each year indicate significant difference at 0.05 level of significance; [‡]Least significant difference (LSD) values provided for P ≤ 0.05; [§]NS, non-significant.

Table 5. Sugarbeet root yield and quality parameters as affected by N management under undrained and tile-drained conditions in 2012 and 2013.

N fertilizer	Root Yield[†]		Sucrose loss to molasses[†]		Net sugar[†]	
	Undrained	Tile-drained	Undrained	Tile-drained	Undrained	Tile-drained
2012	----------$Mg·ha^{-1}$----------		---------------------------------------%---------------------------------			
Control	43.8 ± 1.9 a	51.3 ± 5.4 a	1.62 ± 0.04 a	1.68 ± 0.05 b	16.6 ± 0.4 a	16.9 ± 0.6 a
Urea146	47.0 ± 4.7 a	48.6 ± 4.0 a	1.74 ± 0.09 a	1.86 ± 0.12 ab	16.9 ± 0.8 a	16.6 ± 0.2 a
Urea146 + NP	44.7 ± 3.1 a	45.5 ± 5.3 a	1.75 ± 0.07 a	1.95 ± 0.12 a	17.1 ± 0.4 a	16.5 ± 0.2 a
Urea180	45.6 ± 1.3 a	48.8 ± 5.8 a	1.89 ± 0.06 a	2.03 ± 0.08 a	16.1 ± 0.3 a	15.6 ± 0.2 a
LSD (P ≤ 0.05)[‡]	NS[§]	NS	NS	0.21	NS	NS
2013						
Control	39.4 ± 1.6 a	36.7 ± 2.2 a	1.54 ± 0.14 a	1.69 ± 0.20 a	14.5 ± 0.4 a	14.4 ± 0.4 a
Urea146	40.1 ± 0.5 a	41.0 ± 1.1 a	1.58 ± 0.05 a	1.81 ± 0.12 a	14.3 ± 0.5 a	13.5 ± 0.6 a
Urea146 + NP	39.6 ± 0.6 a	37.0 ± 1.7 a	1.74 ± 0.08 a	1.67 ± 0.08 a	13.8 ± 1.6 a	14.3 ± 0.2 a
Urea180	36.7 ± 1.0 a	35.8 ± 3.8 a	1.73 ± 0.10 a	1.68 ± 0.12 a	14.0 ± 0.1 a	13.9 ± 0.5 a
LSD (P ≤ 0.05)[‡]	NS[§]	NS	NS	NS	NS	NS

[†]Values are means ± standard error (n = 4); Different lowercase letters within a column within each year indicate significant difference at 0.05 level of significance; [‡]Least significant difference (LSD) values provided for P < 0.05; [§]NS, non-significant.

during reproductive growth stages of crops can dramatically reduce yields [43]. In the current study, the crops experienced drought conditions during the growing periods in both years. Therefore, soil water deficit was likely the most limiting factor affecting crop yields due to abnormally dry growing conditions. In addition, water stress could have also lowered N demand in the crops. Consequently, over the two study years, the expected benefit of tile drainage to improve crop yields was likely minimized, regardless of N management.

In 2012, corn yields averaged 7.39 and 7.35 $Mg·ha^{-1}$ under undrained and tile-drained conditions, respectively, with no significant differences among the N fertilizer management sources (**Table 4**). Lack of yield responses to fertilizer N management practices could be attributed in part to large soil residual N and/or organic N mineralization during the growing seasons [37]. In fact, soil N levels measured during the 2012 growing season even exceeded the amount supplied through N fertilizers (**Figure 2**). Moreover, soil N availabilities among the N fertilizer sources and rates were similar throughout the growing season under both drainage conditions. Given that soil N availability from the N sources were similar, it can be safely presumed that N uptake by corn plants were similar for the N sources, which likely limited yield differences among them. Additionally, the 2012 growing season was abnormally dry and the movement of N into the active root zone might have also been hindered within the dry topsoil that consequently may have led to poor root N uptake [44]. Nevertheless, in 2013, corn yields increased with the application of fertilizer-N over the unfertilized control on average by 2.32 $Mg·ha^{-1}$ (27.9% increment) under the undrained condition and by 2.24 $Mg·ha^{-1}$ (27.2% increment) under the tile-drained condition. But, the yields were similar among the N-fertilized treatments under both undrained and drained conditions, and averaged 8.29 $Mg·ha^{-1}$ and 8.25 $Mg·ha^{-1}$, respectively. These results suggest the necessity of N application regardless of N sources in order to optimize crop yields [18]. However, the lack of yield differences among the N-fertilized treatments, regardless of N source and N rate could be attributed to drought stress during the dry mid-growing period that likely reduced the overall corn yield potential, obscuring the effects of N management [12] [44]. Our results are in line with previous studies that have also reported no responses to the application of N sources and rates containing urease inhibitor, nitrification inhibitor, or slow release polymer coated urea under abnormally dry growing periods [10] [12] [45]. The response of N fertilizers on crop yields are usually substantial in coarse-textured soils with higher crop N demands under irrigated condition, whereas the response of N fertilizers may be limited in fine-textured soils having low crop N demand with limited soil water availability [46].

Similar to corn, N management had no influence on sugarbeet root yield and sucrose concentration in both years (**Table 5**). Across N treatments, in 2012, sugarbeet root yield averaged 45.3 and 48.6 Mg·ha^{-1}, with net sucrose concentrations of 16.7 and 16.4% under undrained and tile-drained conditions, respectively. In 2013, the average sugarbeet root yields were 39.0 and 37.6 Mg·ha^{-1}, and sucrose concentrations were 14.2 and 14.0% under undrained and tile-drained conditions, respectively. Soil residual NO_3-N contents measured in the sugarbeet plots were 93 and 272 kg·ha^{-1} (at 0 - 120 cm soil profile) for 2012 and 2013, respectively (**Table 1**). Such high residual N levels could have significantly contributed crop N needs, and consequently no response to the added N through fertilizers was observed. Instead, the application of N increased impurity (SLM %) in beet roots, particularly under the tile-drained condition in 2012 (**Table 5**). Sugarbeet root impurity increased by 13.8% with Urea146 + NP and by 17.2% with Urea180 compared to unfertilized control under the tile-drained condition in 2012. The reduction in sugarbeet root quality with N application could be due to greater crop N uptake with Urea146 + NP and Urea180 treatments because soil N levels measured during the mid-growing season were significantly higher with these treatments as compared to unfertilized control (**Figure 3**). Previous studies have confirmed that greater soil N availability than required usually lowers purity indexes with depressed sucrose concentrations [47]. Excess N fertilization can stimulate sugarbeet top growth beyond the point where maximum root yields are attained and thereby direct photosynthates into regenerating canopy rather than into the root storage [48].

4. Conclusion

Our experiment showed the need for long term studies of subsurface drainage and N management on crop yields and quality, and N availability and N losses in the RRV. Although contrasting weather patterns occurred during the 2012 and 2013 growing seasons, our study emphasized that across intense wetting and drying cycles, subsurface drainage and N management influenced N availability more than crop yield in the silty clay soil. Prolong periods of soil saturation due to poor internal drainage has typically been the greatest limiting factor affecting crop yields in this soil. And, contrary to the norm, soil water deficit was likely the most limiting factor affecting crop yields over the two study years, due to abnormally dry conditions experienced during the growing seasons. Additionally, inherent soil N mineralization appears to be an important factor controlling crop yields and N losses, and therefore assessment of soil mineral N during crop growth period appears to be important to improve our knowledge on N availability for crops. Application of higher than recommended N rates had no effect on yields, but was only associated with more daily soil N_2O emissions. Conversely, the application of recommended N fertilizer rates along with nitrification inhibitor still holds as a viable N management strategy to increase N availability to crops without compromising yields. Also, nitrification inhibitor might have the potential to conserve soil water NO_3^- concentrations at 60 cm depth, particularly under the tile-drained condition. Apart from the high cation exchange associated with the soil under investigation, incorporation of fertilizer-N into soils mechanically and/or rainfall can considerably restrict NH_3 volatilization from the soil, with a possibility of up to 1.9% loss from the applied-N. The research results may provide important information to growers considering suitable N management under subsurface drainage systems within the RRV.

References

[1]　US Department of Agriculture—Natural Resources Conservation Service (2014) Soil Survey Office, Fargo, ND.

[2]　Kandel, H.J., Brodshaug, J.A., Steele, D.D., Ransom, J.K., Desutter, T.M. and Sands, G.R. (2013) Subsurface Drainage Effects on Soil Penetration Resistance and Water Table Depth on a Clay Soil in the Red River of the North Valley, USA. *Agricultural Engineering International: CGIR Journal*, **15**, 1-10.

[3]　Kladivko, R.L., Willowghby, G.L. and Santini, J.B. (2005) Corn Growth and Yield Response to Subsurface Drain Spacing on Clermont Silt Loam Soil. *Agronomy Journal*, **97**, 1419-1428. http://dx.doi.org/10.2134/agronj2005.0090

[4]　Jin, C.X., Sands, G.R., Kandel, H.J., Wiersma, J.H. and Hansen, B.J. (2008) Influence of Subsurface Drainage on Soil Temperature in a Cold Climate. *Journal of Irrigation and Drainage Engineering*, **34**, 83-88. http://dx.doi.org/10.1061/(ASCE)0733-9437(2008)134:1(83)

[5]　Agehara, S. and Warncke, D.D. (2005) Soil Moisture and Temperature Effects on Nitrogen Release from Organic Nitrogen Sources. *Soil Science Society of America Journal*, **69**, 1844-1855. http://dx.doi.org/10.2136/sssaj2004.0361

[6]　Rochette, P., Tremblay, N., Fallon, E., Angers, D.A., Chantigny, M.H., MacDonald, J.D., Bertrand, N. and Parent, L.E. (2010) N_2O Emissions from an Irrigated and Non-Irrigated Organic Soil in Eastern Canada as Influenced by N Ferti-

lizer Addition. *European Journal of Soil Science*, **61**, 186-196. http://dx.doi.org/10.1111/j.1365-2389.2009.01222.x

[7]   Dessureault-Rompre, J.B., Zebarth, B.J., Burton, D.L., Gregorich, E.G., Goyer, C., Georgallas, A. and Grant, C.A. (2013) Are Soil Mineralizable Nitrogen Replenished during the Growing Season in Agricultural Soil? *Soil Science Society of America Journal*, **77**, 512-524. http://dx.doi.org/10.2136/sssaj2012.0328

[8]   Zebarth, B.J. and Paul, J.W. (1997) Growing Season Nitrogen Dynamics in Manured Soils in South Coastal British Columbia: Implications for a Soil Nitrate Test for Silage Corn. *Canadian Journal of Soil Science*, **77**, 67-76. http://dx.doi.org/10.4141/S96-028

[9]   Awale, R. and Chatterjee, A. (2015) Soil Moisture Controls the Denitrification Loss of Urea-Nitrogen from Silty Clay Soil. *Communications in Soil Science and Plant Analysis*, **46**, 2100-2110. http://dx.doi.org/10.1080/00103624.2015.1069317

[10] Nash, P., Nelson, K. and Motavalli, P. (2015) Reducing Nitrogen Loss with Managed Drainage and Polymer-Coated Urea. *Journal of Environmental Quality*, **44**, 256-264. http://dx.doi.org/10.2134/jeq2014.05.0238

[11] Ginting, D., Kessavalou, A., Eghball, B. and Doran, J.W. (2003) Greenhouse Gas Emissions and Soil Indicators Four Years after Manure and Compost Applications. *Journal of Environmental Quality*, **32**, 23-32. http://dx.doi.org/10.2134/jeq2003.2300

[12] Thapa, R., Chatterjee, A., Johnson, J.M.F. and Awale, R. (2015) Stabilized Nitrogen Fertilizers and Application Rate Influence Nitrogen Losses under Rainfed Spring Wheat. *Agronomy Journal*, **107**, 1-10. http://dx.doi.org/10.2134/agronj15.0081

[13] Drury, C.F., Tan, C.S., Reynolds, W.D., Welacky, T.W., Oloya, T.O. and Gaynor, J.D. (2009) Managing Tile Drainage, Subirrigation, and Nitrogen Fertilization to Enhance Crop Yields and Reduce Nitrate Loss. *Journal of Environmental Quality*, **38**, 1193-1204. http://dx.doi.org/10.2134/jeq2008.0036

[14] Randall, G.W. and Vetsch, J.A. (2005) Nitrate Losses in Subsurface Drainage from a Corn-Soybean Rotation as Affected by Fall and Spring Application of Nitrogen and Nitrapyrin. *Journal of Environmental Quality*, **34**, 590-597. http://dx.doi.org/10.2134/jeq2005.0590

[15] Singh, U., Sanabria, J., Austin, E.R. and Agyin-Birikorang, S. (2011) Nitrogen Transformation, Ammonia Volatilization Loss, Nitrate Leaching in Organically Enhanced Nitrogen Fertilizers Relative to Urea. *Soil Science Society of America Journal*, **76**, 1842-1854. http://dx.doi.org/10.2136/sssaj2011.0304

[16] Norman, R.J., Wilson Jr., C.E. and Slaton, N.A. (2003) Soil Fertilization and Mineral Nutrition in US Mechanized Rice Culture. In: Smith, W. and Dilday, R.H., Eds., *Rice: Origin, History, Technology, and Production*, John Wiley & Sons Inc., Hoboken, 331-411.

[17] Al-Kanani, T., MacKenzie, A.F. and Barthakur, N.N. (1991) Soil Water and Ammonia Volatilization Relationships with Surface-Applied Nitrogen Fertilizer Solutions. *Soil Science Society of America Journal*, **55**, 1761-1766. http://dx.doi.org/10.2136/sssaj1991.03615995005500060043x

[18] Maharjan, B., Venterea, R.T. and Rosen, C. (2014) Fertilizer and Irrigation Management Effects on Nitrous Oxide Emissions and Nitrate Leaching. *Agronomy Journal*, **106**, 703-714. http://dx.doi.org/10.2134/agronj2013.0179

[19] Omonode, R.A. and Vyn, T.J. (2013) Nitrification Kinetics and Nitrous Oxide Emissions When Nitrapyrin Is Coapplied with Urea-Ammonium Nitrate. *Agronomy Journal*, **105**, 1475-1486. http://dx.doi.org/10.2134/agronj2013.0184

[20] Parkin, T.B. and Hatfield, J.L. (2013) Influence of Nitrapyrin on N_2O Losses from Soil Receiving Anhydrous Ammonia. *Agriculture, Ecosystems and Environment*, **136**, 81-86. http://dx.doi.org/10.1016/j.agee.2009.11.014

[21] Soil Survey Staff (2014) Keys to Soil Taxonomy. 12th Edition, USDA-Natural Resources Conservation Service, Washington DC.

[22] Franzen, D.W. (2010) North Dakota Fertilizer Recommendation Tables and Equations. Publication No. SF-882 (Revised), NDSU Extension Service, North Dakota State University, Fargo. https://www.ndsu.edu/fileadmin/soils/pdfs/sf882.pdf

[23] Maynard, D.G., Kalra, Y.P. and Crumbaugh, J.A. (2008) Nitrate and Exchangeable Ammonium Nitrogen. In: Carter, M.R. and Gregorich, E.G., Eds., *Soil Sampling and Methods of Analysis*, Canadian Society of Soil Science, CRC Press, Boca Raton, 71-80.

[24] Blake, G.R. and Hartge, K.H. (1986) Bulk Density. In: Klute, A., Ed., *Methods of Soil Analysis, Part 1*, ASA and SSSA, Madison, 363-367.

[25] Combs, S.M. and Nathan, M.V. (1998) Soil Organic Matter. In: Brown, J.R., Ed., *Recommended Chemical Soil Test Procedures for the North Central Region*, NCR Publication No. 221, Missouri Agricultural Experiment Station SB 1001, University of Missouri, Columbia, 53-58.

[26] Thomas, G.W. (1996) Soil pH and Acidity. In: Sparks, D.L., Ed., *Methods of Soil Analysis, Part 3*, ASA, SSSA, Madison, 475-490.

[27] Chapman, H.D. (1965) Cation-Exchange Capacity. In: Black, C.A., Ed., *Methods of Soil Analysis, Part 2: Chemical and Microbiological Properties*, ASA, Madison, 891-900.

[28] Elliott, E.T., Heil, J.W., Kelly, E.F. and Monger, H.C. (1999) Soil Structural and Other Physical Properties. In: Robertson, G.P., *et al.*, Eds., *Standard Soil Methods for Long Term Ecological Research*, Oxford University Press, Inc., New York, 74-88.

[29] Frank, K., Beegle, D and Denning, J. (1999) Phosphorus. In: Brown, J.R., Ed., *Recommended Chemical Soil Test Procedures for the North Central Region*, NCR Publication No. 221, Missouri Agricultural Experiment Station SB 1001, University of Missouri, Columbia, 21-30.

[30] Warncke, D. and Brown, J.R. (1998) Potassium and Other Basic Cations. In: Brown, J.R., Ed., *Recommended Chemical Soil Test Procedures for the North Central Region*, NCR Publication No. 221, Missouri Agricultural Experiment Station SB 1001, University of Missouri, Columbia, 31-33.

[31] Parkin, T.B. and Venterea, R.T. (2010) Chamber-Based Trace Gas Flux Measurements. In: Follet, R.F., Ed., *Sampling Protocols*, USDA-ARS, Washington DC, 3.1-3.39. http://www.ars.usda.gov/research/GRACEnet

[32] Jantalia, C.P., Halvorson, A.D., Follett, R.F., Alves, B.J.R., Polidoro, J.C. and Urquiaga, S. (2012) Nitrogen Source Effects on Ammonia Volatilization as Measured with Semi-Static Chambers. *Agronomy Journal*, **104**, 1595-1603. http://dx.doi.org/10.2134/agronj2012.0210

[33] NDAWN (2015) North Dakota Agricultural Weather Network Center. http://ndawn.ndsu.nodak.edu

[34] Liebig, M.A., Tanaka, D.L. and Gross, J.R. (2010) Fallow Effects on Soil Carbon and Greenhouse Gas Flux in Central North Dakota. *Soil Science Society of America Journal*, **74**, 358-365. http://dx.doi.org/10.2136/sssaj2008.0368

[35] Bateman, E.J. and Baggs, E.M. (2005) Contribution of Nitrification and Denitrification to N_2O Emissions from Soils at Different Water-Filled Pore Space. *Biology and Fertility of Soils*, **41**, 379-388. http://dx.doi.org/10.1007/s00374-005-0858-3

[36] Gu, J., Nicoullaud, B., Rochette, P., Grossel, A., Henault, C., Cellier, P. and Richard, G. (2013) A Regional Experiment Suggests That Soil Texture Is a Major Control of N_2O Emissions from Tile-Drained Winter Wheat Fields during the Fertilization Period. *Soil Biology and Biochemistry*, **60**, 134-141. http://dx.doi.org/10.1016/j.soilbio.2013.01.029

[37] Dell, C.J., Han, K, Bryant, R.B. and Schmidt, J.P. (2014) Nitrous Oxide Emissions with Enhanced Efficiency Nitrogen Fertilizers in a Rainfed System. *Agronomy Journal*, **106**, 723-731. http://dx.doi.org/10.2134/agronj2013.0108

[38] Rochette, P., Angers, D.A., Chantigny, M.H., Gasser, M., MacDonald, J.D., Pelster, D.E. and Bertrand, N. (2013) Ammonia Volatilization and Nitrogen Retention: How Deep to Incorporate Urea? *Journal of Environmental Quality*, **42**, 1635-1642. http://dx.doi.org/10.2134/jeq2013.05.0192

[39] Gioacchini, P., Nastri, A., Marzadori, C., Giovannini, C., Antisari, L.V. and Gessa, C. (2002) Influence of Urease and Nitrification Inhibitors on N losses from Soils Fertilized with Urea. *Biology and Fertility of Soils*, **36**, 129-135. http://dx.doi.org/10.1007/s00374-002-0521-1

[40] Holcomb III, J.C., Sullivan, D.M., Horneck, D.A. and Clough, G.H. (2011) Effect of Irrigation Rate on Ammonia Volatilization. *Soil Science Society of America Journal*, **75**, 2341-2347. http://dx.doi.org/10.2136/sssaj2010.0446

[41] Griggs, B.R., Norman, R.J., Wilson Jr., C.E. and Slaton, N.A. (2007) Ammonia Volatilization and Nitrogen Uptake for Conventional and Conservation Tilled Dry-Seeded, Delayed-Flood Rice. *Soil Science Society of America Journal*, **71**, 745-751. http://dx.doi.org/10.2136/sssaj2006.0180

[42] Martin, S.S. (2001) Growing Sugarbeet to Maximize Sucrose Yield. In: Wilson, R.G., Ed., *Sugarbeet Production Guide, Bulletin ECO*1-156, Nebraska Coop. Ext., University of Nebraska, Lincoln, NE, 3-8.

[43] Cakir, R. (2004) Effect of Water Stress at Different Development Stages on Vegetative and Reproductive Growth of Corn. *Field Crops Research*, **89**, 1-16. http://dx.doi.org/10.1016/j.fcr.2004.01.005

[44] Ray, J.D., Heatherly, L.G. and Fritschi, F.B. (2005) Influence of Large Amounts of Nitrogen on Nonirrigated and Irrigated Soybean. *Crop Science*, **46**, 52-60. http://dx.doi.org/10.2135/cropsci2005.0043

[45] Mckenzie, R.H., Middleton, A.B., Pfiffner, P.G. and Bremer, E. (2010) Evaluation of Polymer-Coated Urea and Urea Inhibitor for Winter Wheat in Southern Alberta. *Agronomy Journal*, **102**, 1210-1216. http://dx.doi.org/10.2134/agronj2009.0194

[46] Abalos, D., Jeffery, S., Sanz-Cobena, A., Guardia, G. and Vallejo, A. (2014) Meta-Analysis of the Effect of Urease and Nitrification Inhibitors on Crop Productivity and Nitrogen Use Efficiency. *Agriculture, Ecosystems and Environment*, **189**, 136-144. http://dx.doi.org/10.1016/j.agee.2014.03.036

[47] Halvorson, A.D. and Hartman, G.P. (1974) Long-Term Nitrogen Rates and Sources Influence Sugarbeet Yield and Quality. *Agronomy Journal*, **67**, 389-393. http://dx.doi.org/10.2134/agronj1975.00021962006700030027x

[48] Anderson, F.N. and Peterson, G.A. (1988) Effect of Incremental Nitrogen Application on Sucrose Yield of Sugarbeet. *Agronomy Journal*, **80**, 709-712. http://dx.doi.org/10.2134/agronj1988.00021962008000050002x

Permissions

All chapters in this book were first published in OJSS, by Scientific Research Publishing; hereby published with permission under the Creative Commons Attribution License or equivalent. Every chapter published in this book has been scrutinized by our experts. Their significance has been extensively debated. The topics covered herein carry significant findings which will fuel the growth of the discipline. They may even be implemented as practical applications or may be referred to as a beginning point for another development.

The contributors of this book come from diverse backgrounds, making this book a truly international effort. This book will bring forth new frontiers with its revolutionizing research information and detailed analysis of the nascent developments around the world.

We would like to thank all the contributing authors for lending their expertise to make the book truly unique. They have played a crucial role in the development of this book. Without their invaluable contributions this book wouldn't have been possible. They have made vital efforts to compile up to date information on the varied aspects of this subject to make this book a valuable addition to the collection of many professionals and students.

This book was conceptualized with the vision of imparting up-to-date information and advanced data in this field. To ensure the same, a matchless editorial board was set up. Every individual on the board went through rigorous rounds of assessment to prove their worth. After which they invested a large part of their time researching and compiling the most relevant data for our readers.

The editorial board has been involved in producing this book since its inception. They have spent rigorous hours researching and exploring the diverse topics which have resulted in the successful publishing of this book. They have passed on their knowledge of decades through this book. To expedite this challenging task, the publisher supported the team at every step. A small team of assistant editors was also appointed to further simplify the editing procedure and attain best results for the readers.

Apart from the editorial board, the designing team has also invested a significant amount of their time in understanding the subject and creating the most relevant covers. They scrutinized every image to scout for the most suitable representation of the subject and create an appropriate cover for the book.

The publishing team has been an ardent support to the editorial, designing and production team. Their endless efforts to recruit the best for this project, has resulted in the accomplishment of this book. They are a veteran in the field of academics and their pool of knowledge is as vast as their experience in printing. Their expertise and guidance has proved useful at every step. Their uncompromising quality standards have made this book an exceptional effort. Their encouragement from time to time has been an inspiration for everyone.

The publisher and the editorial board hope that this book will prove to be a valuable piece of knowledge for researchers, students, practitioners and scholars across the globe.

List of Contributors

Jessique Ghezzi, Anastasios Karathanasis, Chris Matocha, Jason Unrine and Yvonne Thompson
Department of Plant and Soil Sciences, University of Kentucky, Lexington, KY, USA

Christine Noe
Department of Geography, University of Dar es Salaam, Dar es Salaam, Tanzania

Ifeanyichukwu O. Onor
College of Pharmacy, Xavier University of Louisiana, New Orleans, USA

Gabriel I. Onor Junior
Department of Stem Cell and Regenerative Biology, Harvard University, Cambridge, USA

Murty S. Kambhampati
Department of Natural Sciences, Southern University at New Orleans, New Orleans, USA

P. C. Mishra, S. K. Sahu, A. K. Bhoi and S. C. Mohapatra
Department of Environmental Sciences, Sambalpur University, Jyoti Vihar, Sambalpur, Odisha, India

Xiujiao Xu, Jianglong He, Yu Li, Zhaoxi Fang and Shaohui Xu
Department of Environment Science, Qingdao University, Qingdao, China

Liliana Inés Picone, Cecilia Videla and Roberto Héctor Rizzalli
1Facultad Ciencias Agrarias, Unidad Integrada Balcarce Instituto Nacional Tecnología Agropecuaria (INTA) UNMdP, Balcarce, Argentina

Calypso Lisa Picaud
AgroParisTech, Institut des sciences et industries du vivant et de l'environnement, 16 rue Claude Bernard, Paris, France

Fernando Oscar García
International Plant Nutrition Institute (IPNI), Programa Cono Sur de Latinoamérica, Buenos Aires, Argentina

Anthony Z. Sangeda, Frederick C. Kahimba, Reuben A. L. Kashaga, Ernest Semu and Christopher P. Mahonge
Sokoine University of Agriculture, Morogoro, Tanzania

Francis X. Mkanda
Sustainable Land Management Project, Regional Commissioner's Office, Moshi, Tanzania

Fidelis Ifeakachuku Achuba and Patrick Nwanze Okoh
Department of Biochemistry, Delta State University, Abraka, Nigeria

John F. Kessy
Department of Forest Economics, Sokoine University of Agriculture, Morogoro, Tanzania

X. W. Cai, Y. Shao and Z. M. Lin
College of Earth Science, Guilin University of Technology, Guilin City, China

Angelo Indelicato
Dragages Hong Kong Ltd., Hong Kong, China

Richard Y. M. Kangalawe
Institute of Resource Assessment, University of Dar es Salaam, Dar es Salaam, Tanzania

Christine Noe
Department of Geography, University of Dar es Salaam, Dar es Salaam, Tanzania

Felician S. K. Tungaraza
Department of Sociology and Anthropology, University of Dar es Salaam, Dar es Salaam, Tanzania

Godwin Naimani
Department of Statistics, University of Dar es Salaam, Dar es salaam, Tanzania

Martin Mlele
Alpha and Omega Consulting Group Limited, Dar es Salaam, Tanzania

Joel A. Mercado-Díaz, William A. Gould, Grizelle González
USDA Forest Service, International Institute of Tropical Forestry, Jardín Botánico Sur, 1201 Calle Ceiba, Río Piedras, Puerto Rico

Christophe Calvaruso
INRA, UR1138 "Biogeochemistry of Forest Ecosystems", Centre INRA of Nancy, Champenoux, France
"Radiation Physics" Laboratory, University of Luxembourg, Campus Limpersberg, Luxembourg, Luxembourg
EcoSustain, Environmental Engineering Office, Research and Development, Kanfen, France

Christelle Collignon
INRA, UR1138 "Biogeochemistry of Forest Ecosystems", Centre INRA of Nancy, Champenoux, France
INRA, UMR1136 INRA-Nancy University "Interactions Tree-Microorganisms", Centre INRA of Nancy, Champenoux, France

Antoine Kies
"Radiation Physics" Laboratory, University of Luxembourg, Campus Limpersberg, Luxembourg, Luxembourg

Marie-Pierre Turpault
INRA, UR1138 "Biogeochemistry of Forest Ecosystems", Centre INRA of Nancy, Champenoux, France

Fidelis Ifeakachuku Achuba
Department of Biochemistry, Delta State University, Abraka, Nigeria

Sebastian Vogel
University of Tübingen c/o German Archaeological Institute, Berlin, Germany

Michael Märker
Heidelberg Academy of Sciences and Humanities c/o University of Tübingen, Tübingen, Germany
Department of Earth Sciences (DST), University of Florence, Florence, Italy

Stephen Mutimba and Richard Kibulo
Camco Advisory Services (K) Ltd., Nairobi, Kenya

Francis X. Mkanda
Sustainable Land Management Project, Kilimanjaro, Tanzania

Ally-Said Matano
Lake Victoria Basin Commission Secretariat, Kisumu, Kenya
School of Environment and Earth Sciences, Maseno University, Maseno, Kenya

Canisius K. Kanangire
Lake Victoria Basin Commission Secretariat, Kisumu, Kenya

Douglas N. Anyona and Paul O. Abuom
School of Environment and Earth Sciences, Maseno University, Maseno, Kenya

Frank B. Gelder
Probe International, Inc., Auckland, New Zealand

Gabriel O. Dida and Ayub V. O. Ofulla
School of Public Health and Community Development, Maseno University, Maseno, Kenya

Philip O. Owuor
Department of Chemistry, Maseno University, Maseno, Kenya

Bente Foereid
Environment and Climate Division, NIBIO, Ås, Norway

John F. Kessy
Department of Forest Economics, Sokoine University of Agriculture, Morogoro, Tanzania

Abiud Kaswamila
Department of Geography, University of Dodoma, Dodoma, Tanzania

Saman M. K. Rasheed
Department of Soil and Water Sciences, Faculty of Agricultural Sciences, Sulaimani University, Sulaimani, Iraq

Rakesh Awale and Amitava Chatterjee
Soil Science, North Dakota State University, Fargo, ND, USA

Hans Kandel and Joel K. Ransom
Plant Sciences, North Dakota State University, Fargo, ND, USA

www.ingramcontent.com/pod-product-compliance
Lightning Source LLC
Chambersburg PA
CBHW080459200326
41458CB00012B/4027